Lecture Notes in Earth System Sciences 137

Editors:

P. Blondel, Bath
J. Reitner, Göttingen
K. Stüwe, Graz
M.H. Trauth, Potsdam
D. Yuen, Minneapolis

Founding Editors:

G. M. Friedman, Brooklyn and Troy
A. Seilacher, Tübingen and Yale

For further volumes:
http://www.springer.com/series/772

Ilmari Haapala
Editor

From the Earth's Core to Outer Space

Editor
Ilmari Haapala
Huvilakuja 2
02730 Espoo
Finland

ISBN 978-3-642-25549-6   e-ISBN 978-3-642-25550-2
DOI 10.1007/978-3-642-25550-2
Springer Heidelberg Dordrecht London New York

Library of Congress Control Number: 2012931329

© Springer-Verlag Berlin Heidelberg 2012
This work is subject to copyright. All rights are reserved, whether the whole or part of the material is concerned, specifically the rights of translation, reprinting, reuse of illustrations, recitation, broadcasting, reproduction on microfilm or in any other way, and storage in data banks. Duplication of this publication or parts thereof is permitted only under the provisions of the German Copyright Law of September 9, 1965, in its current version, and permission for use must always be obtained from Springer. Violations are liable to prosecution under the German Copyright Law.
The use of general descriptive names, registered names, trademarks, etc. in this publication does not imply, even in the absence of a specific statement, that such names are exempt from the relevant protective laws and regulations and therefore free for general use.

Printed on acid-free paper

Springer is part of Springer Science+Business Media (www.springer.com)

Preface

From the Earth's Core to Outer Space is an extended and revised version of the book *Maan ytimestä avaruuteen* edited by I. Haapala and T. Pulkkinen and published in 2009 in Finnish in the series *Bidrag till kännedom av Finlands natur och folk, No. 180,* of the Finnish Society of Sciences and Letters, Helsinki. That book was based on lectures given in a symposium dealing with timely research topics in geosciences and arranged in January 2008 in Helsinki to celebrate the centennial anniversary of the Finnish Academy of Sciences and Letters. The current version has been written to international readers. The articles have been strongly revised, some of them completely reformulated, and four new articles (Chap. 5 by H. O'Brien and M. Lehtonen, Chap. 12 by M. Viitasalo, Chap. 13 by J. Karhu, and Chap. 14 by A.E.K. Ojala), and three appendices (Geological time table, Layered structure of Earth's interior, Layers of Earth's atmosphere) have been added to widen and deepen the content of the book. The themes of the book are: Earth's Evolving crust, Changing Baltic Sea, Climate Change, and Planet Earth, Third Stone from the Sun.

I am grateful to all authors who, in addition to their official work, have found time to write the articles, and to the reviewers, who in most cases are other authors of the book: Pasi Eilu, Eero Holopainen, Pentti Hölttä, Hannu Huhma, Kimmo Kahma, Juhani Kakkuri, Juha Karhu, Veli-Matti Kerminen, Emilia Koivisto, Annakaisa Korja, Hannu Koskinen, Marita Kulmala, Markku Kulmala, Raimo Lahtinen, Martti Lehtinen, Matti Leppäranta, Wolfgang Maier, Pentti Mälkki, Irmeli Mänttäri, Satu Mertanen, Heikki Nevanlinna, Mikko Nironen, Pekka Nurmi, Hugh O'Brien, Antti Ojala, Risto Pellinen, Markku Poutanen, Tuija Pulkkinen, Tapani Rämö, Juhani Rinne, Jouni Räisänen, Heikki Seppä, and Timo Vesala. Especially, I would like to thank Professor Tuija Pulkkinen, who acted as coeditor of the Finnish version, but retreated from the editorship of the current volume because of her increased new duties at the Finnish Meteorological Institute and, since the beginning of 2011, at Aalto University.

Espoo   Ilmari Haapala

From the Earth's Core to Outer Space

Revised Proceedings of the Centennial Year Symposium (2008) of the Finnish Academy of Sciences and Letters

Edited by
Ilmari Haapala
Emeritus Professor of Geology and Mineralogy University of Helsinki, Finland

From the Earth's Core to Outer Space is an extended and revised version of the book *Maan ytimestä avaruuteen* that was edited by Ilmari Haapala and Tuija Pulkkinen and published in 2009 in the series *Bidrag till kännedom of Finlands natur och folk, No. 180*, Finnish Society of Sciences and Letters.

Contents

1 Introduction 1
   Ilmari Haapala and Tuija Pulkkinen

Part I Earth's Evolving Crust

2 Paleo-Mesoproterozoic Assemblages of Continents: Paleomagnetic
   Evidence for Near Equatorial Supercontinents 11
   S. Mertanen and L.J. Pesonen

3 Seismic Structure of Earth's Crust in Finland 37
   Pekka Heikkinen

4 Evolution of the Bedrock of Finland: An Overview 47
   Raimo Lahtinen

5 Craton Mantle Formation and Structure of Eastern Finland
   Mantle: Evidence from Kimberlite-Derived Mantle Xenoliths,
   Xenocrysts and Diamonds 61
   Hugh O'Brien and Marja Lehtonen

6 Metallic Mineral Resources in Finland and Fennoscandia:
   A Major European Raw-Materials Source for the Future 81
   Pekka A. Nurmi and Pasi Eilu

7 Isotopic Microanalysis: In Situ Constraints on the Origin
   and Evolution of the Finnish Precambrian 103
   O. Tapani Rämö

8 Fennoscandian Land Uplift: Past, Present and Future 127
   Juhani Kakkuri

Part II Changing Baltic Sea

9 Ice Season in the Baltic Sea and Its Climatic Variability 139
   Matti Leppäranta

10 Baltic Sea Water Exchange and Oxygen Balance 151
   Pentti Mälkki and Matti Perttilä

11 Marine Carbon Dioxide 163
   Matti Perttilä

12 Impact of Climate Change on Biology of the Baltic Sea 171
   Markku Viitasalo

Part III Climate Change

13 Evolution of Earth's Atmosphere 187
   Juha A. Karhu

14 Late Quaternary Climate History of Northern Europe 199
   Antti E.K. Ojala

15 Aerosols and Climate Change 219
   Markku Kulmala, Ilona Riipinen, and Veli-Matti Kerminen

16 Enhanced Greenhouse Effect and Climate Change
   in Northern Europe 227
   Jouni Räisänen

17 Will There Be Enough Water? 241
   Esko Kuusisto

Part IV Planet Earth, Third Stone from the Sun

18 Trends in Space Weather Since the Nineteenth Century 257
   Heikki Nevanlinna

19 Space Weather: From Solar Storms to the Technical Challenges
   of the Space Age 265
   Hannu Koskinen

20 Space Geodesy: Observing Global Changes 279
   Markku Poutanen

21	**Destination Mars**	295
	Risto Pellinen	
22	**In Search of a Living Planet**	309
	Harry J. Lehto	

Appendix 1: Geological Time 329

Appendix 2: Layered Structure of Earth's Interior 331

Appendix 3: Layers of Earth's Atmosphere 333

Index 335

Chapter 1
Introduction

Ilmari Haapala and Tuija Pulkkinen

The year 2008 marked the 100th anniversary of the Finnish Academy of Sciences and Letters. On the occasion, the disciplinary groups of the Academy of Sciences and Letters organized a series of mini-conferences focused on timely research topics (see http://www.acadsci.fi/100y.htm). The Group of Geosciences organized two events: the year was opened with a symposium entitled *From the Earth's Core to Outer Space* (January 9–11) and, during the spring and summer the *Exhibition of Geoscientific Expeditions* was open to the general public at the University of Helsinki Museum Arppeanum. This book is based on a collection of articles originating from the presentations given at the symposium. Precursors of many of these articles were published earlier in Finnish (Haapala and Pulkkinen 2009).

1.1 Planet Earth

According to present understanding, the Earth was formed as one of the Solar System planets about 4.57 billion years ago by accretion of material from an inhomogeneous disk-shaped gas and dust cloud that encircled the proto-Sun (Valley 2006, Committee on Grand Research Questions in Solid-Earth Sciences 2008). Earth-like planets close to the Sun formed when the minerals, metals, and dust particles accreted to larger aggregates and combined to form larger objects

---

I. Haapala (✉)
Department of Geosciences and Geography, University of Helsinki, P.O. Box 64, FI-00014 Helsinki, Finland
e-mail: ilmari.haapala@helsinki.fi

T. Pulkkinen
School of Electrical Engineering, Aalto University, P.O. Box 13000, FI-00076 Aalto, Finland
e-mail: tuija.pulkkinen@aalto.fi

whose gravitational attraction gathered other smaller particles. This process led to formation of planetesimals with a diameter of several kilometers. Through gravitational forces and impacts these grew into to the actual planets we now know as Mercury, Venus, Earth, and Mars. Farther from the Sun, in the cooler parts of the Solar nebula, the gaseous giant planets (Jupiter, Saturn, Neptune) formed. Their icy moons contain ice formed of water, ammonium, methane and nitrogen.

The Earth's iron-nickel core was formed when accretion was still in progress: heat produced by radioactive decay and accretion melted iron and nickel, which as heavy drops and large molten patches sank through the partially molten protoplanet to its core. Rock material consisting mainly of magnesium and iron silicates formed a thick mantle around its core. Numerous meteorite impacts and heat from radioactive decay melted the mantle to such extent that molten lava covered the entire surface of the protoplanet. As the impacts thinned out, the molten lava crystallized to form the first basaltic crust, which in later geological processes has been replaced by a crust that in continental areas consist mainly of feldspars- and quartz-bearing rocks.

Formation of the Moon may be a consequence of an impact of Mars-sized object about 2.48 billion years ago (Halliday 2008). The impact blasted numerous pieces of rock material and dust to orbit the Earth, this material later accreted to form the Moon.

There are several theories regarding the birth of the oceans and the atmosphere, and consensus is yet to be reached. At the time of the planet's formation, an early atmosphere made of hydrogen and helium rapidly escaped to the space. The first proper atmosphere was probably formed when water, carbon dioxide, nitrogen and other volatile compounds were either degassed from the solid-semisolid mantle or from the molten lava ocean, or boiled out from lavas during volcanic eruptions, forming a gaseous layer above the crust. Even today, volcanic eruptions release gases comprising 50–80% water vapor supplemented by carbon dioxide, nitrogen, sulfur oxide, hydrogen sulfides, and, small traces of carbon monoxide, hydrogen and chlorine. This shows that water and other volatile compounds are still stored in the inner parts of the Earth. Furthermore, some, perhaps significant amounts of water originated from outside the Earth and were aquaired via impacts of comets and meteorites even after the planet formation.

As the Earth cooled, water vapor condensed at the bottom of craters and valleys, which led to the formation of early oceans. At this time, the atmosphere contained mostly carbon dioxide and nitrogen along with small amounts of water vapor, argon, sulfur oxides, hydrogen sulfides and other gases. After the formation of the hydrosphere, the amount of carbon dioxide decreased as part of it dissolved into seawater as carbonate ions, and part precipitated as carbonate minerals. Chemical erosion of exposed rocks consisting of silicate minerals absorbed much of the carbon dioxide of the atmosphere. The eroded material dissolved in running water and was eventually discharged to the seas.

The first primitive life forms on Earth may have appeared already about four billion years ago, possibly in seafloor hydrothermal vent environment. Blue-green algae, cyanobacteria, appeared about 3.5 billion years ago; layers of algae formed stromatolite structures that have been found in early Archean shallow sea carbonate sediments (Golding and Glikson 2010). Cyanobacteria produced free oxygen for the atmosphere and hydrosphere through the photosynthesis reaction ($H_2O + CO_2 = CH_2O + O_2$). This process created the prerequisites of development of life forms that depend on atmospheric oxygen. Atmospheric oxygen reached its present concentration slowly and stepwise, by the end of the Precambrian supereon.

The layered structure and the uneven distribution of the Earth's intrinsic heat led to mantle-scale convection flows and initiation of plate tectonics. Plate tectonics is a unifying theory that naturally explains the relative motions of the continents, changes in the shapes of the oceans, formation of mountain belts, occurrence of earthquakes, disribution of volcanoes, and many other geological processes.

Since its formation, the Earth has been under continuous changes controlled by celestial mechanics and diverse internal processes. Reactions have taken place and still occur within and between Earth's different layers, powered by geothermal heat, the Sun, and meteorite impacts. Formation of the solid Earth, hydrosphere, and atmosphere has laid ground for the evolution of the biosphere. Balance between many different factors is critical, and even small changes in one of them may easily shift the system from one state to another.

1.2 Themes of the Book

The symposium *From the Earth's Core to Outer Space* comprised 26 presentations by leading Finnish geoscientists on timely topics central to society and the environment. The presentations were divided into four conceptual themes of research:

1. Earth's Evolving Crust (chair Ilmari Haapala)
2. Changing Baltic Sea (chair Pentti Mälkki)
3. Climate Change (chair Timo Vesala)
4. Planet Earth, Third Stone from the Sun (chair Tuija Pulkkinen)

This compilation is composed of 22 articles, most of them based on presentations given at the symposium. The articles are grouped into four parts to comply with the four themes.

1.2.1 Part I

Part I, Earth's Evolving Crust, starts with a paper by Satu Mertanen and Lauri Pesonen. Based on updated paleomagnetic data, this paper presents an extensive synthesis of the drift history of the lithospheric plates showing how these movements have, several times during the geological history of the Earth, led to amalgamation of different continents to supercontinents and to their subsequent breakup.

Pekka Heikkinen presents a summary of the internal structure and thickness of the Earth's crust in Finland and Fennoscandia, based on deep seismic soundings that utilize both the refraction and reflection methods.

To resolve the origin and evolution of rock units in deeply eroded, flat and soil-covered shield areas is a challenging task for geologists. Based on geological, geophysical, geochemical, and isotopic studies, Raimo Lahtinen presents an interpretation of the origin and evolution of the Finnish Precambrian. The oldest part of the bedrock, the Archean continental crust in eastern Finland, consists dominantly of granitoid-migmatite complexes and volcano-sedimentary belts and was formed, for the major part, 2.85–2.62 Ga (billion years) ago. Lahtinen concludes that subduction-related processes were operating already at 2.75 Ga as some volcano-sedimentary belts and plutonic rocks were formed within the wide Mesoarchean–Neoarchean basement of Karelia, eastern Finland. Subsequent evolution included stages of Paleoproterozoic rifting with associated magmatism and sedimentation, the collisional-type Lapland-Kola orogeny, the extensive and composite 1.92–1.79 Ga Svecofennian orogeny, and the bimodal A-type rapakivi granite magmatism at 1.67–1.54 Ga.

With focus on mantle processes, Hugh O'Brien and Marja Lehtonen present a comprehensive review on the origin and evolution of Earth's mantle beneath Archean cratons. This is followed by a review of recent studies of the mantle beneath the Karelian craton, based on detailed petrological, geochemical and isotopic studies of mantle xenoliths recovered from kimberlites and lamproites that intruded the crust in eastern Finland.

Pekka Nurmi and Pasi Eilu describe the state of the art of metallic mining industry in Finland and Scandinavia, present an updated geological review of the important ore deposits and discuss future developments. Mining industry is strongly growing in Finland, and it is estimated that the output of metallic mines will increase from four million tons in the early 2000s to 70 million tons in 2020. The volume and range of types of mineral deposits in Finland, and Fennoscandia as a whole, reflect the long and complex geological history of the crust in this area.

Tapani Rämö introduces, through several examples from the Finnish Precambrian, the opportunities offered by modern isotopic microanalysis in revealing the origin and evolution of the bedrock in Finland.

Juhani Kakkuri's article describes the history and current state of research concerning the land uplift in Fennoscandia during the Holocene. He also estimates how the melting of the ice sheet covering Greenland would change the sea global level.

1.2.2 Part II

Part II focuses on the Baltic Sea and commences with a paper by Matti Leppäranta on the climatological variability of the Baltic Sea and its gulfs. Leppäranta elaborates on winter ice conditions in different parts of the Baltic Sea and their significance to human activity. Changes in ice conditions are examined during the past 100 years over this time the global temperature has increased by about 1°C. Anticipating a 2–4°C temperature rise in the next 100 years, the author estimates the future conditions in the Baltic ice cover: By the year 2100, the Baltic Sea would freeze one month later than at present, the ice would melt about two weeks earlier, and on average the ice cover is 30 cm thinner. On an average winter 100 years from now, only the Gulf of Bothnia and the eastern end of the Gulf of Finland would develop a solid ice cover.

In an article discussing water exchange and oxygen content of the Baltic Sea, Pentti Mälkki and Matti Perttilä examine pulses of saline seawater through the Kattegat strait from the North Sea to the Baltic Sea. They also discuss the effects of Atlantic water exchange to salinity and oxygen concentration in different parts of the Baltic, in particular the conditions within deep basins. Strong saline pulses have become increasingly rare in recent decades, which has led to permanent anoxic conditions in the deep basins of the central Baltic.

As an example of the strong coupling between hydrosphere and atmosphere, Matti Perttilä discusses the carbon cycle in general, and carbon dioxide reactions within the oceans in particular. Perttilä concludes that the oceans, as major sinks of carbon dioxide, have considerably slowed down the increase of atmospheric carbon dioxide, and thus the measurable effects of the climate change.

Markku Viitasalo's article deals with the impact of the climate change on the biology of the Baltic Sea. The complex effects of changing climatic factors to the physics, chemistry and biology of the Baltic Sea are visualized graphically.

1.2.3 Part III

In Part III, Climate Change, Juha Karhu reviews the current knowledge of the evolution of the Earth's atmosphere through the geological history of the planet, with emphasis in the greenhouse gases (carbon dioxide, methane, water) and oxygen. Ancient atmospheres were anoxic and rich in greenhouse gases. Rise of atmospheric oxygen at 2.4 Ga ago produced a drastic environmental change with a wide range of consequences to weathering, atmospheric and oceanic chemistry, and biosphere. Oxygenation of the atmosphere progressed stepwise and reached near-modern levels at the end of the Precambrian.

Antti Ojala discusses the observed natural climate changes over geological timescales and elaborates on their reasons, such as orbital forcing, solar forcing, volcanic activity, concentration of greenhouse gases in the atmosphere, or atmospheric and oceanic circulation. Emphasis is in the long-term Quaternary glacial-interglacial cycles in Eurasia and in the short-term Holocene climate fluctuations in northern Europe, including the historical Medieval Climate Anomaly and Little Ice Age.

Ojala's article demonstrates that the Earth's climate has changed even dramatically during its history. For current setting, however, Markku Kulmala et al. state that *Climate change is probably the most crucial human-driven environmental problem: the humankind has changed the global radiative balance by changing the atmospheric composition.* Their article focuses on the formation of aerosols, their interactions between the atmosphere and the biosphere, significance of aerosols in radiation balance, and due climate change effects.

Jouni Räisänen discusses present climate scenarios. These predict that the climate in southern Finland changes to resemble that in Central Europe today. Thus, warming of the climate in Finland will be much greater than the global average. Radical measures are required to reduce the warming rate, as even keeping the emissions at the present level will increase the carbon dioxide concentration in the atmosphere during the decades to come. Esko Kuusisto examines practical solutions concerning both Finnish and global freshwater reservoirs.

1.2.4 Part IV

Part IV theme deals with the near space above the atmosphere (magnetosphere and ionosphere) and beyond. The ionosphere and magnetosphere affect conditions on Earth and possibly have a bearing on long-term changes in the climate. While the most significant effects on Earth arise from the solar radiation, the Sun also emits a particle flux that fills the Solar system with a fully ionized plasma. The interaction of solar wind with the Earth's intrinsic magnetic field gives rise to a variety of space weather effects in the near-Earth magnetosphere as well as the Aurora Borealis that form at a roughly 100-km altitude in the ionosphere. While the auroras are beautiful to view, the electric currents and charged particle fluxes associated with them may cause disturbances in technological systems both in space and on ground as well as impose a health risk for humans in space and high-altitude aircraft.

Heikki Nevanlinna examines the periodicities in auroral occurrence and disturbances in the geomagnetic field and their dependence on the solar activity. Even if the major periodicity is the 11-year solar cycle, there are hints of also longer-term periods, which may predict lower level of solar activity and thus calmer space weather in the next few decades. Hannu Koskinen discusses the space weather effects that arise from Solar Coronal Mass Ejections. Given the increasing dependence on electric power grids and satellite assets (satellite TV and telephone services, GPS navigation, etc.) it would be vital to increase the accuracy of space weather predicting, but the level of scientific knowledge of the associated processes still pose a significant challenge.

Underlining the significance of the use of space to solid Earth science, Markku Poutanen summarizes detailed space geodetic measurements as a proxy for the Earth's surface and motions of the continents, and elaborates on challenges associated with pertinent data interpretation.

Moving from our home planet to further out in the Solar System, Risto Pellinen discusses physical conditions on Mars. The latest measurements conducted by Mars rovers show that there indeed is water ice on the surface of Mars. Thus there has been (a however subtle) chance for development of Earth-like life forms also on Mars. The article also highlights the difficulties associated with space research: successes and failures alternate in missions that take a decade to carry through. Nevertheless, several space organizations plan to take humans out of Earth's orbit to Mars around the year 2030. In the final contribution of this volume, Harry Lehto discusses one of the basic questions of life: are we alone in the Universe, and if not, how could we observe life elsewhere?

1.3 Epiloque

The 2008 Symposium *From the Earth's Core to Outer Space* was tailored primarily for Finnish scientists and the general public, whereas this revised proceedings volume is directed more to geoscientists and environmental scientists in other countries in Europe and elsewhere. Our intentions were to provide a good snapshot of the Finnish geoscientific and environmental research in a variety of fields that are vital to the future of our living planet. We also hope that the book demonstrates the close relations and interconnections between the different disciplines of geosciences as well as the need for inter disciplinary research, scientific discussion and debate.

References

Committee on Grand Research Questions in the Solid-Earth Sciences (2008) Origin and Evolution of Earth: Research Questions for a Changing Planet. The National Academies Press, Washington DC. http://books.nap.edu/12161

Golding SD, Glikson M (eds) (2010) Earliest Life on Earth: Habitas, Environments and Methods of Detection, DOI 10.1007/978.90481-879-2, Springer

Haapala I, Pulkkinen T (eds) (2009) Maan ytimestä avaruuteen. Bidrag till kännedom av Finlands natur och folk 180:1–246

Halliday AN (2008) A young Moon-forming giant impact at 70–110 million years accompanied by late-stage mixing, core formation and degassing of the Earth. Phil Trans R Soc A 366:4163–4181

Valley JW (ed.) (2006) Early Earth. Elements 2 (4): 201–233

Part I
Earth's Evolving Crust

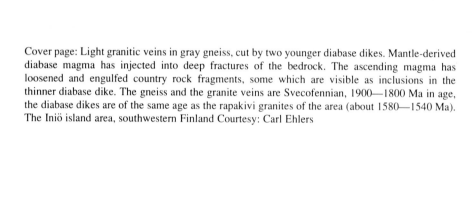

Cover page: Light granitic veins in gray gneiss, cut by two younger diabase dikes. Mantle-derived diabase magma has injected into deep fractures of the bedrock. The ascending magma has loosened and engulfed country rock fragments, some which are visible as inclusions in the thinner diabase dike. The gneiss and the granite veins are Svecofennian, 1900—1800 Ma in age, the diabase dikes are of the same age as the rapakivi granites of the area (about 1580—1540 Ma). The Iniö island area, southwestern Finland Courtesy: Carl Ehlers

Chapter 2
Paleo-Mesoproterozoic Assemblages of Continents: Paleomagnetic Evidence for Near Equatorial Supercontinents

S. Mertanen and L.J. Pesonen

2.1 Introduction

According to plate tectonic theory, the continents move across the Earth's surface through time. The hypothesis of plate tectonics and formation of supercontinents was basically developed already at 1912 by Alfred Wegener who proposed that all the continents formed previously one large supercontinent which then broke apart, and the pieces of this supercontinent drifted through the ocean floor to their present locations. According to the current plate tectonic model, the surface of the Earth consists of rigid plates where the uppermost layer is composed of oceanic crust, continental crust or a combination of both. The lower part consists of the rigid upper layer of the Earth's mantle. The crust and upper mantle together constitute the lithosphere, which is typically 50–170 km thick. This rigid lithosphere is broken into the plates, and because of their lower density than the underlying asthenosphere, they are in constant motion. The driving force for the plate motion are **convection currents** which move the lithospheric plates above the hot astenosphere. Convection currents rise and spread below divergent plate boundaries and converge and descend along the convergent plate boundaries. At converging plate boundaries the rigid plates either pass gradually downwards into the astenosphere or when two rigid plates collide, they form mountain belts, so called orogens.

S. Mertanen (✉)
Geological Survey of Finland, South Finland Unit, FI-02151 Espoo, Finland
e-mail: satu.mertanen@gtk.fi

L.J. Pesonen
Division of Geophysics and Astronomy, Department of Physics, University of Helsinki, FI-00014 Helsinki, Finland
e-mail: lauri.pesonen@helsinki.fi

Supercontinent is a large landmass formed by the convergence of multiple continents so that all or nearly all of the Earth's continental blocks are assembled together. Their role is essential in our understanding of the geological evolution of the Earth. Rogers and Santosh (2003) presented that continental cratons began to assemble already by 3 Ga or possibly earlier. They proposed that during Archean time there existed two supercontinents, *Ur* (ca. 3 Ga, comprising Antarctica, Australia, India, Madagascar, Zimbabve and Kaapvaal cratons) and *Arctica* (ca. 2.5 Ga including the cratons of the Canadian shield and the Aldan and Anabar cratons of the Siberian shield) which were followed by a slightly younger supercontinent, *Atlantica* (including Amazonia, Congo-São Francisco, Rio de la Plata and West Africa cratons), that was formed during the early Paleoproterozoic at ca. 2.0 Ga. According to Rogers and Santosh (2003) these three ancient continental assemblies may have remained as coherent units during most of the Earth's history until their breakup of the youngest supercontinent Pangea at about 180 Ma ago. The existence of these supercontinents will be explored in this paper. Based on present geological knowledge, during the Paleo-Mesoproterozoic era there have been at least *two* times when all of the continental cratons were fused into one large supercontinent, and several other times when more than one craton were accreted to form smaller blocks (Rogers and Santosh 2003, 2004; Bleeker 2003). A larger continental assembly, *Nena* (including cratons of North America, Greenland, Baltica, Siberia and North China) existed at ca. 2–1.8 Ga and it formed part of the first real supercontinent *Nuna* (Hoffman 1997) which is also called as *Columbia* or *Hudsonland* (e.g. Meert 2002; Rogers and Santosh 2003; Zhao et al. 2004; Pesonen et al. 2003, 2011), where nearly all of the Earth's continental blocks were assembled into one large landmass at ca. 1.9–1.8 Ga (see Reddy and Evans 2009). The Nuna supercontinent started to fragment between 1.6 and 1.2 Ga and finally broke up at about 1.2 Ga. The next large supercontinent was *Rodinia* which existed from ca. 1.1 Ga to 800–700 Ma and comprised most of the Earth's continents (McMenamin and McMenamin 1990; Hoffman 1991). The breakup of Rodinia was followed by formation of the enormous Gondwana supercontinent at around 550 Ma including the present southern hemispheric continents Africa, most of South America and Australia, East Antarctica, India, Arabia, and some smaller cratonic blocks (Fig. 2.1). The present northern continents; Laurentia and Baltica collided at about 420–430 Ma, and formed the Laurussia continent (Fig. 2.1). The youngest and last world-wide supercontinent was *Pangea* that started to form at about 320 Ma when Gondwana, Laurussia, and other intervening terranes were merged together. Figure 2.1 shows the reconstruction at ca. 250 Ma when Pangea started to break apart. This process still continues today, seen for instance as spreading of the Atlantic ocean due to separation of Laurussia continents in the north (separation of North America and Europe) and the Gondwana continents in the south (separation of South America from Africa).

The oldest Precambrian continental assemblies presented above are in many cases based solely on geological evidences. However, geologically based reconstructions can be tested by the paleomagnetic method. In this paper, we will

Fig. 2.1 Pangea supercontinent at ca. 250 Ma (modified from Torsvik et al. 2009, 2010b)

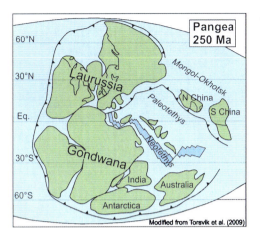

use the paleomagnetic method to reconstruct the Precambrian supercontinents during the time period 2.45–1.05 Ga. In the following, the basic principles of the method are shortly outlined.

2.2 Paleomagnetic Method

Paleomagnetism provides a method to constrain the configurations of cratons that have changed their relative positions through time. The method is based on the assumption that the Earth's magnetic field has always been dipolar and that the magnetic poles coincide as a long term approximation with the rotation axis of the Earth. Consequently, the magnetic field direction shows systematic variation between latitudes so that e.g. vertical geomagnetic field directions occur at the poles and horizontal directions at the equator. Deviations from these existing Earth's magnetic field directions shows that the continents have moved. By measuring the rock's remanent magnetization direction acquired when a magmatic rock cooled below the blocking temperatures of its magnetic minerals, or when magnetic particles were aligned according to the geomagnetic field direction of a sedimentary rock, it is possible to restore the craton back to its original latitude and orientation. The method has two limitations. First, because of the longitudinal symmetry of the Earth's magnetic field, only the ancient paleolatitude and paleo-orientation, but not the paleolongitude, can be defined. This gives the freedom to move the craton along latitude (Fig. 2.2). Second, due to the rapid (in geological time scheme) reversals of the Earth's magnetic field from normal to reversed polarity or vice versa, either polarity of the same magnetization direction can be used. This results to the possibility to place the continent to an antipodal hemisphere with inverted orientation (Fig. 2.2). In all cases, information about the continuations of geological structures between continents is vital in locating the cratons relative to each other.

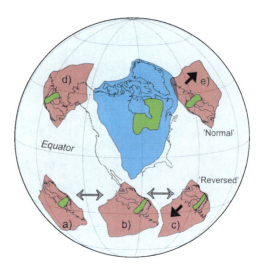

Fig. 2.2 Palaeomagnetic method for making reconstructions used in this paper. Laurentia (*blue*) and Baltica (*red*) are plotted at correct latitude and orientation, based on palaeomagnetic poles. The actual data come from the Superior (Laurentia) and Karelia (Baltica) cratons, marked in green, but for clarity, the whole continents are outlined. Here, Laurentia is kept stationary and Baltica can be moved around it as follows: positions (a), (b) and (c) show that the continent can be moved freely along latitude, but so that the continent retains its orientation. Positions (c) and (e) as well as (a) and (d) demonstrate that the polarity can be chosen between "Normal" and "Reversed" when the continent can be placed upside down on the antipodal hemisphere, depending on the polarity choice. The black arrow shows the antipodal remanence directions. Note that due to spherical projection, the form of the continent varies

2.3 Sources of Data and Cratonic Outlines

In the previous paleomagnetic compilation (Pesonen et al. 2003), the continents were assembled into their Proterozoic positions using the high quality paleomagnetic poles, calculated from the remanent magnetization directions, which were available at that time. Since then, not only have new data been published but also novel, challenging geological models of the continental assemblies during the Proterozoic have been proposed (e.g. Cordani et al. 2009; Johansson 2009; Evans 2009). In this paper, we use the updated (to 2011) paleomagnetic database (Pesonen and Evans 2012), combined with new geological information, to define the positions of the continents during the Paleoproterozoic (2.5–1.5 Ga) and Mesoproterozoic (1.5–0.8 Ga) eras. The data presented here come mainly from the largest continents (Fig. 2.3) which are Laurentia (North America and Greenland), Baltica, Amazonia, Kalahari, Congo, São Francisco, India, Australia, North China and Siberia. The smaller "microcontinents", such as Rio de la Plata, Madagascar or South China are not included due to lack of reliable data from the investigated period 2.45–1.04 Ga (see Li et al. 2008 and references therein). In the following, we use terms such as Laurentia and Baltica for the continents and within

2 Paleo-Mesoproterozoic Assemblages of Continents

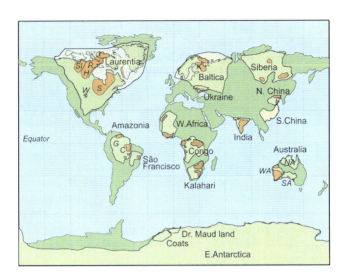

Fig. 2.3 Map showing the continents in their present day geographical positions. Precambrian continental cratons (partly overlain by younger rock sequences) are outlined by yellow shading. The exposed Archean rocks are roughly outlined by orange color. The following continents are used in the reconstructions or discussed in text: Laurentia, Baltica, Siberia, North China, India, Australia, Kalahari, Congo, West Africa, Amazonia and São Francisco. In addition, the Precambrian continents not used in present reconstructions, Ukraine, South China, East Antarctica, Dronning Maud Land and Coats Land are shown. The Archean cratons are marked as follows: for Laurentia Superior (S), Wyoming (W), Slave (Sl), Rae (R), and Hearne (H); for Baltica Karelia (K); for Australia North Australia (NA) (Kimberley and Mc Arthur basins), West Australia (WA) (Yilgarn and Pilbara cratons), and South Australia (SA) (Gawler craton); and for Amazonia Guyana Shield (G) and Central Amazonia (C). Galls projection

each continent those *cratons* (the nuclei of the ancient continents) where the source paleomagnetic data come from are outlined. The Archean to Proterozoic continents consist of individual cratons which may have been drifting, colliding and rifting apart again. Therefore, the consolidation time of the Precambrian continents should be taken into account. For example, most of the poles from Laurentia are derived from rocks within the Superior Province and only a few are derived from other provinces like Slave or Hearne (Fig. 2.3). According to paleomagnetic studies of Symons and Harris (2005), it is possible that the presently assembled Archean terranes of Laurentia did not drift as a coherent continent until at ca. 1,815 Ma to ca.1,775 Ma. Therefore, the data from e.g. Superior craton before 1.77 Ga concerns only that craton. The same is true for Baltica, where Kola and Karelia cratons may have had their own drift histories during Archean-Paleoproterozoic even though they are close to each other within present-day Baltica.

Some cratons, which are now attached with another continent than their inferred original source continent, have been rotated back into their original positions before paleomagnetic reconstruction. For example, the Congo craton is treated together with the São Francisco craton (Fig. 2.3), since geological and paleomagnetic data are consistent that they were united already at least since 2.1 Ga.

2.4 Data Selection

The used paleomagnetic poles come from the updated Precambrian paleomagnetic data compilation that includes the paleopoles from all continents (Pesonen and Evans 2011). The data are graded with the so called Van der Voo (1990) grading scheme (Q-values) that takes into account e.g. statistics of the data, used paleomagnetic methods, isotopic age determinations and tectonism of the studied unit. The highest grade has Q-value 6; we have used data with a minimum value four. In some exceptional cases, however, lower values were accepted. Seven age periods were chosen for reconstructions: 2.45, 1.88, 1.78, 1.63, 1.53, 1.26 and 1.04 Ga. These ages were chosen because paleomagnetic data are available for them from several cratons. In some cases, there are many coeval well-defined paleomagnetic poles from the same craton, and in those cases a mean pole (Fisher 1953) was calculated to be used in the reconstruction. The poles, either individual or mean poles, their ages and other relevant data are listed in Table 2.1.

All original poles are given in Pesonen et al. (2011). The reconstructions are shown in Figs. 2.4, 2.5, 2.6, 2.7, 2.8, 2.9 and 2.10. The main errors with the relative positions of cratons arise from the uncertainty in the pole positions as expressed by the 95% confidence circles of the poles, and from the age difference of poles between different cratons. In some extreme cases when exactly matching data were not available, an age difference of even as high as about 100 Ma was accepted (like e.g. the 2.45 Ga reconstruction where the age of the pole from the Superior craton is ca. 2,470 Ma and that from the Dharwar craton ca. 2,370 Ma, see Pesonen et al. 2012).

2.5 Continental Reconstructions During the Paleo-Mesoproterozoic

2.5.1 Reconstruction at 2.45 Ga

Paleomagnetic data for 2.45 Ga reconstruction (Fig. 2.4) are available from two Nena continental fragments (from Superior craton of Laurentia and Karelia of Baltica) and from two Ur continental fragments (Yilgarn craton of Australia and Dharwar craton of India). At about 2.45 Ga the Nena cratons of Laurentia and Baltica lie near the equator whereas the Ur cratons of Australia and especially India are clearly at high, almost polar (south) latitudes. Although the relative positions of

2 Paleo-Mesoproterozoic Assemblages of Continents

Table 2.1 Mean values of paleopoles used for reconstructions

Continent (Craton)	Age (Ma)	N	Dr (°)	Ir (°)	Plat (°N)	Plon (°E)	A95 (°)	S̄ (°)	Q$_{1-6}$	E-Plat (°N)	E-Plon (°E)	E-Angle (°)
2.45 Ga reconstruction												
Laurentia (Superior)	2,473	1	23.7	43.9	−52.0	239.0	3.3	3.5	6.0	64.0	14.0	96.1
Baltica (Karelia)	2,440	1	312.1	−15.6	9.6	256.2	4.9	7.9	2.0	29.5	317.7	−95.7
Australia (Yilgarn)	2,415	1	248.5	−67.4	−8.0	157.0	8.2	14.9	4.0	0	247.0	82.0
India (Dharwar)	2,367	1	91.4	−83	−17.8	243.4	16.8	24.3	6.0	14.0	313.4	−112.8
1.88 Ga reconstruction												
Laurentia (Superior)	1,880	1	258.6	59.1	28.7	216.0	8.2	14.6	5.0	0	126	61.3
Baltica (Karelia)	1,880	8	341.7	35.1	43.7	232.2	3.5	10.1	2.5	11.5	317.2	−47.3
Amazonia	1,880	3	158.5	−5.4	−68.3	32.4	10.9	14.1	2.3	65.6	147.4	54.2
Australia (WA)	1,850	1	34.7	23.6	45.2	40.0	1.8	11.2	5.0	13.3	275	−143.6
Siberia (Akitkan)	1,878	1	185.4	−1.9	−30.8	98.7	3.5	6.1	5.0	28.6	82.7	164.3
Kalahari	1,875	8	237.9	66.4	−13.6	190.2	10.3	15.4	5.1	34.3	312.7	97.0
1.78 Ga reconstruction												
Laurentia (Churchill)	1,781	1	177.5	56.4	7.0	277.0	8.0	16.4	5.0	16.3	352	−87.3
Baltica (Karelia)	1,788	3	349.1	33.8	43.9	222.4	11.2	10.9	4.0	0	132.4	46.1
Amazonia	1,789	1	358.1	−45.1	−63.3	298.8	11.4	19.4	5.0	69.4	69.8	85.7
Australia (NA)	1,770	3	91.5	38.3	8.5	25.1	18.3	15.3	5.0	12.5	280.1	−101.8
Kalahari	1,770	1	299.4	54.5	−7.0	159.0	7.1	14.8	5.0	41.9	302.5	125.8
India (Bundelkhand)	1,798	1	253.8	0.1	15.4	173.2	7.9	13.7	5.0	21.5	100.7	81.3
North China	1,769	1	37.0	−4.2	−36.0	67.0	3.0	8.1	5.0	53.0	294.5	−97.9
1.63 Ga reconstruction												
Laurentia (Greenland)	1,622	1	201.7	52.0	4.3	256.8	3.0	3.2	–	8.0	174.3	86.8

(continued)

Table 2.1 (continued)

Continent (Craton)	Age (Ma)	N	Dr (°)	Ir (°)	Plat (°N)	Plon (°E)	A95 (°)	S̄ (°)	Q$_{1-6}$	E-Plat (°N)	E-Plon (°E)	E-Angle (°)
Baltica (Karelia)	1,637	3	23.1	5.1	26.3	182.1	12.0	14.3	3.7	34.2	117.1	79.3
Amazonia	1,640	2	323.8	10.1	53.5	213.6	15.9	10.1	3.0	65	168.6	95.6
Australia (NA)	1,641	5	166.4	49.1	−74.1	183.2	8.0	17.7	4.6	4.6	128.2	167
Kalahari	1,649	1	154.7	−71.5	−8.7	202.0	19.3	20.3	4.0	45.3	352.0	135.4
1.53 Ga reconstruction												
Laurentia (Slave)	1,525	3	191.3	24.8	−16.4	263.5	20.7	14.3	4.3	41.1	230.0	138.4
Baltica (Karelia)	1,538	3	15.9	8.8	29.3	189.7	9.4	15.8	3.3	58.8	174.7	154.2
Amazonia	1,535	3	318.7	−32.1	45.8	179.6	15.9	15.6	3.3	65.9	204.6	−133.9
Australia (NA)	1,525	1	185.2	49	−79.0	110.6	8.4	12.0	4.0	1.4	185.6	−169.4
Siberia (Anabar)	1,513	1	205.8	15.3	−19.2	77.8	4.2	18.3	5.0	2.5	164.3	−109.4
North China	1,503	2	84.0	16.8	10.1	202.2	8.8	21.9	5.0	36.0	149.7	105.1
1.26 Ga reconstruction												
Laurentia (incl. Greenland)	1,256	6	267.8	17.6	6.7	191.4	6.2	13.5	4.5	0	101.4	83.3
Baltica (Karelia)	1,257	13	53.1	−31.7	−0.8	158.2	3.2	10.1	4.0	2.5	70.7	90.9
Amazonia	1,200	1	301.0	−55.6	24.6	164.6	5.5	4.0	4.0	41.8	109.6	92.9
Kalahari	1,240	1	358.1	32.6	47.2	22.3	2.9	8.8	5.0	65.7	52.3	−124.5
West Africa	1,250	1	338.3	−36.6	48.7	206.6	1.9	8.7	4.0	48.3	271.6	−64
Congo-São Fr. (Congo)	1,236	1	107.0	3.0	−17.0	112.7	8.0	10.9	3.0	15.6	123.0	−150
1.04 reconstruction (Rodinia)												
Laurentia	1,050	3	274.0	−4.7	−0.1	180.4	14.6	11.3	4.3	22.9	115.4	100.4
Baltica (Karelia)	1,018	1	355.2	−46.8	−2.1	212.2	13.8	11.3	4.0	4.8	127.2	92.5
Amazonia	1,065	1	20.3	−55.3	49.5	89.3	13.2	16.0	4.0	9.7	206.8	144.2
Australia (WA)	1,073	2	321.2	53.9	18.4	87.03	82.0	15.2	5.0	47.7	112	−141.9
Siberia (Aldan)	1,053	4	156.1	45.2	−16.4	220.3	33.7	12.4	4.0	51.5	60.3	−148.2

Kalahari	1,085	1	348.2	11.7	57.0	3.0	7.0	7.9	2.0	55	67	−59.3
India (Dharwar)	1,026	1	24.4	55.6	10.0	211.4	12.4	18.5	5.0	39.6	21.4	167.2
Congo-São Fr. (São Fr.)	1,011	7	93.6	−77.1	−10.3	290.0	13.5	20.3	4.1	40.3	155	−114.2

Continent (Craton) refers to text and Fig. 2.3. Original poles and references are given in Pesonen et al. (2012). N number of entries used for mean calculation. Dr, Ir refer to mean Declination, Inclination of the characteristic remanent magnetization of a central reference location for each craton/continent. Reference locations: Laurentia: 60°N, 275°E, Baltica: 64°N, 28°E, Australia: North Australia NA: 20°S, 135°E, West Australia WA: 27°S 120°E, South Australia SA: 30°S, 135°E, India: 18°N, 78°E, Siberia: 60°N, 105°E, Kalahari: 25°S, 25°E, West Africa: 15°N, 35°E, Amazonia: 0°, 295°E, Congo: 5°S, 23°E, São Francisco (São Fr.): 13°S, 315°E, North China: 40°N, 115°E. Plat, Plon are latitude and longitude of the paleomagnetic pole. A95 is the 95% confidence circle of the pole. Greenland poles have been rotated relative to Laurentia using Euler pole of 66.6°N, 240.5°E, rotation angle −12.2° (Roest and Srivistava 1989). S is the mean Angular Standard Deviation as explained in text. Q_{1-6} is the mean Q of N entries in Appendix 1. E-Plat, E-Plon are the co-ordinates of the single Euler rotation pole. Angle: Euler rotation angle

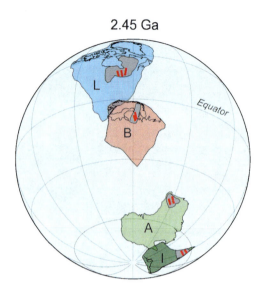

Fig. 2.4 Reconstruction of Archean cratons at 2.45 Ga. Data available from Laurentia (L), Baltica (B), Australia (A) and India (I) (Table 2.1). The Archean cratons Superior (Laurentia), Karelia (Baltica), Yilgarn (West Australia) and Dharwar (India) are shown in grey. Dyke swarms are shown as red sticks and they are: Matachewan dykes (Laurentia), Russian-Karelian dykes (Baltica), Widgiemooltha dykes (Yilgarn) and Dharwar E-W dykes (India). Orthogonal projection

Yilgarn and Dharwar are different to Ur configuration of Rogers and Santosh (2002), the existence of Ur may hold true during the early Paleoproterozoic.

Bleeker (2003) and Bleeker and Ernst (2006) have presented a model of "Superia supercraton" that implies a Superior-Karelia (together with Kola-Hearne-Wyoming blocks) unity at 2.45 Ga, where Karelia is located on the southern margin of the Superior craton. When using a paleomagnetic pole that is not so well-defined from the 2.45 Ga dolerite dykes in Karelia (Mertanen et al. 1999) and the well-defined pole from the 2.45 Ga Matchewan dykes in Superior (Evans and Halls 2010), we end up to a reconstruction shown in Fig. 2.4. This reconstruction is in close accordance with the "Superia" model, when taking into account the error limits of the poles, which allows the cratons to be put closer to each other. It is possible that the previously used pole for Karelia (Mertanen et al. 1999) which clearly separates the two cratons, is actually slightly younger, ca. 2.40 Ga, obtained during cooling of Karelia after heating by the 2.45 Ga magmatism. In the present configuration (Fig. 2.4), the Matachewan and the Karelia dyke swarms become parallel, pointing to a mantle plume centre in the Superia supercraton, as suggested by Bleeker and Ernst (2006).

Dykes of 2.45–2.37 Ga ages exist also in Australia and India as shown in the reconstruction of (Fig. 2.4). The Widgiemooltha swarm (~2.42 Ga; Evans 1968) of the Yilgarn craton (Australia) has a similar trend as the Matachewan-Karelia swarms in this assembly, but its distance to these swarms is more than 90° in latitude (>10,000 km), which does not support a genetic relationship between Australia and Laurentia-Baltica at 2.45 Ga. On the other hand, the E-W trending dykes in the Dharwar craton of India, with an age of 2.37 Ga (Halls et al. 2007; French and Heaman 2010), form a continuation with the Widgiemooltha dyke swarm (Fig. 2.4).

Between 2.40 and 2.22 Ga the Superior, Karelia and Kalahari cratons experienced one to three successive glaciations (Marmo and Ojakangas 1984; Bekker et al. 2001). It is noteworthy that the sequences also contain paleoweathering layers, lying generally on top of the glaciogenic sequences (Marmo et al. 1988). These early Paleoproterozoic supracrustal strata are similar to Neoproterozoic strata that also contain glaciogenic sequences and paleoweathering zones (e.g. Evans 2000). Moreover, in both cases the paleomagnetic data point to low latitudes ($\leq 45°$) during glaciations. Taking Laurentia as an example, it maintained a low latitude position from 2.45 to 2.00 Ga during the time when the glaciations took place (e.g. Schmidt and Williams 1995). If the Superia model of Bleeker and Ernst (2006) is valid, according to which the Karelia and Superior cratons formed a unity during the whole time period from 2.45 to 2.1 Ga, then also Karelia was located at subtropical paleolatitudes of 15–45° at that time (see also Bindeman et al. 2010).

Various models have been presented to explain the fascinating possibility of glaciations near the equator (see Maruyama and Santosh 2008 and references therein). These include the hypothesis of "Snowball Earth" which proposes that the whole Earth was frozen at ca. 2.4–2.2 Ga, possibly resulting from high Earth's orbital obliquity (e.g. Maruyama and Santosh 2008). Eyles (2008) presented that glaciations near the equator could be due to high elevations by tectonic processes. In paleomagnetic point of view one explanation could be remagnetization, or the enhanced non-dipole nature of the geomagnetic field (see Pesonen et al. 2003 and references therein).

In addition to the development of nearly coeval glaciogenic sequences and paleoweathering zones during the Paleoproterozoic, Laurentia, Baltica, Australia and India experienced another rifting episode at ca. 2.20–2.10 Ga as evidenced by widespread mafic dyke activity (e.g. Vuollo and Huhma 2005, Ernst and Bleeker 2010, French and Heaman 2010) and passive margin sedimentations (Bekker et al. 2001). It is possible that this rifting finally led to breakup of Laurentia-Baltica and possibly also Australia-India. Based on geological evidence, Lahtinen et al. (2005) proposed that the breakup of Laurentia-Baltica took place as late as 2.05 Ga ago.

2.5.2 Reconstruction at 1.88 Ga

The period 1.90–1.80 Ga is well known in global geology as widespread orogenic activity. Large amounts of juvenile crust were added to the continental margins, and black shales, banded iron formations (BIFs), evaporites as well as shallow marine phosphates were deposited in warm climatic conditions (Condie et al. 2001).

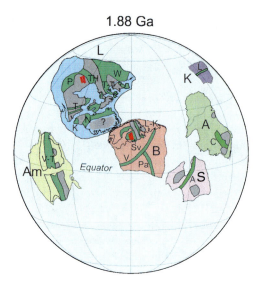

Fig. 2.5 Reconstruction of cratons and orogenic belts (*green*) at 1.88 Ga. The Archean cratons are shown in gray. Data available from Laurentia (L), Baltica (B), Amazonia (Am), Siberia (S), Australia (A) and Kalahari (K). The ca. 1.90–1.80 Ga orogenic belts are shown in dark green and they are: in Laurentia Nagssugtoqidian (N), Ketilidian (K), Torngat (T), Trans-Hudson (TH), Penokean (P), Woopmay (W), and Taltson-Thelon (T-T); in Baltica Lapland-Kola (L-K) and Svecofennian (Sv); in Amazonia Ventuari-Tapajos (V-T); in Siberia Akitkan (A); in Australia Capricorn (C); and in Kalahari Limpopo (L)

These deposits support the existence of a supercontinent at low to moderate latitudes at ca. 1.88 Ga (see Pesonen et al. 2003 and references therein).

Reliable poles of the age of about 1.88 Ga are available from three Nena cratons (Baltica, Laurentia and Siberia), from two Ur cratons (Australia and Kalahari), and from one Atlantica craton (Amazonia). The reconstruction is shown in Fig. 2.5. All continents have moderate to low latitudinal positions with the exception of Kalahari which seems not to belong to this "Early Nuna" landmass. The proposed Ur continent is thus not supported due to significant separation between Australia and Kalahari. Likewise, based on dissimilarity of paleomagnetic data on ca. 2.0 Ga units from Amazonia and Congo-São Francisco, D'Agrella-Filho et al. (2011) argue that neither Atlantica supercontinent ever existed.

The assembly of Laurentia and Baltica cratons at 1.88 Ga, together with Australia and Siberia, marks the onset of development of the supercontinent Nuna although the final amalgamation may have occurred as late as ~1.53 Ga (see later). The position of Baltica against Laurentia is rather well established as the paleomagnetic data from Baltica are available from several Svecofennian 1.88–1.87 Ga gabbros. However, the age of magnetization is somewhat uncertain because the paleomagnetic data from Baltica come from slowly cooled plutons, in which the magnetization may block a few years later compared to the crystallization age. The uncertainty of the position of Laurentia is due to complexity related to the paleomagnetic pole of the 1.88 Ga Molson dykes (Halls and Heaman 2000). In the 1.88 Ga reconstruction (Fig. 2.5) the relative position of Laurentia (Superior craton) and Baltica (Karelia craton) departs significantly from that at 2.45 Ga (Fig. 2.4), consistent with separation of Laurentia from Baltica at about 2.15 Ga. The data further suggest that a considerable latitudinal drift and rotation from 2.45 to 1.88 Ga took place for Laurentia but much less for Baltica.

The model in Fig. 2.5 provides the following scenario to explain the ca. 1.90–1.80 Ga orogenic belts in Laurentia and Baltica. After rifting at 2.1 Ga the cratons of both continents drifted independently until ~1.93 Ga. Subsequently, the Laurentia cratons collided with Baltica cratons causing the Nagssugtoqidian and Torngat orogens in Laurentia and the Lapland–Kola orogen in Baltica. It is likely that collision between Laurentia and Baltica caused intra-cratonic orogenic belts (e.g. between Superior and Slave in Laurentia and between Kola and Karelia in Baltica). Simultaneously, in Baltica, accretion and collision of several microcontinents to the Karelia continental margin may also have taken place (Lahtinen et al. 2005). The complexity of these collisions is manifested by the anastomosing network of 1.93–1.88 Ga orogenic belts separating the Archean cratons in Baltica and Laurentia (Fig. 2.5). The same seems to have happened also in other continents, like in Australia, Kalahari, and Amazonia.

In addition to the above mentioned collisions within Baltica and Laurentia, a collision of Laurentia-Baltica with a "third continent" may be responsible for at least some of the 1.93–1.88 Ga orogenic belts (Pesonen et al. 2003). Candidates for this "third continent" include Amazonia, North China, Australia, Siberia and Kalahari. Each of these have 1.93–1.88 Ga orogenic belts: the Trans China orogen in China, the Capricorn orogen in Australia, the Ventuari-Tapajos orogen in Amazonia, the Akitkan orogen in Siberia and the Limpopo belt in Kalahari (Geraldes et al. 2001; Wilde et al. 2002). Although the data from Amazonia are not of the best quality (quality factor Q only 2–3), Amazonia was probably not yet part of the 1.88 Ga Laurentia-Baltica assembly (Fig. 2.5).

2.5.3 Reconstruction at 1.78 Ga

Reliable paleomagnetic data (Table 2.1) at 1.78 Ga come from Laurentia, Baltica, North China, Amazonia, Australia, India and Kalahari (Fig. 2.6). These continents remained at low to intermediate latitudes during 1.88–1.77 Ga. The 1.78 Ga configuration of Baltica and Laurentia differs from that at 1.88 Ga. This difference is mostly due to rotation of Laurentia relative to more stationary Baltica. The considerable rotation of Laurentia may reflect poor paleomagnetic data, but it is also possible that there was a long-lasting accretion to the western margin of the closely situated Laurentia-Baltica cratons. This may have included relative rotations along transform faults between the accreting blocks (Nironen 1997) until their final amalgamation at ca. 1.83 Ga. Support for the continuation of Laurentia-Baltica from 1.83 to 1.78 Ga comes from the observations that the

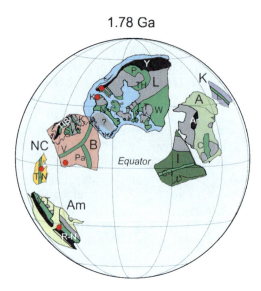

Fig. 2.6 The reconstruction of continents at 1.78 Ga. Data available from Laurentia (L), Baltica (B), North China (NC), Amazonia (Am), India (I), Australia (A) and Kalahari (K). The Archean cratons are shown as grey shading (see Figs. 2.3 and 2.4). The 1.90–1.80 Ga orogenic belts (*green*) in Laurentia, Baltica, Australia, Amazonia and Kalahari are the same as in Fig. 2.5. In North China: Trans-North China orogen (T-N); in India: Central Indian tectonic zone (C-I). The ca. 1.8–1.5 Ga orogenic belts (*black*) are in Laurentia Yavapai (Y); in Baltica Transscandinavian Igneous Belt (TIB); in Amazonia Rio Negro-Juruena (R-N); in Australia Arunta (Ar). The 1.78–1.70 Ga rapakivi granites are shown as *red circles*

geologically similar Trans Scandinavian Igneous (TIB) belt in Baltica and the Yavapai/Ketilidian belts of Laurentia (e.g. Karlström et al. 2001; Åhäll and Larson 2000) become laterally contiguous when reconstructed according to paleomagnetic data of the age of 1.83, 1.78 Ga and 1.25 Ga (see Buchan et al. 2000; Pesonen et al. 2003; Pisarevsky and Bylund 2010).

The configuration of Laurentia, Baltica, North China and Amazonia in the "Early Nuna" configuration at 1.78 Ga is similar with that of Bispo-Santos et al. (2008) where the North China craton is placed between Amazonia and Baltica. This location of North China probably lasted only for a short time period. If the Trans-North China orogen was formed at 1,850 Ma, possibly representing the same orogenic event as the orogens in Baltica and Amazonia, it probably drifted apart from Amazonia-Baltica after 1.78 Ga, as already suggested by Bispo-Santos et al. (2008).

The paleomagnetic data from Amazonia at 1.78 Ga shows that it was located in the southern hemisphere (Fig. 2.6). The 2.0–1.8 Ga Ventuari-Tapajos and 1.8–1.45 Ga Rio Negro-Juruena orogenic belts of Amazonia are coeval with the 1.9–1.8 Ga Svecofennian orogenic belt and ca. 1.8–1.7 Ga TIB and 1.7–1.6 Ga Kongsbergian-Gothian belts of Baltica, and with the corresponding 1.8–1.7 Ga Yavapai and 1.7–1.6 Ga Mazatzal and Labradorian belts in Laurentia (Zhao et al. 2004). Accordingly, Amazonia, North China, Baltica and Laurentia may have formed a united

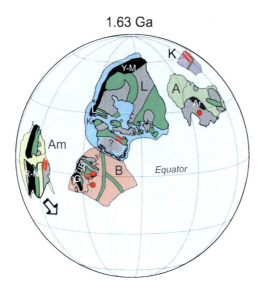

Fig. 2.7 The reconstruction of continents at 1.63 Ga. Data available from Laurentia (L), Baltica (B), Amazonia (Am), Australia (A) and Kalahari (K). The 1.8–1.5 Ga orogenic belts (*black*) are in Laurentia Yavapai-Mazatzal (Y-M), Labradorian (L), and Ketilidian (K); in Baltica Gothian (G) and Transscandinavian Igneous Belt (TIB); in Amazonia Rio Negro-Juruena (R-N); in Australia Arunta (Ar). For other belts, see Figs. 2.5 and 2.6. The SE pointing arrow shows the possible direction of placing Amazonia below Baltica. The 1.63 Ga rapakivi intrusions and related dykes are shown as *red circles* and sticks, respectively

continent with a joint western active margin. This is supported by geological reasoning about the continuity of Amazonia-Baltica-Laurentia (e.g. Åhäll and Larson 2000; Geraldes et al. 2001 and references therein) which favours the idea that all these coeval belts are accretional and were formed during Cordilleran type subduction and arc-accretion from west onto a convergent margin. However, taking into account the possible existence of North China between Baltica and Amazonia at 1.78 Ga, and the reconstruction at 1.63 Ga (Fig. 2.7) where Amazonia is clearly apart from Baltica, it is possible that Amazonia may have been separated from Laurentia-Baltica until 1.53 Ga (Figs. 2.7 and 2.8). This is discussed in the following chapters.

At 1.78 Ga (Fig. 2.6) Australia is located slightly apart from Laurentia to let it be together with its possible Ur counterparts India and Kalahari. Karlström et al. (2001) stressed that geological data of the 1.80–1.40 Ga belts from Laurentia-Baltica landmass (such as Yavapai – Ketilidian and TIB belts) continue into the 1.8–1.5 Ga Arunta belt of eastern Australia. This is paleomagnetically possible: if we take into account the error of pole of 18.3° (Table 2.1), we can shift Australia upwards (Fig. 2.6), which would bring the assembly of Baltica-Laurentia-Australia close to the one suggested by Karlström et al. (2001).

Several episodes of rapakivi magmatism are known during the Paleo-Mesoproterozoic (e.g. Rämö and Haapala 1995, Vigneresse 2005). The

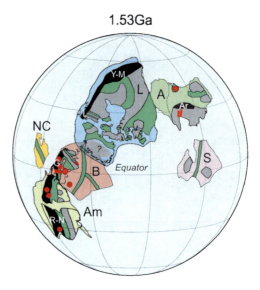

Fig. 2.8 Paleomagnetic reconstruction at 1.53 Ga. Data available from Laurentia (L), Baltica (B), Amazonia (Am), North China (NC), Siberia (S) and Australia (A). The ca. 1.55–1.50 Ga rapakivi intrusions and related dykes are shown as red circles and sticks, respectively. For other explanations, see Figs. 2.5, 2.6 and 2.7

1.77–1.70 Ga and the slightly younger 1.75–1.70 Ga rapakivi-anorthosites are known in Laurentia, Ukraine (part of Baltica), North China and Amazonia (Fig. 2.6). Due to their sparse occurrence at 1.78 Ga they cannot be used to test the paleomagnetic reconstruction, but as will be shown in 1.53 Ga reconstruction (Fig. 2.8), the occurrence of younger rapakivi granites can give some hints for the continuity of the cratons.

2.5.4 Reconstruction at 1.63 Ga

As previously described, the current geological models for Laurentia, Baltica and Amazonia favour the scheme that the post-1.83 Ga orogenic belts were formed along the joint western margin by prolonged subduction and arc-accretions. For the 1.63 Ga reconstruction, paleomagnetic data are available from Laurentia, Baltica, Amazonia, Australia and Kalahari (Fig. 2.7, Table 2.1).

The exact paleomagnetic data places Amazonia onto the same latitude as Baltica and into a situation where the successively younging orogenic belts in Baltica have a westerly trend, in the same sense as in Amazonia. Therefore, in Pesonen et al. (2003) Amazonia was shifted some 25° southeast which was within maximum error of data from both continents. This configuration formed the previously proposed elongated continuation to the 1.78–1.63 Ga Laurentia-Baltica assembly. In that configuration, the successive orogenic belts show a westward younging trend in all three continents (e.g. Åhäll and Larson 2000; Geraldes et al. 2001). However, here the Amazonia craton at 1.63 Ga has been kept in the position defined by the paleomagnetic pole as such (Fig. 2.7), because, as discussed below, there is still the possibility that the final docking of Amazonia took place later than 1.63 Ga.

2.5.5 Reconstruction at 1.53 Ga

Reliable paleomagnetic data at 1.53 Ga come from Laurentia, Baltica, Amazonia, Australia, North China and Siberia (Fig. 2.8). The Laurentia-Baltica assembly at 1.53 Ga differs only slightly from the 1.78 and 1.63 Ga configurations. Therefore, taking into account the uncertainties in the poles, we believe that the previously proposed Laurentia-Baltica unity (where the Kola peninsula is adjacent to present southwestern Greenland) still holds at 1.53 Ga (see also Salminen and Pesonen 2007; Lubnina et al. 2010).

In this reconstruction the successively younging 1.88 Ga to ~1.3 Ga orogenic belts in Laurentia, Baltica and Amazonia are now continued as described in the context of 1.78 Ga and 1.63 Ga reconstructions. However, because in the 1.63 Ga reconstruction (Fig. 2.7) Baltica and Amazonia were still separated at 1.63 Ga, when using the most strict paleomagnetic data, it is possible that the final amalgamation between Baltica and Amazonia took place as late as between 1.63 Ga and 1.53 Ga. Likewise, comparison of reconstructions at 1.78 Ga, 1.63 Ga and 1.53 Ga (Figs. 2.6, 2.7, and 2.8) reveals that North China was still moving with respect to Baltica and Amazonia during 1.78–1.53 Ga. Consequently, by using paleomagnetic data alone, we suggest that the formation of Nuna supercontinent was still going on at ca. 1.53 Ga.

One of the major peaks of rapakivi-anorthosite pulses took place during 1.58–1.53 Ga (Rämö and Haapala 1995; Vigneresse 2005). The coeval occurrences of bimodal rapakivi granites and anorthosites associated with mafic dyke swarms in Baltica and Amazonia during 1.58–1.53 Ga further supports the close connection between these continents in the Nuna configuration.

The position of Australia in (Fig. 2.8), on the present western coast of Laurentia is consistent with the previous reconstructions, thus suggesting that Australia was also part of the Nuna supercontinent. The occurrence of ca. 1.60–1.50 Ga rapakivi intrusions in Australia further supports the idea that Australia was in close connection with Laurentia-Baltica and Amazonia at 1.53 Ga.

Paleomagnetic data could allow Siberia to be in contact with northern Laurentia at 1.53 Ga (Fig. 2.8). However, Pisarevsky and Natapov (2003) noted that almost all the Meso-Neoproterozoic margins in Siberia are oceanic margins and therefore a close connection between Siberia and Laurentia is not supported by their relative tectonic settings. Also in recent Laurentia-Siberia reconstructions (Wingate et al. 2009; Lubnina et al. 2010) the two continents have been left separate although they would become parts of the Nuna supercontinent at ~1.47 Ga. Possibly there was a third continent between Laurentia and Siberia at ~1.53 Ga (see Wingate et al. 2009; Lubnina et al. 2010). Laurentia and Baltica probably remained at shallow latitudes from 1.50 to 1.25 Ga (Buchan et al. 2000). Preliminary comparisons of paleomagnetic poles from the ca. 1.7–1.4 Ga red beds of the Sibley Peninsula (Laurentia) and Satakunta and Ulvö sandstones (Baltica) (e.g. Pesonen and Neuvonen 1981; Klein et al. 2010), support the 1.53 Ga reconstructions within the uncertainties involved. Buchan et al. (2000) implied that the paleomagnetic data from the ca. 1.3 Ga Nairn anorthosite of Laurentia suggest that it remained at low latitudes during ca. 1.40–1.30 Ga, also consistent with a low latitude position of Laurentia at that time.

In some studies (e.g. Rogers and Santosh 2002, 2004; Zhao et al. 2004), the breakup of Nuna supercontinent is regarded to have started by continental rifting already at ca. 1.6 Ga, the timing corresponding with the widespread anorogenic magmatism in most of its constituent continents. This rifting is considered to have continued until the final breakup at about 1.3–1.2 Ga, marked by the emplacement of ca. 1.26 Ga dyke swarms and associated basaltic extrusions in Laurentia, Baltica, Australia and Amazonia (e.g. Zhao et al. 2004). However, we suggest that the separation of Laurentia and Baltica probably occurred much later (even as late as ~1.12 Ga) when a number of rift basins, graben formation and dyke intrusions occurred globally (see below).

2.5.6 Reconstruction at 1.26 Ga

Figure 2.9 shows the assembly of Laurentia, Amazonia, Baltica, West Africa, Kalahari and Congo/São Francisco at ca. 1.26 Ga. These continents are all located at low to intermediate latitudes. The configuration of Laurentia-Baltica is similar to the previous reconstructions during 1.78–1.53 Ga. The Kalahari, Congo-São Francisco and West Africa cratons form a unity slightly west from Amazonia-Baltica-Laurentia. Although the relative position of Baltica-Laurentia at this time is roughly the same as during 1.78–1.53 Ga, the whole assembly has been rotated 80° anticlockwise and drifted southwards.

The 1.26 Ga assembly of Baltica-Laurentia is supported by geological data. For example, as shown previously, the 1.71–1.55 Ga Labradorian-Gothian belts will be aligned in this configuration. As suggested by Söderlund et al. (2006), the ages of the 1.28–1.23 Ga dolerite sill complexes and dike swarms in Labrador, in SW Greenland and in central Scandinavia (Central Scandinavian Dolerite Group, CSDG) are best explained by long-lived subduction along a continuous Laurentia-Baltica margin (see Fig. 2.9). Consequently, the rifting model with separation of Laurentia and Baltica at ca. 1.26 Ga, as presented previously in Pesonen et al. (2003) is not valid any more. It is worthwhile to note that the 1.26 Ga dyke activity is a global one and is well documented in several other continents (see Ernst et al. 1996). Unfortunately, reliable paleomagnetic data from 1.26 Ga dykes are only available from Laurentia and Baltica.

Fig. 2.9 Reconstruction of continents at 1.26 Ga. Data available from Laurentia (L), Baltica (B), Amazonia (Am), West Africa (WA), Congo-São Francisco (C-Sf) and Kalahari (K). The ca. 1.26 Ga dyke swarms in Laurentia and Baltica are shown as red sticks. Kimberlite occurrences of about this age are shown as yellow diamonds. For explanation, see Figs. 2.5, 2.6, 2.7 and 2.8

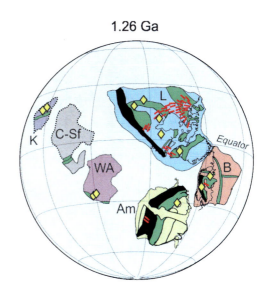

In (Fig. 2.9), the ca. 1.25–1.20 Ga kimberlite pipes are plotted on the 1.26 Ga reconstruction. The kimberlite pipes seem to show a continuous belt crossing the whole Laurentia up to Baltica, making then a ~90° swing and continuing from Baltica to Amazonia. However, the coeval kimberlites in Kalahari and West Africa seem to form clusters rather than a belt. We interpret the kimberlite belt to support the proximity of Laurentia, Baltica and Amazonia although the underlying geological explanation for it remains to be solved (Pesonen et al. 2005; Torsvik et al. 2010a and references therein).

2.5.7 Reconstructions 1.04 Ga: Amalgamation of Rodinia

Baltica and Laurentia probably still formed a unity at 1.25 Ga, but after that, possibly as late as after 1.1 Ga, Baltica was separated from Laurentia and started its journey further south (Fig. 2.10). The southerly drift of Baltica between 1.25 and 1.05 Ga is associated with a ca. 80° clockwise rotation and ca. 15° southward movement. This rotation, suggested already by Poorter (1975), is supported by coeval paleomagnetic data from dolerite dykes in northern Baltica and from the Sveconorwegian orogen of southwestern Baltica (Table 2.1). In this Rodinia model (Fig. 2.10), the Sveconorwegian belt appears continuous with the Grenvillian belt of Laurentia.

After the course of drift and rotations of the detached continents during about 1.10–1.04 Ga, almost all of the continents were amalgamated at ~1.04 Ga to form the Rodinia supercontinent (Fig. 2.10). Unlike most Rodinia models (e.g. Hoffman 1991; Li et al. 2008; Johansson 2009), the new paleomagnetic data of Amazonia

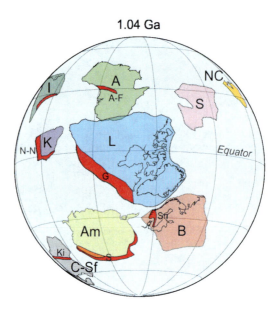

Fig. 2.10 Reconstruction of continents at 1.04 Ga showing the Rodinia configuration. Data available from Laurentia (L), Baltica (B), Amazonia (Am), Congo/SãoFrancisco (C-Sf), Kalahari (K), India (I), Australia (A), Siberia (S) and North China (NC). The Grenvillian age orogenic belts are shown in red and they are: in Laurentia Grenvillian (G), in Baltica Sveconorwegian (Sn), in Amazonia Sunsas (S), in Congo-São Francisco Kibaran (Ki), and in Kalahari Natal-Namagua (N-N). The orange belt in Amazonia marks the possible first collisional location after which the continent was rotated, the red belt was formed in a subsequent collision. For explanation, see text

places the Grenvillian Sunsas-Aguapei belt to be oceanward and not inward (see also Evans 2009). One possible scenario to explain this position is that the Grenvillian collisions occurred episodically, including rotations and strike slip movements (e.g. Fitzsimons 2000; Tohver et al. 2002; Pesonen et al. 2003; Elming et al. 2009). We suggest that during the first collisional episode between 1.26 and 1.1 Ga Amazonia collided with Laurentia on its southwestern border. This collision produced a piece of the inward (against Laurentia's SW coast) pointing Sunsas orogenic belt. Subsequently, Amazonia must have been rotated ~140° anticlockwise swinging the older part of the Sunsas belt to an oceanward position (Fig. 2.10). The second collision by Amazonia, now with Baltica took place at ~1.05 Ga (Fig. 2.10) producing the younger part of the Sunsas-Aguapei belt. The two-phase collisional scenario of Amazonia could explain the oceanward position relative to Laurentia, provided that the Sunsas-Aguapei belt has two segments of variable ages. The same observation may also concern the Namagua-Natal belt in Kalahari, which is also oceanward (Fig. 2.10).

Australia and India are located to the southwest of the present western coast of Laurentia (Fig. 2.10), the space between them occupied by East Antarctica, which formed part of the Gondwana continent. The position and orientation of Siberia

(Fig. 2.10) is somewhat different from that at 1.53 Ga (Fig. 2.8) indicating that Siberia may have been separated from Laurentia during ca. 1.50–1.10 Ga.

The 1.04 Ga time marks the final assembly of Rodinia with possible minor adjustments taking place during 1.04–1.0 Ga. This scenario predicts that late Grenvillian events should have occurred in NW Baltica, in Barentia (Svalbard) and in eastern coast of Greenland due to the collision of Baltica with NE Laurentia (Fig. 2.10). Different scenarios to describe the continent-continent collisions and the formation of Rodinia are presented by Li et al. (2008), Pisarevsky et al. (2003) and Evans (2009).

2.6 Conclusions

1. In this paper we present reconstructions of continents during the Paleo-Mesoproterozoic eras as based on updated global paleomagnetic data. The new data suggest that continents were located at low to intermediate latitudes for much of the period from 2.45 to 1.04 Ga. Sedimentological latitudinal indicators are generally consistent with the proposed latitudinal positions of continents with the exception of the Early Proterozoic period where low-latitude continental glaciations have been noted.
2. The data indicate that two large supercontinents (Nuna and Rodinia) existed during the Paleo-Mesoproterozoic. The configurations of Nuna and Rodinia depart from each other and also from the Pangea assembly. The tectonic styles of their amalgamations are also different reflecting changes in size and thickness of the cratonic blocks, and in the thermal conditions of the mantle with time.
3. The present paleomagnetic data implies that Nuna supercontinent was possibly assembled as late as ~1.53 Ga ago. The configuration of Nuna is only tentatively known but comprises Laurentia, Baltica, Amazonia, Australia, Siberia, India and North China. We suggest that the core of the Nuna was formed by elongated huge Laurentia-Baltica-Amazonia landmass. Australia was probably part of Nuna and in juxtaposition with the present western margin of Laurentia. A characteristic feature of Nuna is a long-lasting accretion tectonism with new juvenile material added to its margin during 1.88–1.4 Ga. These accretions resulted in progressively younging, oceanward stepping orogenic belts in Laurentia, Baltica and Amazonia. The central parts of Nuna, such as Amazonia and Baltica, experienced extensional rapakivi-anorthosite magmatism at ca. 1.65–1.3 Ga. The corresponding activity in Laurentia occurred slightly later. Global rifting at 1.25 Ga, manifested by mafic dyke swarms, kimberlite belts, sedimentary basins, and graben formations took place in most continents of Nuna.
4. The Rodinia supercontinent was fully amalgamated at ca. 1.04 Ga. Rodinia comprises most of the continents and is characterized by episodical Grenvillian continent-continent collisions in a relatively short time span.

Acknowledgement The new Precambrian paleomagnetic database used in this work would not be possible without the years lasting co-operation with David Evans (Yale University).

References

Åhäll K-I, Larson SÅ (2000) Growth-related 1.85–1.55 Ga magmatism in the Baltic Shield. Review addressing the tectonic characteristics of Svecofennian, TIB 1-related and Gothian events. Geol Fören Stockholm Förh 122:193–206

Bekker A, Kaufman AJ, Karhu JA, Beukes NJ, Swart QD, Coetzee L, Eriksson KA (2001) Chemostratigraphy of the Paleoproterozoic Duitschland Formation, South Africa: implications for coupled climate change and carbon cycling. Am J Sci 301:261–285

Bindeman IN, Schmitt AK, Evans DAD (2010) Limits of hydrosphere-lithosphere interaction: origin of the lowest known $\delta^{18}O$ silicate rock on Earth in the Paleoproterozoic Karelian rift. Geology 38:631–634

Bispo-Santos F, D'Agrella-Filho MS, Pacca IIG, Janikian L, Trindade RIF, Elming S-Å, Silva MAS, Pinho REC (2008) Columbia revisited: paleomagnetic results from the 1790 Ma Colider volcanic (SW Amazonian Craton, Brazil). Precambrian Res 164:40–49

Bleeker W (2003) The late Archean record: a puzzle in ca. 35 pieces. Lithos 71:99–134

Bleeker W, Ernst R (2006) Short-lived mantle generated magmatic events and their dyke swarms: the key unlocking Earth's paleogeographic record back to 2.6 Ga. In: Hanski E, Mertanen S, Rämö T, Vuollo J (eds) Dyke swarms - time markers of crustal evolution. Balkema Publishers, Rotterdam, pp 1–24

Buchan KL, Mertanen S, Park RG, Pesonen LJ, Elming S-Å, Abrahamsen N, Bylund G (2000) Comparing the drift of Laurentica and Baltica in the Proterozoic: the importance of key palaeomagnetic poles. Tectonophysics 319:167–198

Condie KC, Des Marais DJ, Abbot D (2001) Precambrian superplumes and supercontinents: a record in black shales, carbon isotopes, and paleoclimates? Precambrian Res 106:239–260

Cordani UG, Teixeira W, DÁgrella-Filho MS, Trindade RI (2009) The position of the Amazon Craton in supercontinents. Gondwana Res 115:396–407

D'Agrella-Filho MS, Trindade RIF, Tohver E, Janikian L, Teixeira W, Hall C (2011) Paleomagnetism and 40Ar/39Ar geochronology of the highgrade metamorphic rocks of the Jequié block, São Francisco craton: Atlantica, Ur and beyond. Precambrian Res 185:183–201

Elming S-Å, D'Agrella-Filho MS, Page LM, Tohver E, Trindade RIF, Pacca IIG, Geraldes MC, Teixeira W (2009) A palaeomagnetic and 40Ar/39Ar study of Late Precambrian sills in the SW part of the Amazonian craton: Amazonia in the Rodinia reconstruction. Geophys J Int 178:106–122

Ernst R, Bleeker W (2010) Large Igneous provinces (LIPs), giant dyke swarms, and mantle plumes: significance for breakup events within Canada and adjacent regions from 2.5 Ga to the Present. Can J Earth Sci 47:695–739

Ernst RE, Buchan KL, West TD, Palmer HC (1996) Diabase (dolerite) dyke swarms of the world: first edition. Geological Survey of Canada Open File 3241, Scale 1:35,000,000 map and report 104 p

Evans ME (1968) Magnetization of dikes: a study of the paleomagnetism of the Widgiemooltha dike suite, western Australia. J Geophys Res 73:3261–3270

Evans DAD (2000) Stratigraphic, geochronological, and paleomagnetic constraints upon the Neoproterozoic climatic paradox. Am J Sci 300:347–433

Evans DAD (2009) The palaeomagnetically viable, long-lived and all-inclusive Rodinia supercontinent reconstruction. Geol Soc London Spec Publ 327:371–404

Evans DAD, Halls HC (2010) Restoring Proterozoic deformation within the Superior craton. Precambian Res 183:474–489

Eyles N (2008) Glacio-epochs and the supercontinent cycle after 3.0 Ga: tectonic boundary conditions for glaciation. Palaeogeogr Palaeoclimatol Palaeoecol 258:898–129

Fisher RA (1953) Dispersion on a sphere. Proc R Soc Lond A217:295–305

Fitzsimons ICW (2000) Grenville-age basement provinces in east Antarctica: evidence for three separate collisional orogens. Geology 28:879–882

French JE, Heaman LM (2010) Precise U-Pb dating of Paleoproterozoic mafic dyke swarms of the Dharwar craton, India: implications for the existence of the Neoarchean supercraton Sclavia. Precambrian Res 183:416–441

Geraldes MC, Van Schmus WR, Condie KC, Bell S, Texeira W, Babinski M (2001) Proterozoic geologic evolution of the SW part of the Amazonian craton in Mato Grosso state, Brazil. Precambrian Res 111:91–128

Halls HC, Heaman LM (2000) The paleomagnetic significance of new U-Pb age data from the Molson dyke swarm, Cauchon Lake area, Manitoba. Can J Earth Sci 37:957–966

Halls HC, Kumar A, Srinivasan R, Hamilton MA (2007) Paleomagnetism and U-Pb geochronology of easterly trending dykes in the Dharwar craton, India: feldspar clouding, radiating dyke swarms and the position of India at 2.37 Ga. Precambrian Res 155:47–68

Hoffman PF (1991) Did the breakout of Laurentia turn Gondwana inside out? Science 252:1409–1412

Hoffman PF (1997) Tectonic genealogy of north America. In: van der Pluijm BA, Marshak S (eds) Earth structure: an introduction to structural geology and tectonics. McGraw-Hill, New York, pp 459–464

Johansson Å (2009) Baltica, Amazonia and the SAMBA connection – 1,000 million years of neighbourhood during the Proterozoic? Precambrian Res 175:221–234

Karlström KE, Åhäll K-I, Harlan SS, Williams ML, McLelland J, Geissman JW (2001) Long-lived (1.8–1.0 Ga) convergent orogen in southern Laurentia, its extension to Australia and Baltica, and implications for refining Rodinia. Precambrian Res 111:5–30

Klein R, Pesonen LJ, Mertanen S, Kujala H (2010) Paleomagnetuc study of Satakunta sandstone, Finland. In: Heikkinen P, Arhe K, Korja T, Lahtinen R, Pesonen LJ, Rämö T (eds) Lithosphere 2010 – Sixth symposium on the structure, composition and evolution of the lithosphere in Finland. Programme and Extended Abstracts, Helsinki, Finland, 27–28 Oct 2010. Institute of Seismology, University of Helsinki, Report S-55, pp 33–35

Lahtinen R, Korja A, Nironen M (2005) Paleoproterozoic tectonic evolution. In: Lehtinen M, Nurmi P, Rämö OT (eds) Precambrian geology of Finland – key to the evolution of the Fennoscandian shield. Elsevier, Amsterdam, pp 481–532

Li ZX, Bogdanova SV, Collins AS, Davidson A, Waele BD, Ernst RE, Fitzsimons ICW, Fuck RA, Gladkochub DP, Jacobs J, Karlstrom KE, Lu S, Natapov LM, Pease V, Pisarevsky SA, Thrane K, Vernikovsky V (2008) Assembly, configuration, and break-up history of Rodinia: a synthesis. Precambrian Res 160:179–210

Lubnina NV, Mertanen S, Söderlund U, Bogdanova S, Vasilieva TI, Frank-Kamenetsky D (2010) A new key pole for the east European craton at 1451 Ma: palaeomagnetic and geochronological constrains from mafic rocks in the lake Ladoga region (Russian Karelia). Precambrian Res 183:442–462

Marmo J, Ojakangas R (1984) Lower Proterozoic glaciogenic deposits, eastern Finland. Geol Soc Amer Bull 95:1055–1062

Marmo J, Kohonen J, Sarapää O, Äikäs O (1988) Sedimentology and stratigraphy of the lower Proterozoic Sariola and Jatuli Groups in the Koli-Kaltimo Area, eastern Finland. In: Sedimentology of the Precambrian formations in eastern and northern Finland: Proceedings of IGCP 160 Symposium at Oulu, 21–22 Jan 1986. Geological Survey Finland Special Paper 5, pp 11–28

Maruyama S, Santosh M (2008) Models on Snowball Earth and Cambrian explosion: a synopsis. Gondwana Res 14:22–32

McMenamin MAS, McMenamin DL (1990) The emergence of animals: the Cambrian breakthrough. Columbia University Press, New York

Meert JG (2002) Paleomagnetic evidence for a Paleo-Mesoproterozoic supercontinent Columbia. Gondwana Res 5:207–215

Mertanen S, Halls HC, Vuollo JI, Pesonen LJ, Stepanov VS (1999) Paleomagnetism of 2.44 Ga mafic dykes in Russian Karelia, eastern Fennoscandian shield – implications for continental reconstructions. Precambrian Res 98:197–221

Neuvonen KJ, Korsman K, Kouvo O, Paavola J (1981) Paleomagnetism and age relations of the rocks in the main sulphide ore belt in central Finland. Bull Geol Soc Finland 53:109–133

Nironen M (1997) The Svecofennian orogen: a tectonic model. Precambrian Res 86:21–44

Pesonen LJ, Evans DAD (2012) Paleomagnetic data compilation – Precambrian era (in prep.)

Pesonen LJ, Neuvonen KJ (1981) Paleomagnetism of the Baltic shield – implications for Precambrian tectonics. In: Kröner A (ed) Precambrian plate tectonics. Elsevier, Amsterdam, pp 623–648

Pesonen LJ, Elming S-Å, Mertanen S, Pisarevsky S, D'Agrella-Filho MS, Meert JG, Schmidt PW, Abrahamsen N, Bylund G (2003) Palaeomagnetic configuration of continents during the Proterozoic. Tectonophysics 375:289–324

Pesonen LJ, O'Brien H, Piispa E, Mertanen S, Peltonen P (2005) Kimberlites and lamproites in continental reconstructions – implications for diamond prospecting. In: Secher K, Nielsen MN (eds) Extended abstracts of the workshop on Greenland's diamond potential. Geological Survey of Denmark and Greenland, Ministry of the Environment, Copenhagen, pp 7–9

Pesonen LJ, Mertanen S, Veikkolainen T (2012) Paleo-Mesoproterozoic supercontinent - a paleomagnetic view. Geophysica 48 (in print)

Pisarevsky S, Bylund G (2010) Paleomagnetism of 1780–1770 Ma mafic and composite intrusions of Småland (Sweden): implications for the Mesoproterozoic supercontinent. Am J Sci 310:1168–1186

Pisarevsky SA, Natapov LM (2003) Siberia and Rodinia. Tectonophysics 375:221–245

Pisarevsky SA, Wingate MTD, Powell CMcA, Johnson S, Evans DAD (2003) Models of Rodinia assembly and fragmentation. Geol Soc London Spec Publ 206:35–55

Poorter RPE (1975) Palaeomagnetism of Precambrian rocks from southeast Norway and south Sweden. Phys Earth Planet Int 10:74–87

Rämö OT, Haapala I (1995) One hundred years of rapakivi granite. Miner Petrol 52:129–185

Reddy SM, Evans DAD (2009) Palaeoproterozoic supercontinents and global evolution: correlations from core to atmosphere. Geol Soc London Spec Publ 323:1–26

Roest WR, Srivistava SP (1989) Sea-floor spreading in the Labrador Sea: a new reconstruction. Geology 17:1000–1003

Rogers JJW, Santosh M (2002) Configuration of Columbia, a Mesoproterozoic supercontinent. Gondwana Res 5:5–22

Rogers JJW, Santosh M (2003) Supercontinents in Earth history. Gondwana Res 6:357–368

Rogers JJW, Santosh M (2004) Continents and supercontinents. Oxford University Press, Oxford

Salminen J, Pesonen LJ (2007) Paleomagnetic and rock magnetic study of the Mesoproterozoic sill, Valaam island, Russian Karelia. Precambrian Res 159:212–230

Schmidt PW, Williams GE (1995) The Neoproterozoic climatic paradox: equatorial palaeolatitude for Marinoan glaciation near sea level in south Australia. Earth Planet Sci Lett 134:107–124

Söderlund U, Elming S-Å, Ernst RE, Schissel D (2006) The central Scandinavian dolerite group – protracted hotspot activity or back-arc magmatism? Constraints from U-Pb baddeleyite geochronology and Hf isotopic data. Precambrian Res 150:136–152

Symons DTA, Harris MJ (2005) Accretion history of the Trans-Hudson orogen in Manitoba and Saskatchewan from paleomagnetism. Can J Earth Sci 42:723–740

Tohver E, van der Pluijm A, Van der Voo R, Rizzoto G, Scandolara JE (2002) Paleogeography of the Amazon craton at 1.2 Ga: early Grenvillian collision with the llano segment of Laurentia. Earth Planet Sci Lett 199:185–200

Torsvik TH, Burke K, Steinberger B, Webb SJ, Ashwal LD (2010a) Diamonds sampled by plumes from the core-mantle boundary. Nature 466:352–355

Torsvik TH, Rousse S, Labails C, Smethurst MA (2009) A new scheme for the opening of the South Atlantic Ocean and dissection of an Aptian Salt Basin. Geophys J Int 177:1315–1333

Torsvik TH, Rousse S, Smethurst MA (2010b) Reply to comment by D. Aslanian and M. Moulin on 'a new scheme for the opening of the south Atlantic ocean and the dissection of an Aptian salt basin'. Geophys J Int 183:29–34

Van der Voo R (1990) The reliability of paleomagnetic data. Tectonophysics 184:1–9

Vigneresse JL (2005) The specific case of the mid-prototerozoic rapakivi granites and associated suite within the context of the Columbia supercontinent. Precambrian Res 137:1–34

Vuollo J, Huhma H (2005) Paleoproterozoic mafic dikes in NE Finland. In: Lehtinen M, Nurmi PA, Rämö OT (eds) Precambrian geology of Finland – key to the evolution of the Fennoscandian Shield. Elsevier, Amsterdam, pp 193–235

Wilde SA, Zhao G, Sun M (2002) Development of the North China Craton during the late archaean and its final amalgamation at 1.8 Ga; some speculations on its position within a global palaeoproterozoic supercontinent. Gondwana Res 5:85–94

Wingate MTD, Pisarevsky SA, Gladkochub DP, Donskaya TV, Konstantinov KM, Mazukabzov AM, Stanevich AM (2009) Geochronology and paleomagnetism of mafic igneous rocks in the Olenek uplift, northern Siberia: implications for Mesoproterozoic supercontinents and paleogeography. Precambrian Res 170:256–266

Zhao G, Suna M, Wilde SA, Li S (2004) A Paleo-Mesoproterozoic supercontinent: assembly, growth and breakup. Earth Sci Rev 67:91–123

Chapter 3
Seismic Structure of Earth's Crust in Finland

Pekka Heikkinen

3.1 Introduction

The single most important scientific observation of the Earth's crust was done about 100 years ago. When studying the earthquake that occurred in Kupa valley on the 8th of October in 1909, Andrija Mohorovicic found out that the first seismic waves at distances larger than 300 km from the epicenter arrived earlier than expected. His interpretation was that at about the depth of 54 km there was a boundary at which the velocity of the seismic P-waves suddenly increased from 5.4 km/s to the value of 7.7 km/s. This boundary between the crust and the mantle of the Earth is still called the Mohorovicic boundary, or MOHO for short, and still this boundary is defined more or less in the same way, a boundary where the P-wave velocity reaches the value more than 7.6 km/s.

One feature that is unique to our planet is the compositionally differentiated continental crust. It covers about 40% of surface of the Earth and is distinctly different from the crust beneath the oceans. The oceanic crust is young, less than 180 million years, whereas some parts of the continental crust can be more than one order of magnitude older. Fennoscandia is a typical Precambrian shield area where the age of the rocks ranges from 500 million years to over 3,500 million years. The continental crust is also thicker – 20–70 km – than the oceanic crust which has depth varying from 5 to 15 km. The continental crust is thickest beneath the active orogens, like the Himalayas and thinnest beneath rift zones like the East African rift.

The average thickness of the continental crust is about 41 km (Christensen and Mooney 1995). Even the deepest boreholes reach only the surface layers of the crust. For example, the deepest borehole in Finland in Outokumpu reaches only the depth of 2.5 km. Besides surface studies, we need geophysical methods which help

P. Heikkinen (✉)
Institute of Seismology, University of Helsinki, P.O. Box 68, FI-00014 Helsinki, Finland
e-mail: pekka.j.heikkinen@helsinki.fi

us to penetrate deeper into the Earth's crust and get information from the geological structures lying tens of kilometers beneath our feet. Present understanding of the Earth's crust is mostly based on geophysical measurements, especially seismic surveys. In this paper, a short overview is given on what we now know about the crustal structure in Finland, based on the seismic surveys done in our territory. It is obvious that all the seismic studies done over the past decades cannot be covered here and likewise, the results from other fields of solid earth geophysics are not included. In Finland we have excellent gravity- and aeromagnetic maps as well as comprehensive data of electromagnetic studies, and all these fields would deserve a treatise of their own.

3.2 Controlled Source Seismic Surveys

Most of the information we have on the upper part of the Earth is based on the results from the controlled source seismic measurements. In broad terms these studies can be divided into two groups, refraction and reflection surveys. In refraction surveys, the main goal is to determine the distribution of the velocities of the seismic waves from the travel times. Typically, the interval between recording points is about 1–3 km and the source points, usually explosions, are at intervals of few tens of kilometers. The largest distances between sources and receivers can be several hundreds of kilometers. In this way it is possible to see the waves that have propagated in the mantle, just like Adrian Mohorovicic did. Reflection surveys are like large scale echo soundings. When a seismic wave propagating downwards in Earth's crust hits an interface at which the acoustic properties of the crust change rapidly, part of the wave reflects back and can be recorded on the surface. In reflection surveys the recording interval is typically 50 m and the wave length is one order of magnitude smaller than in refraction surveys and the resolution correspondingly one order of magnitude better. Unfortunately, also the costs increase respectively. However, both types of measurements are necessary, as they give a complementary view of the Earth's crust. Velocity distributions obtained from refraction surveys give basic information on the composition of the crust whereas near vertical reflections map the boundaries between different geological units, i.e., the tectonic architecture of the crust.

In Finland the SVEKA81 and BALTIC profiles, shot in 1981 and 1982, were the first proper deep refraction/wide angle reflection profiles (Fig. 3.1). Work continued in Lapland in 1985 with the POLAR profile, in SW Finland in 1991 with SVEKA91 and in southern Finland in 1994 with the FENNIA profile (see Luosto 1997 and references therein). In conjunction with the POLAR profile, a small near vertical reflection test was conducted, but proper reflection surveys started with the BABEL project in 1989 in the Gulf of Bothnia and the Baltic Sea (BABEL Working Group 1990). A major step forward was the deep reflection program FIRE 2001–2003 (Kukkonen et al. 2006). In the FIRE program, over 2,000 km of deep reflection data

3 Seismic Structure of Earth's Crust in Finland

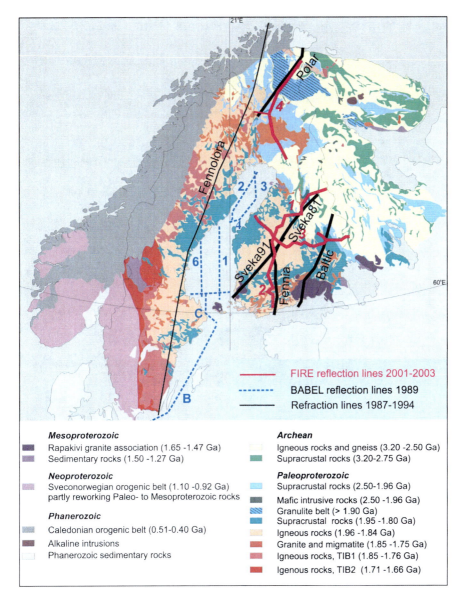

Fig. 3.1 Seismic profiles in Finland 1979–2003. Solid black lines are refraction profiles. *Blue hatched lines* mark near vertical reflection lines of the BABEL project. *Red lines* are reflection lines of the FIRE project (map after Lahtinen et al. 2005)

were acquired, and almost whole Finland was covered. Although the onshore surveys are much more expensive that marine reflection surveys, they have the advantage that the results can be directly compared to known results from other geophysical and geological studies.

3.3 Thickness and Velocity Distribution of the Crust

Figure 3.2 shows velocity models of the refraction profiles in Finland (Luosto 1997). Typically the P-wave velocity at the top of the crust is 5.8–5.9 km/s. The velocity increases with depth to over 7 km/s at the base of the crust, mainly due to the general compositional changes when the crust becomes more basic. Below MOHO in the mantle the P-wave velocity is typically 8.0–8.4 km/s. The most surprising results of the refraction surveys came already from SVEKA 81, when it turned out that the crust in Finland can be as thick as 60 km, a value typical to active orogens. Later surveys have confirmed this observation.

In Finland the crust can be divided into three layers based on the P-wave velocity: upper, middle and lower crust, where the velocities are about 5.9–6.4, 6.6–6.7 and 6.9–7.5 km/s, respectively. A major change occurs along Lake Ladoga – Bothnian Bay zone at the boundary between Archean and Proterozoic crust. In the Archaean, northeast of this zone, the thickness varies between 40 and 48 km, whereas southwest of the boundary, underneath the Proterozoic Svecofennian orogen, the thickness occasionally reaches a value of 60 km. (Fig. 3.3). During

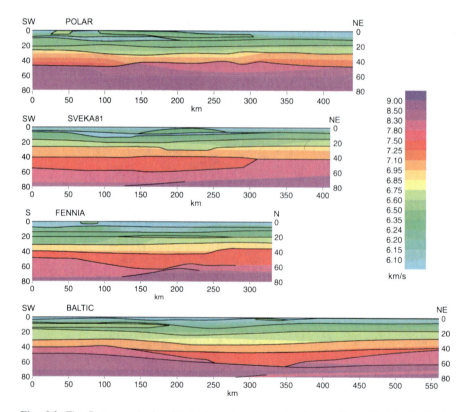

Fig. 3.2 The P-wave velocity distribution from the refraction profiles POLAR, FENNIA, SVEKA81 and BALTIC. *Red colors* correspond to high velocities, *blue* to low velocities. The *color bar* in the *right side* of the figure shows the velocity scale

3 Seismic Structure of Earth's Crust in Finland

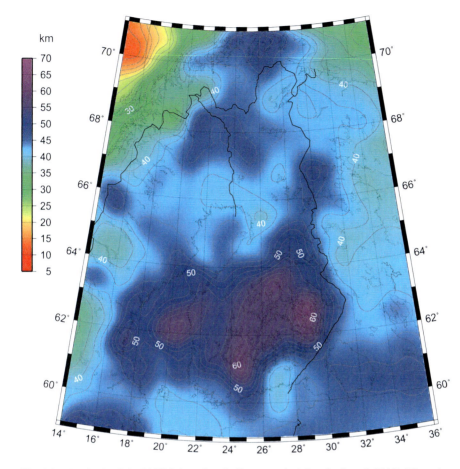

Fig. 3.3 The depth of the MOHO boundary in Fennoscania (after Grad et al. 2009). The color coded depth is on the lower left corner of the figure

the millions of years after the orogenic activity, the thickness of the crust should get back to normal values due to erosion and isostatic balancing. In this collapse of the orogen, the crust also should get its typical layered structure shown by the refraction surveys. An exceptional feature in the Proterozoic southern and central Finland is that the lower crust can be divided into two layers: the lower velocity upper part which is similar to normal lower crust with velocities of about 6.8–6.9 km/s and the high velocity lower part where the P-wave velocity values range from 7.0 to 7.4 km/s. Thickness of the lower layer correlates directly with the thickness of the crust. Where the crust is thinner, like underneath the POLAR profile and in the SW and NE ends of the BALTIC-profile, there is no high velocity layer at the base of the crust. When compared with the laboratory measurements, we can see (Kuusisto et al. 2006), that the high velocity lower part of the crust consists of basic rocks which have densities close to the density of the upper mantle. This would explain the stability of the thick crust.

3.4 Reflectivity and Architecture of the Crust

A more detailed image of the crustal structures can be obtained with the use of seismic reflection method. The reflected waves image sudden changes in the properties of the crust. The changes have to occur over a short distance compared with the seismic wave length used in reflection survey, typically 100–300 m. Thus the gradual changes caused by increase of pressure and temperature do not produce strong reflections. The tectonic processes during which the blocks of the crust move with respect to each other, however, can produce sharp acoustic contrasts which can be seen as seismic near vertical reflections, giving us a chance to map the size and location of the different parts of the crust, in short, the architecture of the crust.

The near vertical sections, both from the BABEL and FIRE profiles, show that the Precambrian crust in Fennoscandia has reflective boundaries from the top of the crust to its base. Many of the reflections indicate structures that we did not expect from previous geological or geophysical studies. The crust seems to be composed of blocks that are distinct with respect of their reflectivity and which have sizes varying from tens of kilometers to a couple of hundred kilometers. Reflectivity in the lower crust is more diffuse than in the upper crust where the reflections are sharp. This change seems to occur in the depth range of 10–15 km and could be caused by changes in the rheological properties of the crust (Kukkonen et al. 2006).

The results of the seismic reflection surveys have been crucial in understanding the geological evolution of the Fennoscandian Shield (Korja and Heikkinen 2005, Korja and Heikkinen 2008, Lahtinen et al. 2005). In the seismic reflection sections from southern and central Finland, we see a variety of structures that are mainly related to the 1.76–1.96 billion years old Svecofennian orogen. When combined with other geophysical and geological studies, these results show that in these orogenic processes the crust has been composed of smaller Proterozoic fragments, microcontinents, island arcs and intervening sedimentary basins. The collision process starts with subduction in which the oceanic crust is thrusting under continental crust, much in the same way as today the Pacific plate is thrusting under the Andes. Traces of these old processes can be seen in the lower crust along the BABEL-3 and BABEL-4 profiles (Fig. 3.4, BABEL Working Group 1990). The structures produced by the subsequent continental collisions where parts of the crust were overthrusted on top of each other, followed be an extensional collapse, can be seen along FIRE-3 as east dipping reflective zones.

Present understanding is (Daly et al. 2006, Patison et al. 2006) that the northern part of the Fennoscandian Shield beneath FIRE-4 is a product of the Lapland-Kola orogeny, during which pieces of the old Achaean crust were welded together. Relatively small amount of new crust was produced when compared with the Svecofennian orogeny. This is also displayed in the velocity models (Fig 3.2), as well as in the near vertical reflection sections (Fig. 3.4). The thickness of the crust beneath the NE end of the profile FIRE-4 and the POLAR profile is 42–48 km and the lower crust has no high velocity lower part. In the near vertical reflection section the MOHO can be seen as a clear base of the reflectivity, as expected. In the

3 Seismic Structure of Earth's Crust in Finland

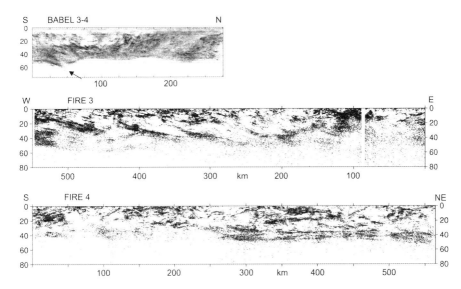

Fig. 3.4 Seismic reflection sections from the profiles BABEL-3-4, FIRE-3 and FIRE-4. *Dark areas* correspond to strongly reflecting parts within the crust, *lighter areas* are weakly reflecting, more homogeneous parts. The remnants of the ancient subduction beneath BABEL 3-4 are marked with an *arrow*

Proterozoic central Finland beneath FIRE-1, which runs along the SVEKA81 profile and where the crustal thickness is about 60 km, such an abrupt change in reflectivity cannot be seen. This is understandable as one would expect that acoustic contrast between the mantle and the high velocity and high density lower crust is lower. Beneath the NE end of the FIRE-4 profile the strong reflectivity is possibly caused by an extension during the collapse of the Lapland-Kola orogen.

The FIRE-1 profile is coincident with the refraction profile SVEKA 81, allowing direct comparison of the velocity distribution and reflectivity. As can be seen in Fig. 3.5, the most distinct common property in these different types of results is the change at the MOHO. In near vertical reflection sections it is seen as the base of reflectivity at about the depth of 60 km. The difference between these images of the crust – velocity distribution vs. reflectivity – does not mean that one method would be more truthful than the other, they simply image different features of the crustal structure.

3.5 Tomographic Studies of the Crust

The accretionary nature of the Proterozoic crust in central and southern Finland, shown clearly in the reflectivity images, should be visible also in the velocity models. This kind of horizontally varying structure is imaged better in 3D velocity models than in standard two-dimensional models like in Fig 3.2. In 1998–1999 the

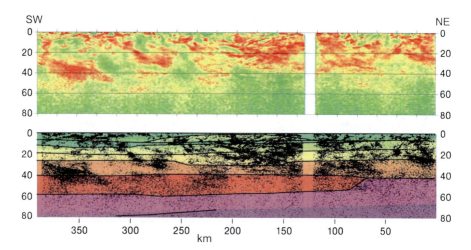

Fig. 3.5 Seismic reflection section along FIRE1 profile (*upper panel*) and coincident velocity model from the SVEKA81 refraction profile (*lower panel*). The color in the reflection section indicates reflectivity: *red areas* are strongly reflective and *green* weakly reflective. In the *lower panel* the black-white reflection section is overlaying the color coded velocity model. The color scale of the *lower panel* is the same as in Fig. 3.2. The vertical scale is in km. The horizontal scale gives the distance from the NE end the FIRE1 profile

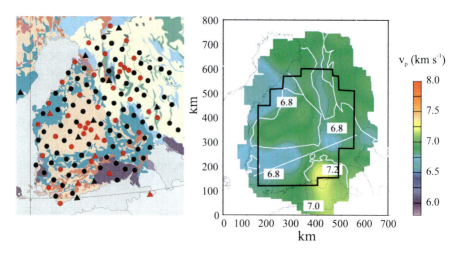

Fig. 3.6 The map on the left shows the location of the seismic stations of the SVEKALAPKO net. Black dots maps short period stations, red broad band stations. On the right is the P-wave velocity distribution at the depth of 20 km (after Hyvönen et al. 2007)

SVEKALAPKO seismic tomography study was conducted in southern and central Finland (Bock and SVEKALAPKO Seismic Tomography Working Group 2001). The main goal of the project was to study the upper mantle using teleseismic earthquakes. The network was recording also large number of local mining explosion and small earthquakes, which were later used to construct 3D velocity models of the crust. Figure 3.6 shows an example of P-wave velocity models obtained by travel time inversion (Hyvönen et al. 2007). The figure displays the horizontal crosscut at the depth of 20 km in southern and central Finland. The velocity distribution shows lateral variation that was expected on the basis of reflection sections. The possible sedimentary basins between microcontinents and island arcs can be distinguished as areas of lower velocity. A distinct high velocity body can be seen in SE Finland beneath the Vyborg rapakivi massive, being most probably a basic intrusion related to the rapakivi granite. One should remember that the horizontal resolution of the tomographic surveys is of the order of the station spacing, in the SVEKALAPKO survey about 50 km.

Thanks to Mohorovicic and generations of geophysicists after him, we have a common theoretical framework – plate tectonics – that helps us to understand the structure and evolution of Earth's crust. The seismic structure of the Earth's crust in Finland shows that the same plate tectonic processes active today in the Earth, were also forming the Fennoscandian shield about 2 billion years ago.

References

BABEL Working Group (1990) Evidence for early proterozoic plate tectonics from seismic reflection profiles in the Baltic Shield. Nature 348:34–38

Bock G, SVEKALAPKO Seismic Tomography Working Group (2001) Seismic probing of archaean and proterozoic lithosphere in Fennoscandia. EOS T Am Geophys Un 82:628–629

Christensen NI, Mooney WD (1995) Seismic velocity structure and composition of the continental crust: a global view. J Geophys Res 100:B9761–B9766

Daly JS, Balagansky VV, Timmerman MJ, Whitehouse MJ (2006) The Lapland-Kola orogen: Palaeoproterozoic collision and accretion of the northern Fennoscandian lithosphere. In: Gee DG, Stephenson RA (eds) European lithosphere dynamics. Geological Society London Memoirs 32:561–578

Grad M, Tiira T, ESC Working Group (2009) The Moho depth map of the European plate. Geophys J Int 176:279–292. doi:10.1111/j.1365-246X.2008.03919.x

Hyvönen T, Tiira T, Korja A, Heikkinen P, Rautioaho E, the SVEKALAPKO Seismic Tomography Working Group (2007) A tomographic crustal velocity model of the central Fennoscandian Shield. Geophys J Int 168:1210–1226

Korja A, Heikkinen P (2005) The Accretionary Svecofennian Orogen – insight from the BABEL profiles. Precambrian Res 136:241–268

Korja A, Heikkinen P (2008) Seismic images of Paleoproterozoic microplate boundaries in the Fennoscandian Shield. In: Condie KC, Pease V (eds) When did plate tectonics begin on planet Earth? Geological Society of America, Special paper 440: 229–248

Kukkonen IT, Heikkinen P, Ekdahl E, Hjelt S-E, Yliniemi J, Jalkanen E, FIRE Working Group (2006) Acquisition and geophysical characteristics of reflection seismic data on FIRE transects, Fennoscandian Shield. In: Kukkonen IT, Lahtinen R (eds) Finnish reflection experiment FIRE 2001–2005 Geological Survey of Finland, Special paper 43:13–43

Kuusisto M, Kukkonen IT, Heikkinen P, Pesonen LJ (2006) Lithological interpretation of crustal composition in the Fennoscandian Shield with seismic velocity data. Tectonophysics 420:283–299

Lahtinen R, Korja A, Nironen M (2005) Palaeoproterozoic tectonic evolution. In: Lehtinen M, Nurmi P, Rämö T (eds) The Precambrian Geology of Finland – key to the evolution of the Fennoscandian Shield. Elsevier, Amsterdam, pp 418–453

Luosto U (1997) Structure of the Earth's crust in Fennoscandia as revealed from refraction and wide-angle reflection studies. Geophysica 33(1):3–16

Patison NL, Korja A, Lahtinen R, Ojala J, the FIRE Working Group (2006) FIRE seismic reflection profiles 4, 4A and 4B: Insights into the crustal structure of Northern Finland from Ranua to Näätämö. In: Kukkonen IT, Lahtinen R (eds) Finnish reflection experiment FIRE 2001–2005 Geological Survey of Finland, Special paper 43:161–222

Chapter 4
Evolution of the Bedrock of Finland: An Overview

Raimo Lahtinen

4.1 Introduction

The Fennoscandian Shield, situated in the north-westernmost part of the East European Craton (Fig. 4.1), is the largest exposed area of Precambrian rocks in Europe and shares many similarities with the ancient shields in Canada, Australia, Brazil and South Africa. The shield constitutes large parts of Finland, NW Russia, Norway, and Sweden and occurs between the Caledonian orogenic belt and the Phanerozoic cover sequences of the East European platform. Within the shield, Archean crust dominates in the east, Paleoproterozoic Svecofennian crust in the centre, and Mesoproterozoic rocks are found in the southwest.

The bedrock of Finland forms the core of the Fennoscandian Shield and can be divided into two main areas; Archean and Proterozoic. The Archean area in Finland includes exposed Archean rocks (23%) and Archean crust overlain by Paleoproterozoic supracrustal and plutonic rocks (30%). The Archean belongs dominantly in the Karelian province but small fragments of the Archean from the Belomorian, Kola and Norrbotten provinces occur also in Finland (Fig. 4.1). The Proterozoic area is composed of the Paleoproterozoic Svecofennian Province (41%) intruded by rapakivi granites (4%). The Svecofennian is further divided into central and southern Svecofennia (Fig. 4.1). Younger Proterozoic to Phanerozoic sedimentary rocks, mafic dikes and alkaline plutonic rocks occur in minor amounts in Finland and are left out from discussion in this paper.

The schematic age distribution of magmatic rocks in the Finnish bedrock (Fig. 4.2) shows two peaks at 2.8 and 2.7 in the Archean and one major peak at 1.9 Ga in the Paleoproterozoic. Sparse older 3.5–3.0 Ga rocks are found from the western part of the Karelian province (Hölttä et al. 2008; Lauri et al. 2010). Cratonic cover comprises rocks in age from 2.50 to 1.92 Ga and a major mafic

---

R. Lahtinen (✉)
Geological Survey of Finland, P.O. Box 96, FI-02151 Espoo, Finland
e-mail: raimo.lahtinen@gtk.fi

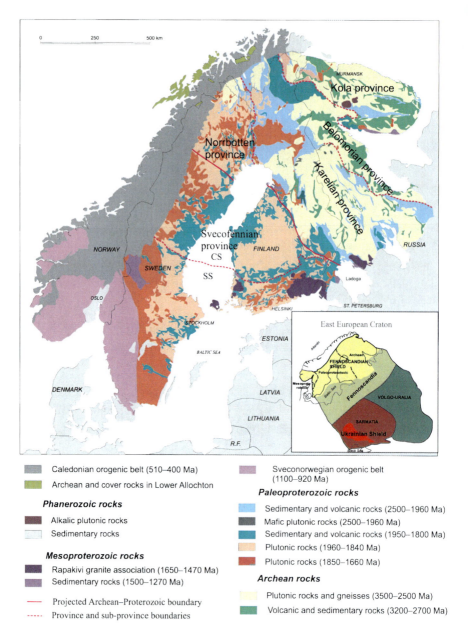

Fig. 4.1 Simplified geological map based on Koistinen et al. (2001) and insert map based on Gorbatschev and Bogdanova (1993) (Subareas: CS – Central Svecofennia; SS – Southern Svecofennia)

igneous activity is recorded at 2.44, 2.22, 2.14–2.10 and 2.06 Ga (Huhma et al. 2011). Interestingly the detrital zircon data from sedimentary rocks in Finland show a major age peak at c. 2.0 Ga, which is almost absent in the igneous age record from

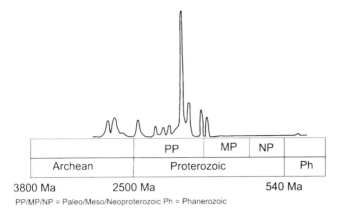

Fig. 4.2 Schematic age distribution of plutonic and volcanic rocks in the bedrock of Finland (Modified after Huhma et al. 2011, Fig. 4.3)

the present erosion level. Few 2.02 and 1.98 Ga ages have been found in Finnish Lapland (Hanski et al. 2005; Huhma et al. 2011) and 1.98–1.96 Ga age arc magmatism occurs in Kola (Daly et al. 2006). Also the occurrence of hidden c. 2.0 Ga microcontinents in the Svecofennian province has been proposed (Lahtinen et al. 2005). The Svecofennian province in Finland comprises few 1.93–1.92 Ga volcanic and plutonic rocks but otherwise the present erosion level is dominated by 1.90–1.80 Ga magmatic ages with a major peak at 1.89–1.88 Ga (Fig. 4.2). The last major magmatic event after 1.80 Ga in Finland is the emplacement of the voluminous rapakivi granites at 1.65–1.57 Ga.

The evolution of the Finnish bedrock can be divided into four main stages: Archean evolution, Paleoproterozoic rifting stage, Paleoproterozoic orogenic stage and Paleo- to Mesoproterozoic rapakivi granite stage. These stages are briefly discussed below. The main focus is on the evolution of the Finnish bedrock but remarks on the evolution of Fennoscandia as a whole are also given when needed. Referring is mainly to the papers in Lehtinen et al. (2005) and newer publications in order to keep the number of references to minimum.

4.2 Archean Evolution

The Archean Earth was in many ways different from modern Earth and one major pending question is when modern-type plate tectonics have started, already in Archean or later in Proterozoic. Anyhow it is evident that tectonic interactions between segments of the lithosphere and melt-generating mantle processes were affected by secular change caused by the exponential decline of the Earth's radiogenic heat production from Archean to Modern times. The occurrence of ultramafic volcanic rocks, komatiites, combined with the overall felsic nature of the crust, dominated by the voluminous occurrence of tonalites, trondjemites and granodiorites (TTG), and the buoyant and depleted nature of the lithospheric mantle are characteristic features of the Archean lithosphere. Possibly the tectonic evolution in the Archean changes from a secular or step-wise change from plume-dominated and/or atypical subduction processes (e.g., Arndt et al. 2009; Rollinson 2010) towards more Phanerozoic-type subduction-type processes. The Neoarchean ophiolite-like rocks and eclogites in the Belomorian province and the sanukitoid-type plutonic rocks are possibly examples of the first Phanerozoic-style subductions and collisions in the Fennoscandian Shield (Hölttä et al. 2008).

The Archean crust in Finland is an amalgamation of small complexes. These might even be less than 40 km wide and 100 km long (Sorjonen-Ward and Luukkonen 2005; Hölttä et al. 2008). Outside the cratonic core (Fig. 4.3c) Paleoproterozoic rifting and later stacking has affected their size and orientation but it seems that already during the Archean time the accreted complexes have been relatively small. This is also evident from the very limited occurrence of older rocks (3.5–3.1 Ga) along the western edge (Iisalmi and close to Oijärvi in Fig. 4.3a). Only restricted information exists from the Archean rocks in Kilpisjärvi and Inari (Fig. 4.3a) belonging to the Norrbotten and Kola provinces, respectively. The continuation of the Karelian province towards NW and the boundaries to other provinces are also poorly constrained. Anyhow the occurrence of Archean basement windows (e.g., Suomujärvi in Fig. 4.3a) and isotopic composition of cross-cutting plutonic rocks indicate that the concealed area is dominantly underlain by the Archean crust.

The Archean crust in Finland is characterized by TTG-type granitoids and gneisses, of which the latter are often migmatitic. The present isotopic data suggest there are some older (>2.9 Ga) domains within the TTG complexes, which predominantly appear to be 2.8–2.7 Ga in age. Age grouping at 2.83–2.78 Ga and 2.76–2.72 Ga occur in TTG rocks, whereas leucogranitoids and leucosomes in migmatites are typically c. 2.7 Ga (Käpyaho et al. 2007, Huhma et al. 2011). Sanukitoid-type, derived from subduction-related slab dehydration processes from a mantle wedge source, plutonic rocks have ages grouping at 2.74 and 2.70 Ga where the older age group has been found in Ilomantsi and younger age group characterizes other areas of the Karelian province in Finland (Heilimo et al. 2011).

Major volcanic-dominated supracrustal belts are Oijärvi, Kuhmo-Suomussalmi ja Kovero-Ilomantsi (Fig. 4.3a). The Oijärvi and Kuhmo-Suomussalmi volcano-sedimentary belts have variable ages between 2.94 and 2.79 Ga. These volcanic rocks formed in within-plate, probably oceanic, environments, whereas the Ilomantsi volcanic-sedimentary belt is younger (c. 2.75 Ga) and shows arc-type characteristics (Sorjonen-Ward and Luukkonen 2005; Hölttä et al. 2008). The Kovero belt contains 2.75 Ga rocks with small amount of older ca. 2.88 Ga rocks (Huhma et al. 2011). The Nurmes paragneisses (Fig. 4.3a) have an age of deposition close to 2.7 Ga and the MORB-type volcanic intercalations suggest that they were deposited in a back- or intra-arc setting (Kontinen et al. 2007). The Kalpio and Tuntsa areas (Fig. 4.3a) are also dominated by paragneisses.

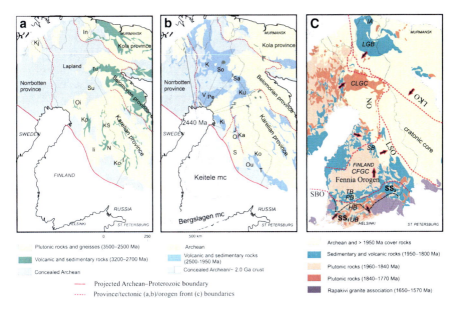

Fig. 4.3 Simplified geological maps based on Koistinen et al. (2001) showing the Precambrian growth of Finland. *Legend 3a*, belts: I – Ilomantsi; K – Kovero; KS – Kuhmo-Suomussalmi; Oi – Oijärvi, areas: Ii – Iisalmi; In – Inari; Kj – Kilpisjärvi; Kp – Kalpio; N – Nurmes; Su – Suomujärvi; Tu – Tuntsa. *Legend 4.3b*, belts: Ka – Kainuu; Ki – Kiiminki; Ku – Kuusamo; Pe – Peräpohja, areas: J – Jormua; K – Kittilä; Ko – Koli; O – Otanmäki; Ou – Outokumpu; Sa – Salla; S – Siilinjärvi; So – Sodankylä; V – Väystäjä; T – Tohmajärvi. *Legend 4.3c*, geological units: CLGC – Central Lapland Granitoid Complex; CFGC – Central Finland Granitoid Complex; LGB – Lapland Granulite Belt; HB – Häme belt; PB – Pirkanmaa belt; PoB – Pohjanmaa belt; SB – Savo Belt; TB – Tampere belt; UB – Uusimaa belt, orogens: LKO – Lapland-Kola orogen; LSO – Lapland-Savo orogen; NO – Nordic ororgen; SBO – Svecobaltic orogen. Dashed line separates areas SS_1 and SS_2 in southern Finland (see text for explanation) (Arrows indicate inferred vergence of thrusting)

Most parts of the Karelian province record metamorphism under upper amphibolite facies and granulite facies conditions, except some supracrustal belts that have peak metamorphic assemblages in the lower or middle amphibolite facies (Hölttä et al. 2008). Metamorphism took place at 2.71–2.62 Ga, partly coeval with emplacement of the youngest granites. An Archean carbonatite, intruded in an anorogenic setting at 2.6 Ga, exemplifies the end of Archean magmatic activity in Finland.

The supracrustal belts are normally relatively shallow (<2–5 km) and often tightly upright folded whereas surrounding migmatite areas are sub-horizontal. Thus, the genetic link between these two is not always clear, and a far-travelled nature for some supracrustal belts is possible. Subduction-related rocks like ophiolites in Belomorian province, sanukitoid-type plutonic rocks and arc-type volcanic and sedimentary rocks, like in Ilomantsi, indicate that most probably subduction-like processes were operating at 2.75 Ga. Also the occurrence of eclogites in the Belomorian province (Volodichev et al. 2004) favours a subduction-related continent-continent-like collision, at least locally. The still open universal question is how the older TTG-crust and supracrustal belts and the depleted Archean lithospheric mantle have formed. Anyhow it seems that the Archean crust in the Karelian province in Finland accreted at 2.75–2.70/2.65 Ga, exhumed at 2.65–2.62 Ga and then stabilized c. 2.6 Ga.

4.3 Paleoproterozoic Rifting Stage

The latest Archean-Paleoproterozoic rifting of the Archean crust in Fennoscandia started with 2.505–2.10 Ga, multiphase, southwest-prograding, intraplate rifting, where the 2.505 Ga ages have been only found in the Kola province (Lahtinen et al. 2008). In Finland, in the Karelian province, major mafic igneous activity occurs at 2.44, 2.22, 2.14–2.10 and 2.06 Ga, with few mafic dykes at 2.3 and at 1.96 Ga (Huhma et al. 2011).

If an Archean supercontinent existed it must have been fragmented during a period of global rifting, which has been proposed to begin at c. 2.45 Ga (Heaman 1997). In Finland we have a scattered occurrence of 2.44 Ga layered intrusions (Fig. 4.3b) originally formed in an ancient major failed rift system (Iljina and Hanski 2005). Volcanic rocks of about same age have been also found in Salla (Fig. 4.3b). The present position and shape of layered intrusions south of Kuusamo belt is due to subsequent block faulting and younger Paleoproterozoic compression (Karinen 2010). One layered intrusion is found from the Norrbotten province in Finland (Fig. 4.3b). If the aulacogen model is correct a major question lies where the rifted margin located at 2.44 Ga. Does the boundary between the Norrbotten and Karelian provinces mark an adjacent continental margin with the proposed aulacogen at 2.45 Ga, or was that margin located more far-away position in the NW?

Rift-basins, uplifts and erosional valleys typical of plume-generated continental flood basalt provinces were formed after 2.44 Ga. Erosion and deep weathering, enhancing the rate of silicate weathering and a considerable consumption of atmospheric CO_2, eventually result in the onset of the Huronian glaciation, (Melezhik 2006). The best examples of this stage glaciogenic rocks in Finland are the dropstones and diamictites found southeast from Koli (Fig. 4.3b). Deep chemical weathering covered large areas in the Karelian at c. 2.35 Ga and it formed also on Archean basement and glaciogenic rocks. Typically volcanic rocks and sedimentation occurred in terrestrial to shallow marine environments at 2.44–2.35 Ga in Finland indicating that the existing rift-basins were shallow and intra-plate in nature (Laajoki 2005).

The chemical weathering period was followed by erosion and sedimentation of conglomerates and thick successions of quartz arenites, locally feldspathic and sericite- and/or fuchsite-bearing, with few occurrences of amygdaloidal mafic lavas (Hanski and Huhma 2005; Laajoki 2005). Fluvial, delta and paralic type of sedimentation are noticed indicating shallow water depths. The 2.2 Ga layered intrusions and sills, intruding both the basement and sedimentary rocks, are found in the Koli, Kuusamo, Peräpohja and Sodankylä areas (Vuollo and Huhma 2005, Fig. 4.3b). The sills can be 150 km long and 100–400 m thick and are important marker horizons.

Next, ca. 2.1 Ga magmatic event occurs mainly as dike swarms but is regionally extensive and affected the whole Karelian province and its cover in Finland (Vuollo and Huhma 2005). Local shallow marine environments were marked by deposition of diverse stromatolites on extensive carbonate platforms at 2.2–2.1 Ga (e.g., Peräpohja Fig. 4.3b). All carbonate deposits show a large positive $\delta^{13}C$ isotope anomaly during the Lomagundi-Jatuli Event (Karhu 2005; Melezhik et al. 2007). Initial separation of the Archean supercontinent (Kenorland) along the Kola province boundary is roughly dated to 2.1 Ga (Daly et al. 2006). In Finland the thinning of Archean crust is seen as mafic c. 2.1 Ga volcanic rocks and turbidites in Tohmajärvi, and pelitic metasedimentary rocks and mafic-ultramafic volcanic rocks (>2.06 Ga) in central Lapland (Fig. 4.3b, Vuollo and Huhma 2005; Hanski and Huhma 2005).

The termination of the perturbation of the global carbon cycle occurred at ca. 2.06 Ga (Karhu 2005; Melezhik et al. 2007). Bimodal 2.06–2.05 Ga volcanics of alkaline-affinity intercalated with deep-water turbiditic sediments are found in Siilinjärvi and Väystäjä, and correlative alkaline anorogenic-type granitoids close to the c. 2.06 Ga mafic plutonic rocks in Otanmäki (Fig. 4.3b). Similar age komatiites and picrites and a mafic intrusion occur in central Lapland. This ca. 2.06 Ga stage is a good candidate for the break-up of the Archean craton at the western margin of the Karelian province.

No clear examples of subduction-related magmatism between 2.70 and 2.05 Ga has been found in Finland. The 2.02 Ga felsic volcanic rocks in Kittilä (Fig. 4.3b) occur in association with oceanic island arc-type rocks and are the oldest candidates for Paleoproterozoic subduction-related rocks (Hanski and Huhma 2005). Associated continental within-plate character volcanic rocks are possibly related to the continuing craton break-up. Both units are presently located within thrust units that include fragments of serpentinized mantle peridotites. The 1.98 Ga continental-type bimodal alkaline-tholeiitic rocks north of Peräpohja belt (Fig. 4.3b, Hanski et al. 2005) show that rift-magmatism continued further until 1.98 Ga.

Slightly after, the Jormua-Outokumpu ophiolites formed (Fig. 4.3b). They are a unique example of Archean subcontinental lithospheric mantle with a thin veneer of 1.95 Ga oceanic crust (Peltonen 2005). The ophiolites have been linked either to a continental break-up and subsequent formation of a passive margin or to a rift within an already existing passive margin generated at 2.05 Ga. Anyhow the preserved craton-derived sedimentation after 2.1–2.06 Ga is dominated by turbiditic sediments showing deep-water conditions. The Outokumpu, Kainuu and possibly Kiiminki areas and belts comprise thick successions of monotonous <1.94–1.92 Ga turbidites, which are often allochthonous with tectonically enclosed fragments of 1.95 Ga ophiolite bodies. Their material is derived from an orogenic domain (possibly Lapland-Kola Orogen, see below), comprising both Archean and

Paleoproterozoic units, followed by effective mixing during the transport (Lahtinen et al. 2010).

4.4 Paleoproterozoic Orogenic Stage

As discussed above there are still different opinions when and where the Archean craton (Karelian) break up occurred. The 2.02 Ga volcanic rocks in Kittilä and the 1.93–1.92 Ga island arc rocks in the Savo belt (Fig. 4.3b, c) are examples of the oldest yet defined Paleoproterozoic subduction-related oceanic island arc rocks in Finland but older ca. 2.0 Ga lithosphere as buried microcontinents has also been proposed (e.g., Keitele and Bergslagen in Fig. 4.3b). Main part of the central Svecofennia in Finland (Figs. 4.1 and 4.3c) is characterized by 1.89–1.87 Ga granitoids (CFGC), turbiditic sedimentary rocks with WPB-MORB-affinity volcanic intercalations (PB and PoB) and 1.90–1.88 Ga arc-type volcanic rocks (Kähkönen 2005; Nironen 2005). Similar age and type volcanic and sedimentary rocks are also found in the southern Svecofennia (e.g., UB and HB in Fig. 4.3c) but in addition younger ($\leq$1.87 Ga) units of sedimentary and volcanic rocks also occur. The Svecofennian province is characterized by high-T, low-P amphibolite facies metamorphism with local granulite facies domains. Both older (1.89–1.87 Ga) and younger metamorphism at 1.83–1.80 Ga, associated with locally voluminous granites (Figs. 4.1 and 4.3c), are found in the southern Svecofennia whereas only older peak metamorphism has been noticed, e.g., in the Pirkanmaa belt (Fig. 4.3c).

The main Paleoproterozoic orogenic evolution of Fennoscandia resulted in the Lapland-Kola orogen (1.94–1.86 Ga) and the composite Svecofennian orogen (1.92–1.79 Ga). The Lapland-Kola orogen is a continent-continent collision zone with a limited formation of new Paleoproterozoic crust (Daly et al. 2006), whereas the composite Svecofennian orogen is responsible for the main Paleoproterozoic crustal growth of Fennoscandia. The composite Svecofennian orogen has tentatively been divided into four orogens temporally overlapping towards their end phases and named as the Lapland-Savo, Fennian, Svecobaltic and Nordic orogens (Lahtinen et al. 2005, 2009a).

The Lapland-Kola orogen in the north (Fig. 4.3c) is a large and wide orogenic root of a mountain belt mainly comprising dominantly reworked Archean crust (Daly et al. 2006; Lahtinen et al. 2009a). The Lapland granulite belt (LGB in Fig. 4.3c) is composed of sedimentary rocks intruded by ca. 1.91 Ga enderbites and the Inari area (IA in Fig. 4.3c) is characterized by 1.94–1.91 Ga arc-type calc-alkaline plutonic rocks (Tuisku and Huhma 2006). The granulites in the LGB have either formed in the base of a subduction-related arc wedge (Tuisku and Huhma 2006; Daly et al. 2006) or due to the shortening and thickening of a pre-heated, at least partly ensialic, back-arc basin (Lahtinen et al. 2005). Anyhow the LGB and adjacent cover rocks form a large-scale thrust complex with a vergence to SW.

The Lapland–Savo orogen has been divided into northern and southern segments, where the northern segment formed during the collision of the Karelian and Norrbotten Archean cratons (provinces in Fig. 4.3), and the southern segment formed during the collision of the Karelian craton and the Keitele + Bergslagen microcontinent. The accreted terrane in the north is the Kittilä allochthon (Fig. 4.3b) including 2.02 Ga oceanic island-arc affinity rocks, dismembered ophiolitic rocks formed in a suprasubduction zone environment and shallow water mafic volcanic rocks, formed possibly at a passive margin (Hanski and Huhma 2005).

The southern segment of the Lapland-Savo orogen comprises the Central Finland granitoid complex, formed on the 2.1–2.0 Ga Keitele microcontinent, the 1.93–1.92 Ga island arc (Savo belt) and the inverted passive margin of the Karelian craton (Fig. 4.3c). The island arc volcanic rocks in the Savo belt are juvenile without any significantly older component (Lahtinen and Huhma 1997). The concealed Keitele microcontinent is inferred based on the geochemical and isotopical composition of plutonic rocks in the Central Finland granitoid complex (Lahtinen and Huhma 1997; Rämö et al. 2001) and the 2.03–1.97 Ga detritus in the Tampere and Pirkanmaa belt metasedimentary rocks (Lahtinen et al. 2009b). Thick lithosphere with >60 km thick crust (Heikkinen 2012, this volume) characterizes the Archean-Paleoproterozoic collision zone, which is due to the complex and interlayered nature of crust and lithospheric mantle (Peltonen and Brügmann 2006; Lahtinen et al. 2009a).

Subsequent evolution to the c. 1.92 Ga collision, forming the Lapland-Savo orogen, led to rifting and break-up of the Keitele + Bergslagen continent into two continents. At 1.92–1.91 Ga the rift was developed into a subsiding passive margin of the Keitele microcontinent with voluminous turbidite deposition, now seen as graywackes in the Tampere, Pirkanmaa and Pohjanmaa belts (Fig. 4.3c, Lahtinen et al. 2009b). A subduction reversal led to the onset of the Tampere arc volcanism at 1.90 Ga. The subduction terminated when the Bergslagen microcontinent and associated island arc collided with the Keitele microcontinent at c. 1.88 Ga. The major components in Finland, of the resulting Fennia orogen, are the Pohjanmaa, Savo, Tampere, Pirkanmaa, Häme and Uusimaa belts (Fig. 4.3c) but the north–south compression affected also rocks far in the north, e.g., seen as a closure of sedimentary basins in Peräpohja and Kuusamo (Fig. 4.3b).

Numerous ≤ 1.87 Ga quartzites and meta-arkoses (e.g., Bergman et al. 2008) and few lateritic paleosols (Lahtinen and Nironen 2010) have been found in the southern Svecofennia (Figs. 4.1 and 4.3c). The southern Svecofennia in Finland has been tentatively divided into two domains where SS_1 is inferred to be dominated by sedimentary rocks older than 1.88 Ga and SS_2 is inferred to have both older and younger (deposition age ≤ 1.87 Ga) sedimentary rocks (Fig. 4.3c, Lahtinen and Nironen 2010). Exhumation of the Fennian orogen rocks unroofed plutonic rocks and deposited sedimentary material. Strong lateritic weathering and deposition of ultra-mature/mature quartzites took place between 1.87 Ga and 1.85/1.84 Ga followed by rifting at 1.84–1.83 Ga. Basin inversion thrust the pre-rift quartzites and rift-related volcanic and sedimentary rocks over older middle crustal rocks and both older and younger rocks were then metamorphosed and migmatized at 1.82–1.80 Ga. The orogen front of the resulting Svecobaltic orogen (Fig. 4.3c) follows the northernmost occurrence of inferred ≤1.87 Ga supracrustal rocks and/or 1.82–1.80 Ga metamorphism.

The Nordic orogen (Fig. 4.3c) is poorly constrained and includes areas, especially in northern Finland, where c. 1.8 Ga deformation zones and thrusts are found (e.g., Niiranen et al. 2007). The Nordic orogen resulted from far in the northwest occurring continent-continent collision or Andean-type advancing arc, causing retro-arc fold and thrust belts. Late E-W bulk compression has also been inferred in the southern Finland (Skyttä and Mänttäri 2008; Pajunen et al. 2008).

Post-collisional bimodal magmatism at 1.80–1.77 Ga is found, especially in Lapland and southern Finland (Nironen 2005). Magmatism in Finland at 1.77–1.75 Ga is confined to minor occurrence of pegmatites, coeval with cooling of the upper and middle crust. Thus, c. 95% of the present erosional level of the bedrock of Finland was formed and accreted by 1.8 Ga.

4.5 Rapakivi Granite Stage

Magmatic quietness continued more than 100 Ma years in the bedrock of Finland before mafic dikes started to intrude at 1.67–1.65 Ga followed by mafic plutonic rocks and rapakivi granites at 1.65–1.63 Ga in the south-eastern Finland (Figs. 4.1 and 4.3c, Rämö and Haapala 2005). Next major event of the rapakivi granites and associated rocks occurred at 1.58–1.54 Ga in south-western Finland followed by a partly coeval pulse at 1.55–1.53 Ga north of Lake Ladoga in Russia (Fig. 4.1). The rapakivi associations are bimodal where granites have anorogenic A-type characteristics. The alkali feldspar megacrysts in granites are ovoidal and often mantled by a rim of oligoclase-andesine forming one of the characteristic rapakivi textures (Rämö and Haapala 2005).

The rapakivi granite batholiths occur as sheet-like (5–10 km) bodies in the upper crust counter parted by thinned lower crust. Magmatic underplating by mantle magmas and subsequent partial melting of the lower crust is the favoured model for the origin of the bimodal rapakivi granite associations. The parental magmas for mafic rocks in the Finnish rapakivi batholiths derive dominantly from depleted asthenospheric mantle with only minor contribution from subcontinental lithospheric mantle (Heinonen et al. 2010). The c. 100 Ma age span of rapakivi granite batholiths favours that separate pulses of mantle magmas and upwelling occurred. The cause for the mantle perturbations is still somewhat controversial and includes active and passive rifting, orogenic collapse, mantle heterogeneities, deep mantle plumes and inboard extension related to subduction events (Rämö and Haapala 2005 and references therein).

4.6 Summary

The Archean crust in Finland includes some >2.9 Ga old (oldest 3.5 Ga) domains within the 2.8–2.7 Ga granitoid-migmatite complexes. The 2.9–2.8 Ga volcano-sedimentary belts formed in within-plate, probably oceanic, environments, whereas the younger volcanic-sedimentary belt (c. 2.75 Ga) shows arc-type characteristics. The Archean crust in the Karelian province in Finland accreted at 2.75–2.70/2.65 Ga, exhumed at 2.65–2.63 Ga and then stabilized at c. 2.6 Ga.

The next major stage is exemplified by the 2.44 Ga layered intrusions and volcanic rocks possibly formed in an ancient major failed rift system. Remnants of global Huronian glaciation and deep chemical weathering are noticed in eastern Finland. Terrestrial to shallow marine deposition of sedimentary rocks and incipient rifting and mafic magmatism, at 2.22, 2.14–2.10 and 2.06 Ga, characterises the 400 Ma age period from 2.44 Ga to 2.06 Ga. The bimodal magmatism at 2.06 Ga, at the western margin of the Karelian province, was subsequently followed by deposition of deep water sediments. Slightly after, the Jormua-Outokumpu ophiolites formed at 1.95 Ga.

The Svecofennian province in Finland consists of hidden c. 2.0 Ga microcontinents, belts of turbiditic sedimentary rocks, 1.93–1.92 Ga and 1.90–1.88 Ga arc-type volcanic belts and voluminous 1.89–1.87 Ga granitoids. In addition younger (≤1.87 Ga) sedimentary and volcanic rocks and locally voluminous 1.84–1.80 Ga granites are found in the southern Svecofennia. The main Paleoproterozoic orogenic events in the Fennoscandian Shield are the formation of the Lapland-Kola orogen (1.94–1.86 Ga), a continent-continent collision zone with a limited formation of new crust, and the composite Svecofennian orogen (1.92–1.79 Ga. The latter includes amalgamation of microcontinents, sedimentary basins and juvenile arc crust, and is responsible for the main Paleoproterozoic crustal growth of the Fennoscandian Shield. Rocks of the bimodal rapakivi granite association intruded the already stabilized crust in southern Finland and form the last major magmatic input to the bedrock of Finland.

References

Arndt NT, Coltice N, Helmstaedt H, Gregoire M (2009) Origin of Archean subcontinental lithospheric mantle: some petrological constraints. Lithos 109:61–71
Bergman S, Högdahl K, Nironen M, Ogenhall E, Sjöström H, Lundqvist L, Lahtinen R (2008) Timing of palaeoproterozoic intra-orogenic sedimentation in the central Fennoscandian Shield; evidence from detrital zircon in metasandstones. Precambrian Res 161:231–249
Daly JS, Balagansky VV, Timmerman MJ, Whitehouse MJ (2006) The Lapland-Kola orogen: Palaeoproterozoic collision and accretion of the northern Fennoscandian lithosphere. In: Gee DG, Stephenson RA (eds) European lithosphere dynamics. Geol Soc London Mem 32:561–578
Gorbatschev R, Bogdanova S (1993) Frontiers in the Baltic Shield. Precambrian Res 64:3–21
Hanski E, Huhma H (2005) Central Lapland greenstone belt. In: Lehtinen M, Nurmi PA, Rämö OT (eds) Precambrian geology of Finland: key to the evolution of the Fennoscandian Shield. Developments in Precambrian geology, vol 14. Elsevier, Amsterdam, pp 139–194

Hanski E, Huhma H, Perttunen V (2005) SIMS U-Pb, Sm-Nd isotope and geochemical study of an arkosite-amphibolite suite, Peräpohja Schist belt: evidence for ca. 1. 98 Ga A-type felsic magmatism in northern Finland. Geol Soc Finland Bull 77:5–29

Heaman LM (1997) Global mafic magmatism at 2.45 Ga; remnants of an ancient large igneous province? Geology 25:299–302

Heikkinen P (2012) Seismic structure of Earth's crust in Finland. In: Haapala I (ed) From the Earth's core to outer space. Lecture notes in Earth system sciences 137. Springer, Berlin/Heidelberg, pp 37–46

Heilimo E, Halla J, Huhma H (2011) Single-grain zircon U–Pb age constraints of the western and eastern sanukitoid zones in the Finnish part of the Karelian province. Lithos 121:87–99

Heinonen AP, Andersen T, Rämö OT (2010) Re-evaluation of rapakivi petrogenesis: source constraints from the Hf Isotope composition of Zircon in the rapakivi granites and associated mafic rocks of Southern Finland. J Petrol 51:1687–1709

Hölttä P, Balagansky V, Garde A, Mertanen S, Peltonen P, Slabunov A, Sorjonen-Ward P, Whitehouse M (2008) Archaean of Greenland and Fennoscandia. Episodes 31:13–19

Huhma H, O'Brien H, Lahaye Y, Mänttäri I (2011) Isotope geology and Fennoscandian lithosphere evolution. Geol Surv Finland Spec Pap 49:35–48

Iljina M, Hanski E (2005) Layered mafic intrusions of the Tornio-Näränkävaara belt. In: Lehtinen M, Nurmi PA, Rämö OT (eds) Precambrian geology of Finland: key to the evolution of the Fennoscandian Shield. Developments in Precambrian geology, vol 14. Elsevier, Amsterdam, pp 101–138

Kähkönen Y (2005) Svecofennian supracrustal rocks. In: Lehtinen M, Nurmi PA, Rämö OT (eds) Precambrian geology of Finland: key to the evolution of the Fennoscandian Shield. Developments in Precambrian geology, vol 14. Elsevier, Amsterdam, pp 343–406

Käpyaho A, Hölttä P, Whitehouse MJ (2007) U-Pb zircon geochronology of selected Archaean migmatites in eastern Finland. Geol Soc Finland Bull 79:95–115

Karhu J (2005) Paleoproterozoic carbon isotope excursion. In: Lehtinen M, Nurmi PA, Rämö OT (eds) Precambrian geology of Finland: key to the evolution of the Fennoscandian Shield. Developments in Precambrian geology, vol 14. Elsevier, Amsterdam, pp 669–680

Karinen T (2010) The Koillismaa intrusion, northeastern Finland: evidence for PGE reef forming processes in the layered series. Geol Surv Finland Bull 404:176

Koistinen T (comp.), Stephens MB (comp.), Bogatchev V (comp.), Nordgulen Ø (comp.), Wennerström M (comp.), Korhonen J (comp.) (2001) Geological map of the Fennoscandian Shield, scale 1:2 000 000. Espoo: Trondheim: Uppsala: Moscow: Geological Survey of Finland: Geological Survey of Norway : Geological Survey of Sweden: Ministry of Natural Resources of Russia

Kontinen A, Käpyaho A, Huhma H, Karhu J, Matukov DI, Larionov A, Sergeev SA (2007) Nurmes paragneisses in eastern Finland, Karelian craton: provenance, tectonic setting and implications for NeoArchaean craton correlation. Precambrian Res 152:119–148

Laajoki K (2005) Karelian supracrustal rocks. In: Lehtinen M, Nurmi PA, Rämö OT (eds) Precambrian geology of Finland: key to the evolution of the Fennoscandian Shield. Developments in Precambrian geology, vol 14. Elsevier, Amsterdam, pp 279–342

Lahtinen R, Huhma H (1997) Isotopic and geochemical constraints on the evolution of the 1.93–1.79 Ga Svecofennian crust and mantle. Precambrian Res 82:13–34

Lahtinen R, Nironen M (2010) Palaeoproterozoic lateritic paleosol–ultra-mature/mature quartzite–meta-arkose successions in southern Fennoscandia: intra-orogenic stage during the Svecofennian orogeny. Precambrian Res 183:770–790

Lahtinen R, Korja A, Nironen M (2005) Palaeoproterozoic tectonic evolution of the Fennoscandian Shield. In: Lehtinen M, Nurmi PA, Rämö OT (eds) Precambrian geology of Finland: key to the evolution of the Fennoscandian Shield. Developments in Precambrian geology. Elsevier, Amsterdam, pp 418–532

Lahtinen R, Garde AA, Melezhik VA (2008) Palaeoproterozoic evolution of Fennoscandia and Greenland. Episodes 31:20–28

Lahtinen R, Korja A, Nironen M, Heikkinen P (2009a) Paleoproterozoic accretionary processes in Fennoscandia. In: Cawood PA, Kröner A (eds) Earth accretionary systems in space and time. Geol Soc London Spec Publ 318:237–256

Lahtinen R, Huhma H, Kähkönen Y, Mänttäri I (2009b) Paleoproterozoic sediment recycling during multiphase orogenic evolution in Fennoscandia, the Tampere and Pirkanmaa belts, Finland. Precambrian Res 174:310–336

Lahtinen R, Huhma H, Kontinen A, Kohonen J, Sorjonen-Ward P (2010) New constraints for the source characteristics, deposition and age of the 2.1–1.9 Ga metasedimentary cover at the western margin of the Karelian Province. Precambrian Res 176:77–93

Lauri LS, Andersen T, Hölttä P, Huhma H, Graham S (2010) Evolution of the Archaean Karelian Province in the Fennoscandian Shield in the light of U–Pb zircon ages and Sm–Nd and Lu–Hf isotope systematics. J Geol Soc London 167:1–18

Lehtinen M, Nurmi PA, Rämö OT (2005) Precambrian geology of Finland: key to the evolution of the Fennoscandian Shield. Developments in Precambrian geology, vol 14. Elsevier, Amsterdam, p 736

Melezhik VA (2006) Multiple causes of Earth's earliest global glaciation. Terra Nova 18:130–137

Melezhik VA, Huhma H, Condon DJ, Fallick AE, Whitehouse MJ (2007) Temporal constraints n the Paleoproterozoic Lomagundi-Jatuli carbon isotopic event. Geology 35:655–658

Niiranen, T. Poutiainen, M. Mänttäri, I. (2007) Geology, geochemistry, fluid inclusion characteristics, and U-Pb age studies on iron oxide-Cu-Au deposits in the Kolari region, northern Finland. Ore Geology Reviews 30: 75–105

Nironen M (2005) Proterozoic granitoid rocks. In: Lehtinen M, Nurmi PA, Rämö OT (eds) Precambrian geology of Finland: key to the evolution of the Fennoscandian Shield. Developments in Precambrian geology, vol 14. Elsevier, Amstrdam, pp 443–480

Pajunen M, Airo M-L, Elminen T, Mänttäri I, Niemelä R, Vaarma M, Wasenius P, Wennerström M (2008) Tectonic evolution of the Svecofennian crust in southern Finland. In: Tectonic evolution of the Svecofennian crust in southern Finland – a basis for characterizing bedrock technical properties. Geol Surv Finland Spec paper 47: 15–160

Peltonen P (2005) Ophiolites. In: Lehtinen M, Nurmi PA, Rämö OT (eds) Precambrian geology of Finland: key to the evolution of the Fennoscandian Shield. Developments in Precambrian geology, vol 14. Elsevier, Amsterdam, pp 237–278

Peltonen P, Brügmann G (2006) Origin of layered continental mantle (Karelian craton, Finland): geochemical and Re-Os isotope constraints. Lithos 89:405–423

Rämö OT, Haapala I (2005) Rapakivi granites. In: Lehtinen M, Nurmi PA, Rämö OT (eds) Precambrian geology of Finland: key to the evolution of the Fennoscandian Shield. Developments in Precambrian geology, vol 14. Elsevier, Amsterdam, pp 533–562

Rämö OT, Vaasjoki M, Mänttäri I, Elliott BA, Nironen M (2001) Petrogenesis of the post-kinematic magmatism of the central Finland granitoid complex I: radioisotopeconstraints and implications for crustal evolution. J Petrol 42:1971–1993

Rollinson H (2010) Coupled evolution of Archaean continental crust and subcontinental lithospheric mantle. Geology 38:1083–1086

Skyttä P, Mänttäri I (2008) Structural setting of late Svecofennian granites and pegmatites in Uusimaa belt, SW Finland: age constraints and implications for crustal evolution. Precambrian Res 164:86–109

Sorjonen-Ward P, Luukkonen EJ (2005) Archean rocks. In: Lehtinen M, Nurmi PA, Rämö OT (eds) Precambrian geology of Finland: key to the evolution of the Fennoscandian Shield. Developments in Precambrian geology, vol 14. Elsevier, Amsterdam, pp 19–99

Tuisku P, Huhma H (2006) Evolution of migmatitic granulite complexes: implications from Lapland Granulite Belt, Part II: isotopic dating. Geol Soc Finland Bull 78:143–175

Volodichev OI, Slabunov AI, Bibikova EV, Konilov AN, Kuzenko TI (2004) Archaean Eclogites in the Belomorian Mobile belt, Baltic shield. Petrology 12:540–560

Vuollo J, Huhma H (2005) Paleoproterozoic mafic dikes in NE Finland. In: Lehtinen M, Nurmi PA, Rämö OT (eds) Precambrian geology of Finland: key to the evolution of the Fennoscandian Shield. Developments in Precambrian geology, vol 14. Elsevier, Amsterdam, pp 195–236

Chapter 5
Craton Mantle Formation and Structure of Eastern Finland Mantle: Evidence from Kimberlite-Derived Mantle Xenoliths, Xenocrysts and Diamonds

Hugh O'Brien and Marja Lehtonen

5.1 Introduction

The lithosphere is the outer "shell" of the Earth, 15–250 km in thickness, comprised of two parts, the crust and the uppermost mantle physically attached to the crust. It is this part of the Earth that moves as plates in tectonic fashion. In order to get a proper perspective of what the lithosphere "shell" represents, it is necessary to take a step back and look at a whole Earth cross section (Fig. 5.1). The Earth is 6,378 km from the surface to the very center (at the equator, slightly less at the poles). The crust – mantle transition occurs at 30–60 km, the upper – lower mantle transition from 450 to 600 km, the silicate mantle – Fe, Ni metal outer core boundary at about 2,890 km and the outer (liquid) core – inner (solid) core transition at about 5,150 km. The lithosphere represents a very small proportion of the material in this cross section, with oceanic lithosphere and continental lithosphere representing 0.1% and 0.5% of the Earth's mass, respectively (Taylor and McLennan 1995).

These two lithosphere varieties are quite distinctive; oceanic is relatively thin at roughly 30 km, whereas continental ranges up to thicknesses of 250 km or more. Oceanic lithosphere, which underlies all ocean basins, has a relatively simple life, being formed at divergent boundaries such as the mid-Atlantic ridge, and being consumed at convergent boundaries, such as off the west coast of South America. As a consequence of this life cycle, there is no oceanic crust at the Earth's surface that is much older than Jurassic (ca. 200 Ma in age).

Continental lithosphere on the other hand is quite another story, and forms the topic of this chapter. In particular we are concerned with the oldest lithosphere on Earth, consisting of rocks older than 2.5 billion years, formed during the Archean Eon (from arkhaios, Greek for "ancient," from arkhe "beginning"), or even earlier, in the Hadean Eon (from Hades, Greek for "underworld"), the period prior to

H. O'Brien (✉) • M. Lehtonen
Geological Survey of Finland, P.O. Box 96, FI-02151 Espoo, Finland
e-mail: hugh.obrien@gtk.fi

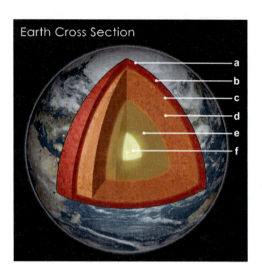

Fig. 5.1 Earth cross section from core to crust showing the major layers of the Earth. (**a**) Crust 0–40 km, (**b**) Upper Mantle 40–410 km, (**c**) Transition Zone 410–660 km, (**d**) Lower Mantle 660–2,890 km, (**e**) Outer Core 2,890–5,150 km, (**f**) Inner Core 5,150–6,378 km (Figure by Adam Dorman, www.adamdorman.com)

ca. 4.0 billion years ago. Archean rocks are not exactly rare, but are nevertheless restricted to about a dozen regions on the planet, areas that have been stable and acting as rigid blocks for most of our planet's history (see Pesonen and Mertanen, this volume, Fig. 2.3). In this chapter, however, we are only peripherally interested in the relatively easy-to-study crustal rocks in these Archean areas, rather the goal is to understand the mantle rocks underneath the continental crust (at a depth of 50–250 km), comprised of olivine-dominated, ultramafic rocks (peridotites).

Directly acquired information about the continental lithospheric mantle in the rock record comes from two main sources: mantle lifted vertically from great depth during major mountain building events (orogenic peridotite massifs) and mantle xenoliths (fragments of mantle rocks) contained in alkaline magmas, also derived from great depth. Each has its own merits and shortcomings in terms of representativeness. Orogenic peridotite massifs are mostly derived from the uppermost parts of the mantle, and very little deep mantle has been discovered at the Earth's surface. Additionally most of the ultramafic rocks in these massifs have been strongly tectonized and highly altered during uplift and emplacement, making original fabrics and mineral/whole rock compositions difficult to ascertain. Xenoliths in alkaline magmas, on the other hand, suffer from the three major problems: (1) They are usually small and are randomly distributed within the host rock, rendering no information about original rock contact relationships (except in rare cases). (2) They are usually rounded and abraded due to mechanical and chemical effects that occur during transportation in the ascending magma, leaving uncertain the extent of host magma interaction. (3) The majority may represent modified mantle along magma conduits, rather than samples representative of the true bulk mantle (see Sect. 5.3.3).

The most important xenolith-containing rock type is kimberlite, a rare rock found as remnants of mostly small (rarely medium-sized) volcanoes. Kimberlite volcanoes formed from very volatile-rich magmas that ascended explosively and rapidly from great depth, and in some cases carried sufficient quantities of diamond to form ore. Much of our knowledge about the structure and composition of the continental lithosphere has to do with the fact that exploration for diamond-bearing kimberlites requires a very large amount of information about the underlying lithospheric mantle to test its diamond prospectivity. The reason such exploration is limited to continental lithosphere is that the formation of diamond requires high pressure (depths of more than about 150 km) and low heat flow (otherwise diamonds turn to graphite), characteristics only found in old and cold Archean continental lithospheric mantle.

In this chapter we start by reviewing how the early Earth first developed into a layered body and how the very earliest rigid blocks of lithosphere were formed. We continue by explaining what is unique about the oldest parts of our planet's outer layer and then provide a review of the latest ideas about how the cratons formed. Finally we turn attention to the Archean portion of Finland, applying the ideas and models presented earlier to explain how these processes may have resulted in the formation of the Karelian craton.

5.2 Hadean Earth

When the Earth formed with the rest of inner planets at 4.567 Ga, it was quite unlike the planet we see today, assembled from a heterogeneous jumble of leftover solar nebulae material. Much of how the Earth is layered today (Fig. 5.1) has to do with a series of events that occurred just after this main accumulation period: (1) Heating due to radioactive decay of abundant highly unstable isotopes and due to continued planetesimal bombardment generated a terrestrial magma ocean (e.g., Ito et al. 2004). (2) Core formation, with the sinking of metallic Ni and Fe to the center of the Earth, commenced shortly thereafter, possibly within the first 30 m.y. after formation of the Earth (Boyet and Carlson 2005). (3) The moon forming event occurred as Mars-sized Theia collided with Earth at about 4.48 Ga (Halliday 2008). This at least partially reset differentiation of the Earth and spewed out the material that collected as our moon. (4) Another or a continuation of the same terrestrial magma ocean crystallized synchronously with the lunar magma ocean (oldest lunar zircon discovered crystallized at 4.417 Ga, Nemchin et al. 2009). (5) Earth moved rapidly to a quiescent stage and the oldest (to date) terrestrial zircons crystallized at 4.4 Ga (detrital grains from Jack Hills sedimentary rocks, Wilde et al. 2001). (6) Late Heavy Bombardment (LHB) in the inner solar system at about 3.9 Ga (well recorded on the moon, e.g., Ryder et al. 2000) may have destroyed much of the earliest formed terrestrial crust. Consequently, the paucity of >4.0 Ga old crust most likely is a problem of preservation; with much of the first generated crust reworked back into the mantle. As an example of remnant crust that managed to survive this destructive period, the 4.0 Ga Acasta gneiss in the Slave craton of NW Canada (e.g., Bowring and Williams 1999) represents some of the oldest crust on our planet.

The time period between Earth's accretion at 4.567 Ga and the oldest extant crustal rocks at ca. 4.0 Ga, is known as the Hadean Eon. Since essentially no terrestrial rock record exists from this period, it is difficult to say when the first significant quantities of felsic (silica- and aluminum-rich) buoyant crust and complementary depleted mantle formed on Earth. Recent discoveries including the Jack Hills 4.4 Ga zircons, and other isotopic and geochemical evidence are beginning to provide tantalizing answers this question, corroborating a record of mantle melting that occurred locally, if not globally, within the first few 100 m.y. of Earth's history. Consequently craton formation may have already begun at this time, but the extent to which present day Earth retains a memory of the Hadean is open to much debate and an area of active current research.

5.3 Archean Cratonic Mantle

5.3.1 *Characteristics*

It can be shown that three main characteristics set cratons apart from other continental lithosphere areas elsewhere on the planet: age, composition and thickness.

5.3.1.1 Age

An age of >2.5 Ga is of course one of the defining characteristics of Archean cratonic rocks, but again, the vast majority of this age information comes from relatively easy to date, zircon-bearing crustal rocks. Ages from cratonic *mantle* samples are, in contrast, rather poorly constrained. This remains true despite the many attempts employing a variety of isotope methods to date mantle rocks in general, and cratonic mantle rocks in particular (e.g., Pearson 1999). The main problem appears to be that at the high ambient temperatures of the mantle, isotopic systems such as Rb–Sr and Sm–Nd simply do not act as closed systems, continuously resetting or partially resetting their isotopic clocks. Consequently most dating of this type provides ages rather close to the host intrusion age. The isotopic system which is the most robust and least affected by external inputs is Re–Os, and today we have a large body of ^{187}Re–^{186}Os data from whole rocks and from sulfides they contain. The results imply that all of the areas of the world judged to be Archean based on their exposed crustal rocks are underlain by Archean mantle, with a strong mode at around 3.0 Ga and maximum ages of around 3.5 Ga (Pearson and Wittig 2008).

Diamonds provide another window into craton mantle genesis and development through time as they are almost exclusively found in Archean areas of the planet. Although it is impossible to date diamond itself; rare mineral inclusions encapsulated in diamonds have been used. These include pyropes, dated by Sm–Nd on composite grain samples (Richardson et al. 1984) and sulfides, dated by Re–Os of single grains (e.g., Pearson 1999). A recent summary of diamond inclusion ages (Gurney et al. 2010) shows that the oldest diamond forming event at 3.1–3.5 Ga produced peridotite-type diamonds likely on a global scale, with good examples from the Slave Craton, Canada, Siberian Craton Russia, and Kaapvaal Craton, southern Africa (see Pesonen and Mertanen, this volume, Fig. 2.2).

5.3.1.2 Thickness

The thickness of the cratonic mantle is best estimated by two main methods, by measuring the PT conditions at which mantle xenoliths last equilibrated in the mantle (see Karelian craton section below) and by the application of geophysical modeling to data provided by large earthquakes. For the latter, there are several methods by which vibrations from large earthquakes can be used to determine physical properties of the rocks through which they pass. The most commonly used method is seismic tomography whereby velocity anomalies are mapped throughout a section of lithosphere. Given that seismic velocities are very sensitive to temperature then the positive anomalies typically observed underlying most cratons are a consequence of a relatively deep and cold mantle root. Seismic velocities are also sensitive to rock composition and it also likely that part of the positive velocity anomalies is due to the highly melt-depleted nature of the bulk of the rocks in Archean cratonic mantle roots.

5.3.1.3 Composition and Mineralogy

Given the tremendous transformation in the Earth from its earliest history to the present, and possibly also changes in the processes or at least the rate of these processes as the planet evolved, it is perhaps not surprising that the ancient Archean lithospheric mantle is not simply older than the rest of the mantle on Earth, it is also compositionally different. Looking at the chemical composition of peridotite samples of various ages in large datasets (e.g., Wittig et al. 2008; Griffin et al. 2009), several clear geochemical differences among age groups can be seen. For example, one obvious difference is that Archean cratonic mantle xenoliths cluster at low Fe compared to Proterozoic (2.5–0.54 Ga) and younger Phanerozoic examples. Archean Mg/(Mg + Fe) values are commensurately high, whereas there is a large range of SiO_2, probably partly due to the introduction of silica-rich, subduction-related fluids or melts that have metasomatized and produced significant secondary orthopyroxene in the overlying lithospheric mantle, particularly on the Kaapvaal craton (Boyd 1989). Based on garnet xenocryst and xenolith data, calculated whole rock CaO wt% and Al_2O_3 wt% values have changed from roughly Archean 1 and 1 through Proterozoic 1.5–2.5 and 2–3 to Phanerozoic 3 and 4 (Griffin et al. 2009). The Phanerozoic values approach those of primitive mantle (i.e., asthenosphere). These secular changes in lithospheric mantle composition argue that the processes by which the cratonic mantle was formed occurred only during the Archean (or earlier), and have not been active in the same way since.

These geochemical differences manifest themselves petrologically as a greater abundance of refractory rock types (those with the most melt removed) constituting the Archean cratonic mantle, particularly the olivine-dominated rock types dunite and harzburgite. Proterozoic and younger mantle rock types are dominated by more fertile, clinopyroxene-bearing lherzolites. Olivine compositions (expressed as the relative abundance of the Mg olivine endmember forsterite Fo) in Archean craton peridotite xenoliths reflect their depleted host rocks, with a rather restricted range from Fo_{89} to Fo_{95} and a strong mode around Fo_{93-94}. This is in strong contrast to the range of Fo_{88} to Fo_{93} with a mode at Fo_{89-90} seen in convecting mantle peridotite samples (Lee 2006).

Consequences for the physical properties of Archean cratonic mantle are many. Most significantly, the relatively low Fe, high Mg composition the Archean cratonic mantle makes it buoyant, and extremely difficult to destroy by sinking into the underlying, more fertile, denser asthenospheric mantle. However, more than buoyancy is required to keep depleted craton mantle from being stirred into the convecting mantle. It must also be relatively non-viscous and rigid, with an inherent strength that makes it difficult to subduct. Viscosity is not so much controlled by mineralogy as it is by the presence or absence of volatiles (Kohlstedt et al. 1996). Consequently the stability and longevity of the Archean cratonic mantle requires not only melt-depleted compositions to remain buoyant, it must also be nearly volatile-free to act rigidly.

5.3.2 Genesis

There are three main models that have been used to explain the formation of depleted mantle rocks that comprise the cratonic mantle, each of these consistent with a specific tectonic environment: stacking of oceanic lithosphere in subduction zones, island arc accretion zones and plumes (Fig. 5.2). All have in common the fact that the strongly depleted nature of the craton mantle peridotites is explained by high degrees of partial melting of an originally more fertile mantle. The main disagreement is whether the partial melting process occurred (or at least was initiated) at high pressure – the plume scenario, or at low pressure – the subduction zone and accretion zone scenarios.

5.3.2.1 Accretion and Stacking of Oceanic Lithosphere

This model suggests that cratonic mantle formed in a subduction zone by the process of continued accretion of slabs of oceanic lithosphere originally formed at mid-ocean ridges. Based on geochemical arguments, a large group of researchers (e.g. Parman et al. 2004; Lee 2006; Pearson and Wittig 2008) have argued that the lithospheric mantle, all of it, including continental and oceanic varieties, must have formed at relatively shallow depths of melting <120 km. A worldwide comparison

5 Craton Mantle Formation and Structure of Eastern Finland Mantle

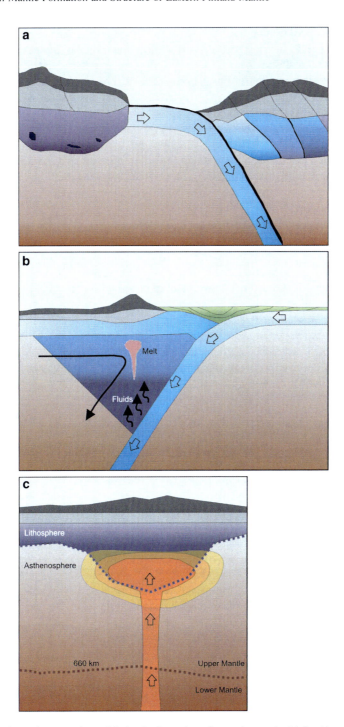

Fig. 5.2 Alternative tectonic models for the formation of cratonic mantle. (a) Stacking of oceanic lithosphere slabs, (b) subduction and wet melting of mantle wedge, (c) plume activity (After Lee (2006) and Arndt et al. (2009))

of xenoliths concludes that uniformly low heavy rare Earth element concentrations in Archean peridotites requires that, if they represent melt residues, then they originally contained little to no garnet, suggesting melting happened at low pressure in the spinel stability field (Wittig et al. 2008). Corroborating evidence for the model comes from the large number of eclogite xenoliths carried by kimberlites that come from the cratonic mantle and are interpreted as high pressure versions of subducted oceanic crust. Diamonds found in some of these xenoliths have "light" carbon (negative $\delta^{13}C$ as low as -30, thought to fingerprint biogenic C) and inclusions of coesite (a high pressure form of quartz) with oxygen isotopic compositions most likely reflecting Earth surface processes (Schulze et al. 2003).

5.3.2.2 Subduction Zone Processes

In this model wet melting in the mantle wedge overlying the subduction zone is used to explain the formation of very refractory harzburgites and dunites. The main problems with this model are that the resulting residues most probably would have arc-type trace element signatures indicative worldwide of arc-related magmatism (e.g., Rb-Ba-enrichment and Ta-Nb-Ti-depletion), and this is generally not the case (Lee 2006). Moreover, it would seem this model would produce more depleted (Fe-poor) residues below less depleted (Fe-rich), forming an unstable buoyancy problem and the need to invert this column to match the stratigraphy observed in most craton mantle roots (Arndt et al. 2009).

5.3.2.3 Mantle Plumes

Rather than subduct ocean crust to generate the Archean cratonic mantle root, Griffin and coworkers (Griffin et al. 2009) point to the Norwegian dunite/harzburgite rocks of the Western gneiss region (Beyer et al. 2006) that formed as high pressure residues (5–6 GPa, op. cit.) as very close analogues of the type of peridotites formed at the base of the lithosphere by plume activity (so-called subcretion). As a corroborating piece of evidence for the plume model, rare diamonds are found that contain inclusions of phases only stable in the lower mantle (e.g., ferropericlase [(Mg, Fe)O], majoritic garnet, stishovite (high pressure SiO_2), and MgSi- and CaSi-perovskite, the latter requiring encapsulation below the 680 km phase transition. Griffin et al. (1999) suggest that diamonds of this type were first carried from the lower mantle and underplated to the craton mantle by ascending plumes. The kimberlites later picked up these diamonds and transported them to the surface during emplacement.

Recently Arndt et al. (2009) revisited the question of which of the three proposed models best fits the current data, and have made a strong case in favor of the plume model. They base this conclusion mostly on the fact that melting at ocean ridges and in subduction zones just does not produce the appropriate composition rocks and ratio of rock types characteristic of the cratonic mantle. They reconfirm that in a hotter Archean mantle, komatiite lavas, which correspond to 30–50% melting of primitive mantle, could represent the melting component extracted in the Archean, leaving a highly depleted residue, although it would necessarily be voluminous volcanism. The problem of too little komatiite existing in Archean greenstone belts (Lee 2006), is suggested to be a preservational issue because such dense rocks would be the least likely to be preserved in ocean basin settings (Arndt et al. 2009).

5.3.3 Metasomatism and Refertilization

No matter how the cratonic mantle actually formed, there is positive evidence that craton mantle roots have been partially to extensively refertilized by later percolating melts and/or fluids. A particularly good example of this type of metasomatic transformation is shown by the dunites from Western Norway (Beyer et al. 2006). The depleted dunites appear to form the host rock for meter-scale thick lenses of garnet lherzolites, where the latter are thought to represent melt channelways along which the original dunites were infiltrated and metasomatized. A similar melt infiltration process has been ascribed to the harzburgite to lherzolite metasomatic transformation suspected for the Lherz massif (Le Roux et al. 2007). A corollary of this process is that potentially all garnet, clinopyroxene and possibly the bulk of orthopyroxene present in continental mantle may have formed through the process of fertilization of dunite, i.e., they can all represent secondary minerals.

Evidence is also building that metasomatic processes control the formation of peridotitic diamonds in the mantle, in this case by the oxidation of asthenosphere-derived $CH_4 \pm H_2O$ fluids as they invade and react with subcontinental mantle peridotites (Malkovets et al. 2007). At the same time this reaction produces metasomatic horizons (subvertical veins?) in the mantle with subcalcic harzburgitic pyropes as one byproduct and explains the fact that this type of garnet is the dominant silicate inclusion in diamond. It is also likely that melts, including kimberlites and lamproites discussed here, are preferentially channelled into pre-existing fractures within the mantle. Given the reactivity of fluids derived from such volatile-rich magmas, it follows that the conduits along which precursor melts have ascended are metasomatized. This leads to the inevitable conclusion that kimberlitic/lamproitic magmas probably are **not** carrying a representative sample of the bulk mantle, but rather the average of a certain metasomatized domain, with the bulk of the mantle lithosphere remaining more refractory than the xenolith record.

5.4 Karelian Craton and Kimberlites

In the remainder of this chapter, we apply the general models and descriptions of cratonic mantle presented above to explain the information that is available on the Karelian cratonic mantle.

5.4.1 Karelian Craton Crust

The Karelian craton is roughly 400,000 km² in size and represents the largest of five Archean blocks that make up the northern and eastern portion of the Fennoscandian shield (Fig. 5.3). To the NE, the Karelian craton is bounded by the Belomorian province, considered to be a mobile belt ranging in age from 2.93 to 2.72 Ga, which has undergone intense polyphase deformation and mid to high pressure metamorphism. It contains a 2.72 Ga ophiolite and eclogite-bearing assemblage, unique on

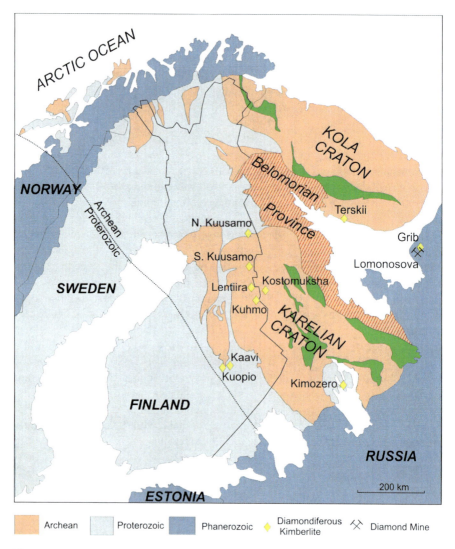

Fig. 5.3 Geologic sketch map showing the location of the Karelian craton and diamond-bearing alkaline rocks described here (Dashed Proterozoic-Archean line is based on Nd isotopes)

the Fennoscandian shield (Volodichev et al. 2004), that has been interpreted as part of a subduction zone complex marking the amalgamation of the Kola and Karelian cratons. To the SW and NW, the Karelian craton is bounded by Paleoproterozoic rocks of the Svecofennian orogen and the Lapland greenstone belt, respectively, and runs under platform cover to the SE.

The Karelian craton and has been divided into three terranes, Western (essentially the part within Finland), Central and Vodlozero (southern third), based on within terrane similarity of structures, lithologies, and ages of granitoids and greenstone belts (Hölttä et al. 2008). All of the mantle samples discussed below come from the Western terrane of the Karelian craton. This terrane hosts the oldest rocks in the Karelian craton (and in fact in Europe), the Siurua trondhjemitic gneisses in the Pudasjärvi granulite belt, with zircon U-Pb ages up to 3.5 Ga (Mutanen and Huhma 2003). The bulk of the terrane is composed of 2.8–2.7 Ga orthogneisses and lesser 2.84–2.79 Ga greenstone belts comprised of oceanic plateau-type komatiitic and tholeiitic basalts and sediments containing banded iron formations (Hölttä et al. 2008).

5.4.2 Eastern Finland Kimberlitic Rocks: The Sampling Tool

Kimberlites and related rocks occur in three major fields in eastern Finland, Kaavi-Kuopio, Lentiira-Kuhmo and Hossa-Kuusamo (Fig 5.3). The Kaavi-Kuopio and Hossa-Kuusamo fields contain archetypal kimberlites, whereas only olivine lamproite-orangeite hybrids are known from the Lentiira-Kuhmo field. The terminology of potentially diamond-bearing alkaline rocks derived from the deep mantle is often problematic. Table 5.1 lists the important mineralogical criteria used to differentiate the various magma types, yet it is clear there is considerable overlap.

5.4.2.1 SW Craton Margin Kaavi-Kuopio Kimberlites

The Kaavi–Kuopio kimberlites are found in two fields, located only ~ 50 and ~ 30 km inboard from the southwestern margin of the craton, respectively (Fig. 5.3). These kimberlites were formed from ultramafic, volatile-charged, incompatible element-rich magmas that represent a mixture of liquid, mantle-derived peridotite and eclogite xenoliths and disaggregated xenoliths (xenocrysts) and megacryst suite minerals such as Ti-pyrope, Mg-ilmenite and Cr-diopside all carried from mantle depths. They also commonly contain xenoliths of lower, mid and upper crustal rocks through which the magmas have traversed. These kimberlites have abundant large, rounded grains (macrocrysts) of olivine, in a matrix of euhderal olivine, monticellite, perovskite, magnesian ulvöspinel-magnetite, Ba-rich phlogopite-kinoshitalite mica, apatite, calcite and serpentine. In addition to having typical kimberlite mineralogies, the Kaavi-Kuopio also have kimberlite major and trace element and Sr, Nd and Pb isotopic compositions similar to others worldwide (O'Brien and Tyni 1999).

Table 5.1 Mineralogical comparison of kimberlites, lamproites, minettes and ultramafic lamprophyres (After Mitchell 1995, Table 1.1; slightly modified)

		Kimberlites	Orangeites	Lamproites	Minettes	Ultramafic Lamprophyres
Mantle	Xenoliths	common	common	rare	---	rare
	Xenocrysts	common	common	rare	rare	rare
Olivine	Macrocrysts	common	common	rare	---	rare
	Xenocrysts	common	common	common	rare	common
Mica	Macrocrysts	common	common, phlogopite	common, phlogopite to Ti-phlogopite	common	common
	Phenocrysts	phlogopite		Ti-phlogopite	phlogopite	phlogopite
	Groundmass	common, phlogopite kinoshitalite	common, tetraferriphlogopite	common, Ti-tetraferriphlogopite	common, Al-biotite	common, Al-biotite
Spinels		abundant, Mg-chromite to Mg-ulvöspinel	rare, Mg-chromite to Ti-magnetite	rare, Mg-chromite to Ti-magnetite	common, Mg-chromite to Ti-magnetite	common, Mg-chromite to Ti-magnetite
Monticellite		common				common
Diopside		---	common, Al- + Ti-poor	common, Al- + Ti-poor	common, Al- + Ti-rich	common, Al- + Ti-rich
Perovskite		common, Sr- + REE-poor	rare, Sr- + REE-rich	rare, Sr- + REE-rich		common, Sr- + REE-poor
Apatite		common, Sr- + REE-poor	abundant, Sr- + REE-rich	common, Sr- + REE-rich	common, Sr- + REE-poor	common, Sr- + REE-poor
Primary Serpentine		abundant	common		rare	
Calcite		abundant	common		rare	common
Sanidine			rare groundmass	common, phenocrysts + groundmass	abundant, groundmass	
K-richterite			rare groundmass	common, phenocrysts + groundmass		
K-Ba-titanates		very rare	common	common		
Zr-silicates		very rare	common	common	very rare	
Mn-ilmenite		rare	common	very rare	common	rare
Leucite			rare pseudomorphs	common, phenocrysts		

--------- = absent | critical characteristic | important characteristic | characteristic of evolved endmember

The kimberlites range from coherent kimberlites in dikes to volcaniclastic kimberlite and volcaniclastic kimberlite breccias formed in steep sided funnel- or carrot-shaped pipes representing diatreme facies remnants of small volcanoes. As in many kimberlite fields hosted by crystalline basement, these are rather small, ranging in size from narrow dikes <1 m wide to elongate bodies 500 × 30 m in size to nearly circular cross-section diatreme-facies pipes up to 4 ha in size. None of the Kaavi-Kuopio pipes appear to have existing crater-facies materials due to erosion of the upper portions of the pipes. Quantities and volume ratios of mantle derived xenocrysts vary considerably among the bodies but, in general, the compositions of these minerals do not show large intra-pipe variations. Many of the pipes are diamondiferous, several having reasonable diamond grades (14–41 ct/100 t; Tyni 1997). Age determinations have been made on four of these kimberlites by ion microprobe analyses of U-Pb in perovskites with measured ages ranging from 589 to 626 Ma, with a suggested age of 600 Ma (O'Brien et al. 2005).

5.4.2.2 Craton Core Kuhmo and Lentiira Olivine Lamproites

Kuhmo and Lentiira olivine lamproite dikes are confined to a N–S zone of faults. The emplacement ages are constrained by phlogopite Ar-Ar analyses from the Kuhmo dikes (1,202 ± 3 Ma; O'Brien et al. 2007). Most of the bodies occur as dikes 0.5–4 m in thickness, extending up to 450 m in length, less commonly as veins or stockworks (500 × 500 m). The most notable feature of these ultramafic rocks is that they contain abundant phlogopite as phenocrysts, microphenocrysts and more rarely as macrocrysts. Groundmass minerals include K-richterite, Mn-rich ilmenite, apatite and perovskite in a calcite + serpentine matrix. Mantle derived xenocrysts are dominantly chrome-spinels zoned to Ti-magnetite, Ti-rich (megacryst-composition) pyropes, less commonly Mg-rich ilmenites and microdiamonds. Notably, mantle-derived Cr-diopside is almost completely absent from the xenocryst suite.

5.4.2.3 North Craton Kuusamo Kimberlites

The Kuusamo kimberlites mark the third, and most recently discovered kimberlite region in Finland, located in the northern part of the Western terrane of the Karelian craton (Fig. 5.3). Six kimberlites occur in two discrete groups. Coherent kimberlites, Kattaisenvaara, Kalettomanpuro and Lampi, consist of serpentinized olivine macrocrysts and phenocrysts in a fine-grained matrix composed of microphenocrysts of phlogopite, perovskite, apatite and spinel. The other three kimberlites, named 47, 45 and 45 South, occur close to each other. Kimberlite 45 contains both hypabyssal and diatreme-facies types, whereas only diatreme material was recovered from 47. The volcaniclastic kimberlite in these small pipes contains abundant pelletal lapilli (lapilli-sized [2–64 mm] pyroclasts, typically with a kernel or seed crystal enveloped in magma, spherical to disc-shaped due to spinning while growing in the eruption column). A seventh discovery is an olivine lamproite dike intersected at Kalettomanpuro containing mantle xenocrysts of picroilmenite, pyrope and chrome diopside. U-Pb analyses of two perovskite fractions from one sample each of the Kattaisenvaara and Kalettomanpuro kimberlites gave weighted mean $^{206}Pb/^{238}U$ ages of 759 ± 15 Ma and 756.8 ± 2.1 Ma, respectively (O'Brien and Bradley 2008).

5.4.3 The Mantle Sample

The two main mantle rock types that kimberlites carry with them to the surface are peridotites, dominated by olivine, and eclogites, composed basically of orange garnet and Al-rich clinopyroxene. Karelian craton kimberlites and lamproites contain a variety of mantle-derived materials either as xenoliths (rocks), or as xenocrysts (separate crystals). Kaavi-Kuopio peridotites range from dunite (olivine + chromite) to harzburgite (by the addition of orthopyroxene) to lherzolite (with the addition of clinopyroxene) and finally to pyroxenites with little olivine. All have equigranular textures lacking evidence of shearing. The harzburgites types are garnet + spinel-bearing derived from 70 to 110 km and garnet-only from >110 km depth (Fig. 5.4a, b). Two main differences can be seen in the peridotite xenolith population at Kuhmo, located in a much more central part of the craton.

Fig. 5.4 Photomicrographs of representative mantle samples for eastern Finland. (**a**) Layer A garnet-spinel harzburgite and (**b**) layer B garnet lherzolite from Kaavi-Kuopio, (**c**) harzburgite with sheared (fluidal porphyroclastic) texture from Kuhmo, (**d**) diamondiferous eclogite from Lahtojoki pipe, Kaavi (Plane polarized light, thin sections are 3 cm wide, eclogite is 1 cm wide)

Here xenoliths are generally clinopyroxene poor. Additionally, a number of peridotites have been recovered from the area that have highly recrystallized textures indicative of shearing in the mantle (Fig. 5.4c), something that one might expect instead near the craton edge at Kaavi-Kuopio. There is also a difference concerning eclogites, which occur in small amounts at Kaavi-Kuopio, and are even more common as xenocrysts. In rare cases the eclogites can be highly diamondiferous (Fig. 5.4d), possibly representing the main diamond source rock at the Lahtojoki kimberlite pipe in Kaavi. In contrast, only one exploration target is known to contain eclogite garnet xenocrysts in the Kuhmo-Lentiira-Kostomuksha area.

Equilibrium pressures and temperatures based on thermodynamic models using major and minor elements in coexisting orthopyroxene, garnet and clinopyroxene have been derived from the Kaavi-Kuopio kimberlite-derived xenolith suite (Kukkonen and Peltonen 1999). Although xenoliths provide us the most reliable information about the mantle, they are relatively rare in kimberlite. Single crystal xenocryst thermobarometry and compositional data require more assumptions (Lehtonen and O'Brien 2009 and references therein), but make up for this uncertainty partly by being vastly more numerous than xenoliths, and partly because we can find them in sediment samples, allowing information on the mantle to be acquired even before the kimberlite source rocks have been discovered.

5.4.4 Eastern Finland Mantle Section

Xenolith and xenocryst PT data indicate that the mantle root of the Karelian craton varies considerably from margin to core (Fig. 5.5). At the margin, in the Kuopio and Kaavi area, the mantle is comprised of at least three distinct layers: (a) a shallow, 60–110 km, garnet-spinel peridotite layer, (b) a variably depleted peridotitic horizon from 110 to 180 km containing diamond-indicative subcalcic (harzburgitic) pyropes, (c) a deep layer, >180 km depth, composed largely of fertile peridotites. Shallow *layer A* contains peridotites that have "ultradepleted" compositions (high Mg, low Al) but yet contains obvious evidence of modal metasomatism (rare phlogopite grains in many xenoliths) and relatively radiogenic Os compositions at a given Al_2O_3 content (Peltonen and Brügmann 2006). The middle *layer B* (at ~110–180 km depth) is the main source of harzburgitic garnet xenocrysts in the Kaavi-Kuopio kimberlites, although it is dominated by lherzolite (Lehtonen and O'Brien 2009). It also has unradiogenic Os isotopic compositions that show good correlations with indices of partial melting imply a melt extraction age of ~3.3 Ga (Peltonen and Brügmann 2006). The underlying *layer C* (at 180–250 km depth) is harzburgitic pyrope-free and represents the main source of Ti-rich pyropes of megacrystic composition. It has an Os isotopic composition more radiogenic compared to layer B, yielding only Proterozoic T_{RD} ages.

The mantle stratigraphy of the craton core, in the Kuhmo, Lentiira and Kostomuksha areas, shows less variation, with no evidence of Layer A (Fig. 5.5).

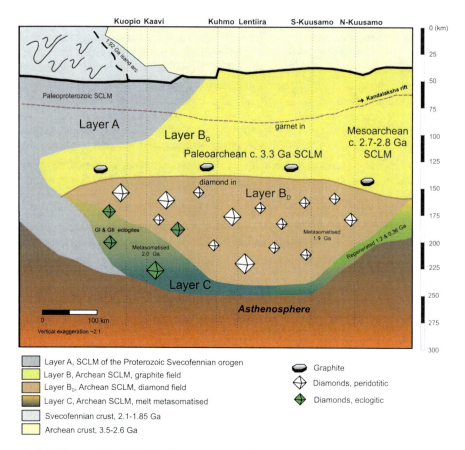

Fig. 5.5 Cross section of the Karelian craton mantle from Kuopio in the SW to N. Kuusamo in the NE (Modified after Lehtonen and O'Brien (2009))

Layer B begins with the lowest temperature pyropes at an inferred depth of 70 km and continues to a depth of about 250 km showing a relatively homogenous distribution of harzburgite and lherzolite pyropes throughout. Differences in craton core layer B compared to craton edge layer B include: (1) Wehrlite garnet is very rare, as is chrome diopside. (2) The harzburgitic to lherzolitic garnet ratio is relatively high implying the craton core layer B is relatively harzburgite-rich. (3) The overall Mg/(Mg + Fe) of the mantle lithosphere in this area is as high as anywhere on the planet and considerably higher than the mantle at Kaavi-Kuopio. Coupled with the rarity of mantle-derived chrome diopside, the implication is that this portion of the mantle underwent high levels of partial melting to produce refractory residua or was assembled originally from extremely melt-depleted source rocks. (4) As described above, eclogites are rare in the craton core based on exploration and hard rock samples. (5) At least within the limitations of the present database there is only a weak indication of a harzburgitite pyrope-free, metasomatically enriched Layer C. However, Ti-rich megacryst composition pyropes are very common, so better evidence for a deep 250–300 km Layer C may become available with further sampling.

The position of the S. Kuusamo area well within the Karelian craton would imply it represents craton core and the distribution and composition of mantle xenocrysts derived from this area confirms the existence of depleted mantle underlying this area. Xenocryst pyrope chemistry demonstrates that lherzolite and harzburgite are roughly equally distributed down to depths of roughly 180 km. Although the data are relatively sparse, we interpret the existence of harzburgite throughout the mantle section, without any obvious layering, to represent lithospheric mantle similar to that in the Kuhmo region, albeit slightly thinner. Bolstering this interpretation is the paucity of eclogite material in exploration and hard rock samples from the area. The existence of abundant harzburgitic rocks, some with ultra-depleted pyrope compositions, and extremely high Mg, implies this is similar Archean mantle as at Kuhmo.

5.4.5 Interpretation

Based on the data presented here, the continental lithosphere thickens from about 220 km at Kuopio to nearly 250 km in the Lentiira–Kuhmo–S. Kuusamo section and then thins again to about 220 km at the N. Kuusamo locality. The latter is in the direction of the Kandalaksha graben where the mantle lithosphere may have already been considerably thinned during the Middle Riphean (c. 1,200 Ma) (Baluev et al. 2000). Geophysical evidence such as teleseismically derived tomographic cross sections for the central Fennoscandian shield published by Sandoval et al. (2004) are remarkably consistent with our interpretation shown in (Fig. 5.5).

Our present understanding of the genesis of the Karelian craton mantle is that, in its original state, it closely resembled the section at Kuhmo-Lentiira-Kuusamo, likely formed by plume activity, high degrees of partial melting and komatiite extraction as per the plume model described in Sect. 5.3.2.3. The timing of this stabilization must be as least as old as the oldest crustal rocks, and we take this to be approximately 3.5 Ga based on U-Pb zircon ages on garnet granulite xenoliths from Lahtojoki (Peltonen et al. 2006) and the Siurua gneiss in the Pudasjärvi granulite belt (Mutanen and Huhma 2003). Surprisingly even at the very craton edge at Kaavi-Kuopio, Layer B mantle gives 3.3 Ga model ages only slightly younger than the oldest 3.5 Ga crustal rocks. However, also at the craton edge is the uppermost mantle layer A, not found elsewhere in the craton cross-section. It is interpreted as Proterozoic arc-wedge mantle produced by subduction processes in a similar fashion to the subduction and wet melting model described in Sect. 5.3.2.2. Metasomatism by slab-derived fluids provided the radiogenic Os it contains. At approximately 1.9 Ga, layer A was accreted to the existing craton core during collision between the Svecofennian arc complex and the Karelian craton. The relative abundance of eclogite component at Kaavi-Kuopio, which is essentially absent in the craton core mantle sample population, suggests eclogitic material was also accreted during this collisional event. Layer C is interpreted as a melt metasomatised equivalent of layer B. This metasomatism most likely occurred at c. 2.0 Ga when the present craton margin formed subsequent to break-up of the proto-craton. Nevertheless, a fragmentary memory of its prior history is contained in a few depleted, fluid-metasomatized pyropes it contains that are characteristic of layer B (see Lehtonen and O'Brien 2009).

5.5 Concluding Remarks

Diamond exploration has lead to the discovery of over 30 kimberlites and olivine lamproites in Finland. Most of these contain at least some diamonds and significant amounts of mantle material; a mantle which shows all indications of being prospective for diamonds. Our knowledge of the geology of the deep mantle root in this part of the world, as elsewhere, has been advanced immeasurably by the need for information of these exploration activities.

References

Arndt N, Coltice N, Helmstaedt H, Gregoire M (2009) Origin of Archean subcontinental lithospheric mantle: some petrological constraints. Lithos 109:61–71

Beyer EE, Griffin WL, O'Reilly SY (2006) Transformation of Archean lithospheric mantle by refertilization: evidence from exposed peridotites in the Western Gneiss Region, Norway. J Petrol 47:1611–1636

Bowring SA, Williams IS (1999) Priscoan (4.00–4.03 Ga) orthogneisses from NW Canada. Contrib Mineral Petrol 134:3–16

Boyd FR (1989) Compositional distinction between oceanic and cratonic lithosphere. Earth Planet Sci Lett 96:15–26

Boyet M, Carlson RW (2005) 142Nd evidence for early (>4.53 Ga) global differentiation of the silicate Earth. Science 309:576–581

Griffin WL, Doyle BJ, Ryan CG, Pearson NJ, O'Reilly SY, Davies R, Kivi K, van Achterbergh E, Natapov LM (1999) Layered mantle lithosphere in the Lac de Gras area, slave craton: composition, structure and origin. J Petrol 40:705–727

Griffin WL, O'Reilly SY, Afonso JC, Begg GC (2009) The composition and evolution of lithospheric mantle: a re-evaluation and its tectonic implications. J Petrol 50:1115–1204

Gurney JJ, Helmstaedt HH, Richardson SH, Shirey SB (2010) Diamonds through time. Econ Geol 105:689–712

Halliday AN (2008) A young Moon-forming giant impact at 70–110 million years accompanied by late-stage mixing, core formation and degassing of the Earth. Phil Trans R Soc A 366:4163–4181

Hölttä P, Balagansky V, Garde A, Mertanen S, Peltonen P, Slabunov A, Sorjonen-Ward P, Whitehouse M (2008) Archean of Greenland and Fennoscandia. Episodes 31(1):13–19

Ito E, Kubo A, Katsura T, Walter MJ (2004) Melting experiments of mantle materials under lower mantle conditions with implications for magma ocean differentiation. Phys Earth Planet Inter 143–144:397–406

Kohlstedt DL, Keppler H, Rubie DC (1996) Solubility of water in the a-phase, b-phase and c-phase of $(Mg, Fe)_2SiO_4$. Contrib Mineral Petrol 123:345–357

Kukkonen IT, Peltonen P (1999) Xenolith-controlled geotherm for the central Fennoscandian shield: implications for lithosphere-asthenosphere relations. Tectonophysics 304:301–315

Le Roux V, Bodinier J-L, Tommasi A, Alard O, Dautria J-M, Vauchez A, Riches AJV (2007) The Lherz spinel lherzolite: refertilized rather than pristine mantle. Earth Planet Sci Lett 259:599–612

Lee C-TA (2006) Geochemical/petrologic constraints on the origin of cratonic mantle. In: Benn K, Mareschal J-C, Condie KC (eds) Archean geodynamics and environments, American geophysical union monograph. AGU, Washington, pp 89–114

Lehtonen ML, O'Brien HE (2009) Mantle transect of the Karelian Craton from margin to core based on P-T data from garnet and clinopyroxene xenocrysts in kimberlites. Bull Geol Soc Finland 81:79–102

Malkovets VG, Griffin WL, O'Reilly SY, Wood BJ (2007) Diamond, subcalcic garnet and mantle metasomatism: kimberlite sampling patterns define the link. Geology 35:339–342

Mitchell RH (1995) Kimberlites, orangeites, and related rocks. Plenum, New York

Mutanen T, Huhma H (2003) The 3.5 Ga Siurua trondhjemite gneiss in the Archean Pudasjärvi Granulite Belt, northern Finland. Bull Geol Soc Finland 75:51–68

Nemchin A, Timms N, Pidgeon R, Geisler T, Reddy S, Meyer C (2009) Timing of crystallization of the lunar magma ocean constrained by the oldest zircon. Nat Geosci 2:133–136

O'Brien HE, Bradley J (2008) New kimberlite discoveries in Kuusamo, northern Finland. 9th International kimberlite conference, Frankfurt, Germany. Extended Abstract 00346

O'Brien HE, Tyni M (1999) Mineralogy and geochemistry of kimberlites and related rocks from Finland. In: Gurney JJ, Gurney JL, Pascoe MD, Richardson SH (eds) Proceedings of the 7th international kimberlite conference, vol 2. University of Cape Town, Cape Town, pp 625–636

O'Brien HE, Peltonen P, Vartiainen H (2005) Kimberlites, carbonatites and alkaline rocks. In: Lehtinen M, Nurmi P, Rämö OT (eds) Precambrian geology of Finland: key to the evolution of the Fennoscandian shield. Elsevier Science B.V, Amsterdam, pp 605–644

O'Brien HE, Phillips D, Spencer R (2007) Isotopic ages of Lentiira-Kuhmo-Kostomuksha olivine lamproite-Group II kimberlites. Bull Geol Soc Finland 79:203–215

Parman SW, Grove TL, Dann JC, de Wit MJ (2004) A subduction origin for komatiites and cratonic lithospheric mantle. South African J Geol 107:107–118

Pearson DG (1999) The age of continental roots. Lithos 48:171–194

Pearson DG, Wittig N (2008) Formation of Archean continental lithosphere and its diamonds: the root of the problem. J Geol Soc London 165:1–20

Peltonen P, Brügmann G (2006) Origin of layered continental mantle (Karelian craton, Finland): geochemical and Re-Os isotope constraints. Lithos 89:405–423

Peltonen P, Mänttäri I, Huhma H, Whitehouse M (2006) Multi-stage origin of the lower crust of the Karelian craton from 3.5 to 1.7 Ga based on isotopic ages of kimberlite-derived mafic granulite xenoliths. Precambrian Res 147:107–123

Richardson SH, Gurney JJ, Erlank AJ, Harris JW (1984) Origin of diamonds in old enriched mantle. Nature 310:198–202

Ryder G, Koeberl C, Mojzsis SJ (2000) Heavy bombardment on the Earth 3.85 Ga: the search for petrographic and geochemical evidence. In: Canup RM, Righter K (eds) Origin of the earth and moon. University of Arizona Press, Tucson, pp 475–492

Sandoval S, Kissling E, Ansorge J, SVEKALAPKO Seismic Tomography Working Group (2004) High resolution body wave tomography beneath the SVEKALAPKO array: II. Anomalous upper mantle beneath the central Baltic shield. Geophys J Int 157:200–214

Schulze DJ, Harte B, Valley JW, Brenan JM, Channer DMDeR (2003) Extreme crustal oxygen isotope signatures preserved in coesite in diamond. Nature 423:68–70

Taylor SR, McLennan SM (1995) The geochemical evolution of the continental crust. Rev Geophys 33:241–265

Tyni M (1997) Diamond prospecting in Finland: a review. In: Papunen H (ed) Mineral deposits: research and exploration, Where do they meet? Proceedings of the 4th SGA meeting, Turku, Finland: . Balkema, Rotterdam, pp 789–791

Volodichev OI, Slabunov AI, Bibikova EV, Konilov AN, Kuzenko TI (2004) Archean eclogites in the Belomorian Mobile belt, Baltic shield. Petrology 12:540–560

Wilde SA, Valley JW, Peck WH, Graham CM (2001) Evidence from detrital zircons for the existence of continental crust and oceans on the Earth 4.4 Gyr ago. Nature 409:175–178

Wittig N, Pearson DG, Webb M, Ottley CJ, Irvine GJ, Kopylova M, Jensen SM, Nowell GM (2008) Origin of cratonic lithospheric mantle roots: a geochemical study of peridotites from the North Atlantic Craton, West Greenland. Earth Planet Sci Lett 274:24–33

Chapter 6
Metallic Mineral Resources in Finland and Fennoscandia: A Major European Raw-Materials Source for the Future

Pekka A. Nurmi and Pasi Eilu

6.1 Introduction

A considerable increase in the world population, more rapid urbanisation and higher material living standards have resulted in an unprecedented demand for metals. It has been forecasted that annually 60–80 million people will move to cities, and that as much as 70% of the population will live in cities by 2030 (World Economic Forum 2010). At the same time, the middle class is becoming larger and more affluent.

Mineral-based materials and products are used either directly or indirectly in almost every aspect of our lives. They are essential in the construction of housing and other buildings, railroads, road networks, power grids and pipelines. Industrial production and the manufacturing of machinery, equipment, vehicles and ICT technology are all largely based upon the utilisation of mineral-based materials. New technologies and environmental challenges have expanded our need for raw materials and mineral-based products even further (e.g., Buchert et al. 2009).

The availability of mineral raw materials is essential for the prosperity of modern societies. For example, in Europe the combined annual turnover of the construction, chemical, automobile, aeroplane, machinery and equipment-manufacturing industries is about 1,300 billion Euros, and they provide employment for 30 million people (Tiess 2010). EU member countries consume 25–30% of the metals produced globally. In contrast, metal mine production within the EU is only about 3% of global production and many important metals are not produced in Europe at all The overall intensity of metal use within the EU is slowly decreasing and, as in other developed countries, recycling is becoming more efficient and new replacement

P.A. Nurmi (✉) • P. Eilu
Geological Survey of Finland, B.O. Box 96, FI-02151 Espoo, Finland
e-mail: pekka.nurmi@gtk.fi

materials are being found. Despite that, it is expected that for the foreseeable future European industries will remain vulnerable to disruptions in the primary metal supply and to market volatility.

Finland's metallic-ore-related mining operations started in the year 1530, when the Ojamo iron mine began operations (Puustinen 2003). Since that date, there have been hundreds of mining operations, but their production up until the 1930s, although being important for the society of the day, was modest. Metal mining on a modern industrial scale began in Finland in the 1930s, though it was not until the 1970s that most of the mines commenced operations. In this period, Finland produced chromium, copper, zinc, gold, nickel, cobalt, iron and vanadium, and mining became an important player in the nations economy. At its peak in 1979, approximately 10 million tons of metal ores were mined in Finland. After a long recession, a new era in Finnish metal mining was initiated in 2008 when two large mines opened for the first time in many years. The volume of mining is expected to grow fast, at least in this decade, and reach a level 10–20 times that of the previous peak.

Finland is located within Fennoscandia, a major Precambrian area of the Earth's crust, which is a traditional producer of mineral raw materials and whose importance as a future source of metals is increasing. In this paper we briefly evaluate the potential of Fennoscandia, with a focus on Finland as a major future source of metals, and use simplified genetic models to describe the most important ore types in Finland. We also discuss the future of mining in Finland.

6.2 What are Mineral Resources?

The Earth's crust is mined where it is enriched with minerals that form ore deposits. Mining occurs where these minerals exist in sufficient amounts at locations where the operations are technically and financially viable. The concept of ore is, therefore, always directly tied to a particular point in time: the benefits minerals bring vary according to current needs, price, their occurrence and quality, and the kinds of technology available for mining. In addition to geological and techno-economic considerations, the mining industry is increasingly subject to the reconciliation of environmental, social and political interests in the use of land.

An ore is a mineral accumulation that contains exploitable metals and minerals, and an ore deposit is a mass of rock from which metals or minerals can be economically harnessed (Australasian Joint Ore Reserves Committee 2004; National Instrument 43-101 2006; Moon and Evans 2006). A mineral resource is a concentration or occurrence of natural, solid, inorganic or fossilized organic material in or on the Earth's crust in such form and quantity and of such a grade or quality that it has reasonable prospects for economic extraction. A mineral reserve is the part of the resource known to be economically feasible for extraction and processing when taking into account the current market, environmental, administrative and social factors (Australasian Joint Ore Reserves Committee 2004; Whateley and Scott 2006).

In this article we also use the term "assumed mineral resources". By this we mean the estimated geological resources which are not supported by large quantities of data, analyses or reporting modes, but which give the impression, in spite of incomplete data, of the potential size of the deposits. In considering the numerical values presented here it is very important to keep in mind the following points: (1) the untapped resources are geological in-situ values; (2) in the majority of instances, the resource estimates are not based on existing industrial standards; (3) also the resources' Euro values are in situ values, which do not take into account any potential mining costs, waste rock dilution, means of enrichment, the mine's profitability, or other factors that might be considered in the calculations. Our estimated tonnage and Euro values are, therefore, in no way to be confused with the results of studies on the economic profitability of the mining projects (cf. Australasian Joint Ore Reserves Committee 2004; National Instrument 43-101 2006).

Accurate information on the full amount of mineral resources in a region is difficult to obtain because the deposits are difficult to find and some may never be found. This is especially the case in glaciated terrains, such as, Finland, where 96% of bedrock is covered by overburden. Evaluation of resources of even a known deposit requires many years of sustained intensive research. Quite often, further investigations reveal new mineral reserves in proximity to or as an extension of even the best-known deposits. It is not uncommon for the lifetime of a mine to dramatically increase size compared to its original assessment. Rising metal prices can make reserves that were previously considered to be of marginal grade into profitable mining operations, and further developments in enrichment and extraction technology can lead to the profitable mining of previously uneconomic deposits. In addition, many previously unknown occurrences, especially in the deeper parts of the bedrock are now coming to light due to the development of exploration techniques and as geological knowledge increases.

6.3 Formation of Ore Deposits

Much of the Earth is affected by plate-tectonic processes that generated the Earth's crust and the ore deposits located therein. Rupture of the Earth's crust and the upper mantle into several individual plates and their movement relative to each other is also the primary reason for ore formation (Fig. 6.1). Rupture of continental plates is associated with felsic and mafic magmatism. These form, for example, basalts, granites, and mafic-ultramafic layered intrusions (Amelin et al. 1995; Eckstrand and Hulbert 2006). Ores are rock types containing economically exploitable amounts of useful minerals. Ores may be produced by all geological processes that form different rock types.

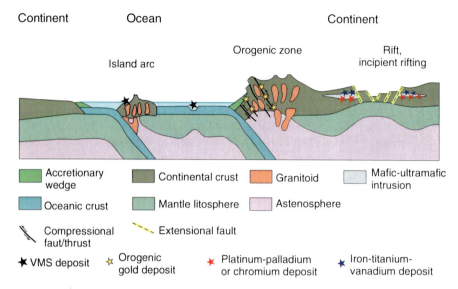

Fig. 6.1 Formation of major ore deposit types found in Finland, as seen in a plate-tectonic context. VMS refers to volcanic copper-zinc-lead deposits. (Modified from Groves et al. 2005)

Magmatism-related geological processes give rise to many types of ore deposits. Mafic and ultramafic magmas usually contain more chromium, iron, titanium, vanadium, nickel, copper and platinum metals than do other magmas. In favourable geological conditions mafic rocks may, for example, contain iron or chromium-rich oxides, or nickel or platinum sulfide ores. Granite magmatism, in turn, can give rise to ores containing copper, molybdenum, uranium and tin. The occurrence of many high-tech metals such as rare earth metals may also be related to granite magmatism.

Volcanism at the plate tectonic spreading centres and volcanic island-arc settings is capable of forming metal sulfide deposits which mainly contain copper, zinc and lead. These have been created by volcanic-related hydrothermal events in submarine environments. They are stratiform accumulations of sulfide minerals that precipitate from hydrothermal fluids on the seafloor in a wide range of ancient and modern geological settings. In modern oceans they are synonymous with sulfurous plumes called black smokers.

Several types of mineralisation are also associated with sedimentary environments and processes. Saline brines and the mixing of different fluids may lead to the formation of large zinc–lead deposits. Major uranium deposits are associated with sandstones as well as ancient placer gold deposits.

When ore minerals are crystallized or deposited at the same time as the host rock, we talk about syngenetic ores, whereas in epigenetic ores ore minerals are crystallized after the formation of the host rock. Regardless of their origins, many ores are deformed and metamorphosed after their initial formation and, in many cases, these later processes have led to the formation of economic resources from originally uneconomic deposits. For example, folding during an orogeny may concentrate metal-rich minerals into economically exploitable ore bodies, or metamorphism may have led to the recrystallization of ore minerals to be more easily recoverable in processing.

6.4 Mineral Resources in Fennoscandia

Fennoscandia comprises Norway, Sweden, Finland, and north-western Russia (Kola Peninsula and Russian Karelia) (Fig. 6.2). The northern and eastern parts are formed by Archean rocks that are over 2,500 million years old, and the central, southern and western parts by younger Proterozoic rocks. The geological variation of Fennoscandia is equally reflected in the volume and the range in types of its mineral deposits. The large number and size, and the extensive commodity range of its mineral deposits, which can be explained by the complex and very long plate-tectonic evolution of the Earth's crust in the region. This is a typical feature of the shield areas in the world, such as those in Western Australia, South Africa and parts of Brazil and Canada, which all are significant metal producers.

The most significant metallic mineral resources in Fennoscandia include: iron, nickel, chromium, copper, zinc, vanadium, palladium and gold. Most of these are hosted by deposits similar to other Precambrian areas of the world. There also are significant uncommon or unique deposit types, such as, the Kiruna-type iron deposits in northern Sweden, the alum shale-hosted uranium-molybdenum-vanadium deposits in Sweden, and the giant alkaline and carbonatite intrusion-hosted rare metal deposits in the Kola Peninsula and Finland. In addition, Fennoscandia hosts the unique Talvivaara and Pechenga nickel, and Outokumpu copper-cobalt deposits, which all contain large metal resources (Weihed et al. 2005, 2008, the Fennoscandian Ore Deposit Database 2010 and references therein).

The Fennoscandian Ore Deposit Database (2010) indicates that very large mineral resources still remain unexploited in the region. Most of these are located in the presently active mines and in large unexploited deposits. Eilu (2011) has estimated that if all the known and assumed Fennoscandian mineral resources were put into production, they would, based on 2007 levels of consumption within the European Union, cover the consumption of chromium, lithium, nickel, rare earth elements, tantalum, titanium, vanadium and probably also of niobium for more than 50 years (Fig. 6.3). Iron ore, cobalt, platinum-group elements and uranium would cover consumption for 10–30 years. These figures indicate the importance of Fennoscandian mineral production and metal resources for manufacturing and other industries within the EU and the nations of Fennoscandia. However, it is important to keep in mind that we are discussing in situ geological resources, not resources defined by present industrial standards.

For the rest of the metals discussed here, the situation is more challenging. There are large numbers of mines and unexploited deposits of many metals, but the known copper, gold, manganese, molybdenum, silver, zinc and zirconium resources could

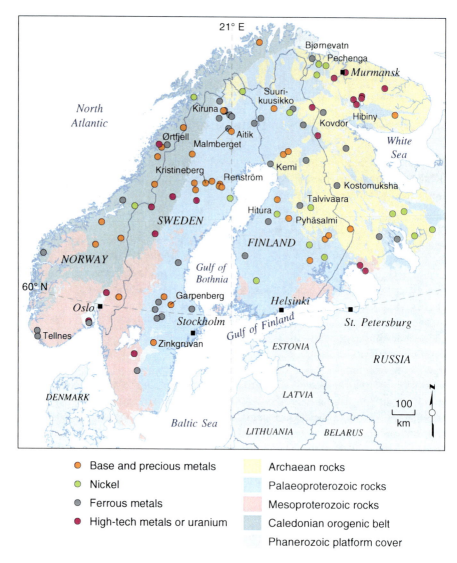

Fig. 6.2 Schematic geological map of Fennoscandia, showing the location of major ore deposits. The largest active metal mines are named

only cover EU demand for a few years; or, for aluminium, lead, tin and tungsten, the resources are quite small compared to EU rates of consumption. Despite this, even the latter two sets of metals include commodities whose production is significant, at least at the regional level, as exemplified by the Suurikuusikko gold mine in Finland and the Zinkgruvan and Skellefte Belt zinc-copper-gold-silver mines in Sweden.

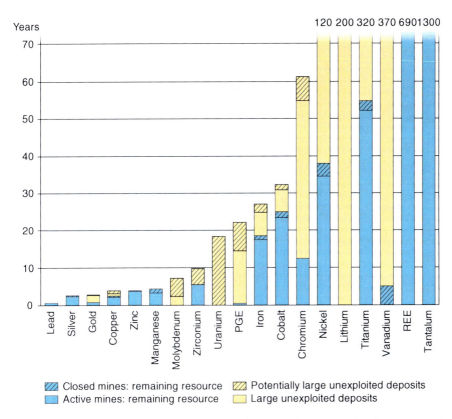

Fig. 6.3 Histogram indicating how many years the presently known mineral resources of Fennoscandia (Finland, Norway, Sweden, and NW Russia) could potentially cover of current EU consumption (in 2007) (Modified from Eilu 2011)

6.5 Mineral Resources in Finland and Their Formation

The metallogenic areas and the main ore deposits of Finland are shown in Fig. 6.4. The metallogenic areas have been defined on a geoscientific basis as regions which include known mineral occurrences of certain deposit types and in which there is a strong potential for new discoveries of those deposit types (Eilu et al. 2009). Some deposits occur outside the defined zones and, naturally, new discoveries are also possible elsewhere. Over forty metallogenic areas have been defined for Finland. As an example, 9% of the land area has good or very good exploration potential for copper and zinc, 9% for nickel and cobalt, and 8% for precious metals. These are vast areas compared to the current areas reserved for mining operations (mining concessions), which comprise only 0.09% of the total land area in Finland.

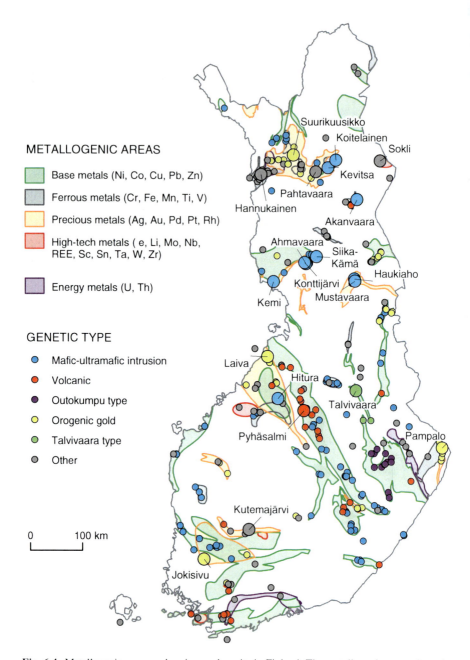

Fig. 6.4 Metallogenic areas and main ore deposits in Finland. The metallogenic areas show the regions where certain ore types occur and where new deposits of that type most probably can be discovered. Active metal mines and large unexploited deposits are named (Data from Eilu et al. 2009 and Fennoscandian Ore Deposit Database 2010)

6 Metallic Mineral Resources in Finland and Fennoscandia

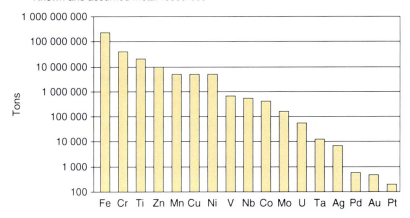

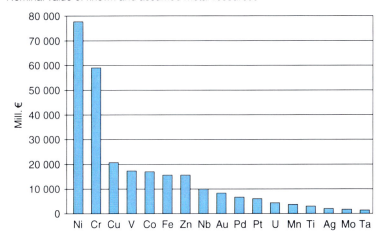

Fig. 6.5 Remaining geological in situ resources (known and assumed, see text for details) of the most important metals in Finnish ore deposits. Data are given for total metal tons (Data from Fennoscandian Ore deposit Database 2010) and for estimated Euro values, based on 10-year average metal prices (2000–2009)

Figure 6.5 shows the remaining metal resources in Finland based on information in the Fennoscandian Ore Deposit Database (2010). These figures contain both known and assumed resources, without taking into account their possible economic feasibility today or in the future. We have also estimated the speculative in situ value of the resources by metal type. These euro figures are mainly based on the average monthly metal prices at the London Metal Exchange (LME) from 2000 to 2009. For metals not listed by the LME, we have used other public sources (Eilu et al. 2007). On the basis of the nominal in situ value of the remaining mineral resources, the most important ore deposits in Finland host significant volumes of nickel, chromium, copper, vanadium, iron, zinc, niobium, gold, platinum and palladium. The diagrams demonstrate the extreme differences in the prices of different metals, from € 50 per ton for iron to € 50 million per ton for platinum.

The most important mineral deposits in Finland can be divided into five genetic types: (1) chromium, platinum metals (PGM), nickel and iron-vanadium ores associated with layered mafic–ultramafic intrusions; (2) copper–zinc ores associated with volcanic rocks; (3) Outokumpu-type copper ores; (4) orogenic gold ores; and (5) the graphitic mica schist-hosted nickel–zinc–copper ore of Talvivaara. Below, we provide simplified genetic models for each ore type and briefly discuss the main deposits in Finland.

6.5.1 Ore Deposits Related to Mafic: Ultramafic Intrusive Rocks

Mafic–ultramafic magmas may intrude into the Earth's crust during both continental rifting and crustal plate collision. Intrusions formed in both settings may contain significant nickel-copper and iron-titanium-vanadium metal ores. Intrusions associated with rifting may also host large chromium and PGM deposits.

Chromium, iron, titanium and vanadium are elements which tend to form oxides at different stages of magma crystallization (Fig. 6.6). These oxide minerals can accumulate in the magma chamber to form ore deposits if crystallization occurs slowly enough. Chromite, the most important mineral containing chromium, tends to accumulate at the base of an intrusion, whereas magnetite and ilmenite, which contain iron, titanium and vanadium, tend to precipitate in the upper portion of the

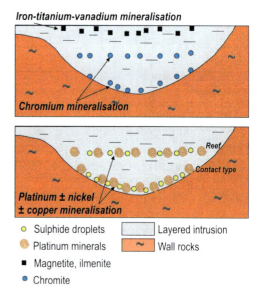

Fig. 6.6 Schematic diagram showing the formation of chromium and iron-titanium-vanadium ores (above), and platinum metal and nickel ores (below) in a mafic–ultramafic layered intrusive system (Petri Peltonen, Pers. comm. 2009)

intrusion. Layered intrusions usually form as a result of several pulses of magma, so stratabound mineral deposits can actually occur anywhere within the intrusion (Eckstrand and Hulbert 2006).

In contrast to the elements mentioned above, nickel, cobalt, copper and platinum metals are "chalcophile" elements, which means that they have a strong tendency to bind to sulfides. If a mafic-ultramafic magma intruding toward the Earth's surface reacts with sulfur-rich rocks or a primarily sulfur-rich magma reacts with silica-rich rocks, a separate sulfide melt may form. In this case, a significant part of the precious metals, nickel, cobalt and copper of the magma will go to the sulfide melt. The sulfide melt will initially form droplets and these can separate from the rest of the magma to form massive or disseminated accumulations during intrusion and crystallization. Platinum metals can also form separate metal droplets in silicate magma. Thus, magma crystallisation may lead to disseminated sulfide minerals in a silicate mass or larger sulfide masses, which both may form economic ores (Fig. 6.6). Because sulfides are heavier than silicates, the ores in most cases occur at the base of an intrusion. Disseminated ore deposits can also occur as thin layers almost anywhere within an intrusion (Iljina and Hanski 2005; Eckstrand and Hulbert 2006).

The mafic–ultramafic intrusions in northern Finland host remarkable resources of chromium, platinum metals, nickel and vanadium (Table 6.1, Figs. 6.4 and 6.5). The Kemi chromium mine (Mutanen 1997; Alapieti et al. 1989) is in one of the approximately 2,400 million-year-old layered intrusions which form a discontinuous belt across southern Lapland, from the Swedish border in the west to Russia in the east. The Kemi mine is the backbone of stainless steel production in northern Europe and has significant resources for the coming decades. There are currently plans to double production. In addition to Kemi, the Koitelainen intrusion further to the north also hosts extensive chromium resources.

The same group of mafic–ultramafic intrusions also contains significant platinum and palladium resources (Iljina and Hanski 2005). The deposits hosted by the Portimo (e.g., Ahmavaara and Konttijärvi, Table 6.1) and Penikat intrusions have recently been evaluated in great detail, but no mining operations have yet been undertaken. The main reasons for the delay in the exploitation of these deposits evidently are the relatively low metal grades and the predominance of palladium over platinum, with the former currently having a much lower price. In addition to the known resources, the intrusions in northern Finland have been assessed to contain a number of yet undiscovered deposits with remarkable resources, even when considered on a global scale (Rasilainen et al. 2010).

The Kevitsa intrusion in central Finnish Lapland (Fig. 6.4) contains large resources of intrusive-hosted nickel, with copper and platinum metals as significant by-products (Table 6.1). Production at Kevitsa is planned to begin in 2012; this will be the first nickel mine in northern Finland and the first producer of platinum metals in Europe.

Table 6.1 The most important metal ore deposits in Finland, including present and closed mines, and unexploited deposits

Name	Main metals	Pre-mining value M€	Pre-mining size (Mt)	Mined (Mt)	Genetic type
Talvivaara	Ni,Co,Zn	75,453	1550.0	27.0	Talvivaara type
Kemi	Cr	30,842	158.8	35.8	Mafic-ultramafic
Koitelainen UC	Cr,V	17,840	70.0		Mafic-ultramafic
Kevitsa	Ni,Cu,Co,Pt,Pd	13,198	208.0		Mafic-ultramafic
Sokli	Nb,Zr,Ta	12,891	250.0		Carbonatite
Akanvaara Cr	Cr,V	11,721	55.1		Mafic-ultramafic
Outokumpu	Co,Cu	6,768	28.5	28.5	Outokumpu type
Ahmavaara	Pd,Pt,Ni,Cu	5,504	187.8		Mafic-ultramafic
Hannukainen	Fe	4,718	202.5	4.5	Iron oxide-copper-gold
Pyhäsalmi	Zn,Cu	3,940	65.1	42.1	Volcanic
Vihanti Zn	Zn,Cu	3,002	37.1	27.9	Volcanic
Otanmäki	V,Fe,Ti	2,845	36.1	25.4	Mafic-ultramafic
Suurikuusikko	Au	2,603	51.9		Orogenic gold
Mustavaara	V,Fe,Ti	2,489	43.4	13.4	Mafic-ultramafic
Siika-Kämä	Pd,Pt,Ni	2,416	43.1		Mafic-ultramafic
Konttijärvi	Pd,Pt,Ni,Cu	1,906	75.2		Mafic-ultramafic
Ruossakero	Ni,Co	1,578	35.6		Mafic-ultramafic
Hitura	Ni,Co	1,490	19.4	15.0	Mafic-ultramafic
Haukiaho	Ni,Pd,Pt,Cu	1,473	27.0		Mafic-ultramafic
Akanvaara V	V	1,459	20.0		Mafic-ultramafic
Kotalahti	Ni,Co	1,222	12.4	12.4	Mafic-ultramafic
Kylylahti	Co,Cu,Ni	1,222	7.9		Outokumpu type
Vuonos Cu	Co,Cu,Zn	1,080	6.7	5.9	Outokumpu type
Sompujärvi	Pt, Pd	1,061	6.7	0.0	Mafic-ultramafic
Laukunkangas	Ni,Co	1,005	12.6	6.7	Mafic-ultramafic

Main metals are given in decreasing order of nominal in situ economic value. Speculative economic euro values (based on 2000–2009 average prices) for the combined mined and existing in situ geological mineral resources are given (see text for details). More detailed data, including metal grades and geological information, can be found in the Fennoscandian Ore Deposit Database (2010)

Order of metals given according to 2000–2009 average metal prices

For the entire Sokli deposit, the main commodity is apatite (phosphate); the combined size of the apatite ore bodies is more than 12,000 million tonnes

M€ mill. Euros, *Mt* mill. Tons

6.5.2 Ore Deposits Related to Volcanic Rocks

Oceanic lithospheric plates form most of the ocean floors. These oceanic plates are traversed by central rift zones thousands of kilometres long, along which new crust is created from mafic magma forcing its way up from the mantle (Fig. 6.7). This igneous activity also generates basaltic volcanoes on the sea floor, causes the circulation of sea water as hydrothermal solutions within the Earth's crust. Additionally, volcanic island arcs are generated when one oceanic plate slides under

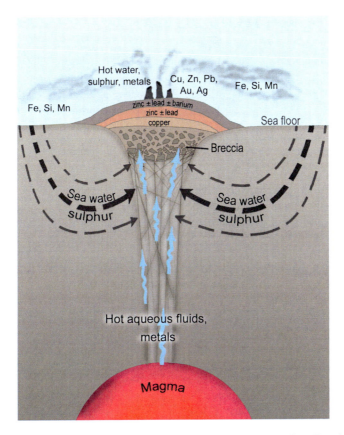

Fig. 6.7 Schematic diagram on the ore forming system related to volcanism (Based on Lydon 1988; Galley et al. 2006)

another, leading to an arc of volcanoes and granitic intrusions at the plate boundary. The movement of a hot circulating fluid through the bedrock dissolves metals from the rock into the fluid. Metals may also originate directly from the magmas (Fig. 6.7). Below the sea floor, in rift and fracture zones, hot, sulfuric, aqueous solutions rise upwards, and metals and sulfides are deposited into and in fissures of the sea floor (Lydon 1988; Galley et al. 2006). Because these types of deposits are closely related to volcanism and are generally in lava and tuff sequences, they are known as volcanic-associated massive sulfide (VMS) deposits.

The largest volcanic-related ore deposits in Finland are the approximately 1,900 million-year-old Pyhäsalmi and Vihanti zinc-copper mines in central western Finland (Table 6.1, Fig. 6.4). Vihanti is already closed, but Pyhäsalmi is active and currently the deepest mine in Europe, with a vertical shaft extending down to 1,500 m. There also is a good exploration potential for additional resources in the Pyhäsalmi–Vihanti area.

6.5.3 Outokumpu-Type Ore Deposits

Outokumpu-type ore deposits are unique. Hence, many hypotheses have been presented on their genesis, the latest by Peltonen et al. (2008), as follows. In the first stage, a hydrothermal system was formed within an ultramafic environment on the ocean floor. This gave rise to copper (± zinc) deposits in the oceanic crust, possibly on the floor of the ocean, in a VMS-like system, but hosted by ultramafic rocks. After this, at the outset of the Svecofennian orogeny, about 1.9 billion years ago, plate tectonic processes pushed the sea floor on top of the continental crust, subjecting the rock sequence to low-temperature hydrothermal alteration. At this point, nickel and cobalt contained within the silicates of the ultramafic rocks were released and bound to the sulfides. At the peak of the Svecofennian orogeny, copper and nickel-bearing ores were mixed, partly mechanically, and partly, perhaps, hydrothermally, resulting in the massive sulfide ores we can see today.

The Outokumpu region has been mined for its major copper, cobalt, zinc, and nickel ores and these mining activities have had a crucial influence on the development of modern mining industry in Finland (Table 6.1, Fig. 6.4). Presently, there are plans to open a new mine in the area. Recent reflection seismic surveys have demonstrated the widespread occurrence of ultramafic intrusions in the near-surface environments, and the potential of this classic area for new ore discoveries remains significant (Kukkonen et al. 2011).

6.5.4 Orogenic Gold Deposits

Most gold occurrences found in Finland can be classified in the category of orogenic gold deposits (Groves 1993; Goldfarb et al. 2001; Eilu et al. 2003). These form at island arc-continent and continent-continent collisional settings during the peak stages of orogenies (Fig. 6.1). Magmas and hot aqueous fluids rise from the deep parts of the crust located under magmatic arcs in these settings (Fig. 6.8). Fluids also originate when volcanic and sedimentary successions are heated and metamorphosed, and water and volatile elements are separated from the rocks. These fluids contain variable amounts of silica, carbon dioxide and sulfur and are, hence, able to extract gold from all rocks on their way towards upper levels of the crust, along major shear and fault zones. Part of the fluids and the gold may originate from deep intrusions and from the mantle of the Earth, although conclusive indications of such origins for the gold are still missing.

The concentration of gold in the hot fluids is low. Hence, formation of an economic deposit requires both large fluid volumes and an effective gold precipitation mechanism. As the auriferous fluids travel upwards, they over-pressurise the local environment and may cause earthquakes. An earthquake results in a sudden decrease in pressure, which causes the destruction of the chemical balance of the

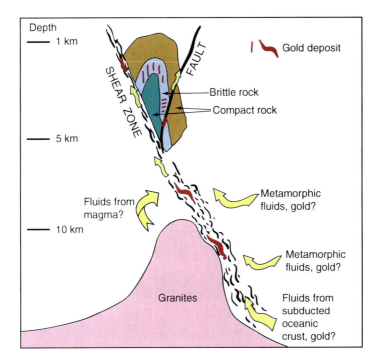

Fig. 6.8 Schematic diagram demonstrating the formation of orogenic gold deposits (After Groves 1993)

fluid and the precipitation of quartz, sulfides and gold as veins. Sulfur compounds which carry the gold may also react with the wall rocks, resulting in further precipitation of gold and sulfides. Both scenarios may produce deposits large enough to be viable for mining.

In Finland, there currently are three small gold mines in operation and another two under construction, in addition to the major Suurikuusikko mine in Lapland (Table 6.1, Fig. 6.4). Suurikuusikko is one of the largest gold deposits in Europe and has potential for discovery of remarkable resources in addition to those presently known.

6.5.5 Talvivaara Nickel Deposit

There are a few exotic nickel-zinc-copper sulfide occurrences associated with the Paleoproterozoic graphitic mica schists in the schist belts covering large areas of eastern Finland. The gigantic Talvivaara deposit is the only deposit that has so far been detected. It has been suggested that the metals and sulfur were precipitated under reducing conditions at the bottom of a large (sea?) water basin simultaneously with carbon-rich mud, silt and sand. This proto-ore was deformed in the

Svecofennian orogeny, giving it its present form and composition (Kontinen et al. 2006; Loukola-Ruskeeniemi and Heino 1996).

Talvivaara is one of the largest metal ore deposits in Europe. It has huge but fairly low-grade resources of nickel, zinc, copper and cobalt (Table 4, Fig. 6.1). The economy of the mine is based on large-scale open-pit mining and a novel, bio-heap leaching technique. The plan is to produce 50,000 t nickel (over 2% of world production), 90,000 t zinc, 15,000 t copper and 1,800 t cobalt annually for the next 50 years. The company is also currently investigating the possibility to recover uranium and manganese from the ore. The uranium concentration in the ore is so low, 0.0017%, that it does not constitute a uranium ore as such. However, the huge tonnage of ore that is processed, combined with effective metal recovery, could evidently lead to a production roughly matching the needs of all the nuclear power reactors in Finland, and Talvivaara could become the largest uranium producer in Europe.

6.6 Discussion

Despite the continuous increase in demand for mineral resources, there is no risk of rapid depletion of raw materials – the Earth is entirely made up of minerals. In addition to identified ore resources, a large number of sub-economic mineral deposits are known to contain huge resources, as demonstrated in the present paper. The sub-economic resources may become economically viable in the future, depending on their geographic location, commodity demand and price fluctuation, and the introduction of new mining and beneficiation technologies (Ericsson 2010; Tilton 2010). Our knowledge of the mineral potential of many areas continues to be deficient, and our level of understanding in areas that have been extensively surveyed is usually limited to the near-surface environment. The success of a new mining operation requires not just adequate mineable reserves, but also the availability of an appropriate technology for ore processing, the existence or possibility to build the infrastructure needed and local political consent for minerals exploitation. Talvivaara, Suurikuusikko and Kevitsa in Finland are good examples of new large-scale mines which are based on recent discoveries, new processing techniques, and pro-mining attitudes of the local community.

Future mining operations will have to be increasingly based on underground operations and large, low-grade, open-pittable deposits. Mining and processing of these ores may be more challenging. Competition for available water and energy is becoming more intense, and anti-mining attitudes of local communities are strengthening, which may prevent mining activity in some areas. Responses to these challenges require innovative technological developments throughout the entire extraction and production chain. Mining also has to earn its social licence to operate. Permit procedures are becoming more demanding, which can lead to delays in production. These factors, when combined with a continuous increase in consumption and demand, raise metal prices, which, in turn, provides incentives for seeking both alternative raw materials and improving recycling efficiency.

Metals and minerals are non-renewable natural resources. However, the lifespan for products derived from them is typically long, on the order of decades, and they can be recycled effectively. At best, recycling can only partially meet the current demand for minerals. For example, more than 80% of copper is recycled but, due to rapidly increasing demand and the fact that the average life of copper products is more than 30 years, recycling covers only one-third of current needs (International Copper Study Group 2010). The increased use of metals can be considerably slowed through careful product planning based on material efficiency and recycling. Price increases, as well as adoption of new and replacement materials, can also lower the demand for traditional raw materials. The overall intensity of metal use, for example in the EU, is slowly decreasing and, as in other developed countries, recycling is becoming more efficient and new replacement materials are being found. Despite that, it is expected that manufacturing industries will, for a long time to come, remain vulnerable to disruptions in the metal supply and to market volatility. In this respect, the large unexploited resources and the good discovery potential for new resources in Finland and Fennoscandia will be increasingly important.

The rapid expansion of information and communications technology, and developments in new energy technologies, have created an entirely new demand for a range of high-tech metals that previously were only used to a small extent. Accordingly, only a relatively small proportion of these metals can be sourced through recycling. To meet the growing demand for these metals, new mines are required, as well as more effective techniques for the recovery of these high-tech metals as by-products of other mining activities.

Europe is heavily or fully reliant upon the import of minerals, so disruptions in availability and supply can pose a significant risk. The European Union has recently defined the most critical and economically important raw materials (Anon 2010). The Finnish bedrock contains significant known deposits of many of the 40 raw materials listed by the European Union, and has considerable potential for the discovery of new resources of critical minerals, as evaluated in Fig. 6.9. The volume and vast range of types of mineral deposits in Finland, and more generally in Fennoscandia, is based on the very long and complex geological history of the Earth's crust in the region.

In recent years, significant new mines have been opened in Finland, while existing mines have recorded increased production, and many more mining projects are in progress. The ore output of metallic mines in Finland is estimated to increase to 70 million tons by 2020 from the modest 4 million tons of the early 2000s (Fig. 6.10). It can also be expected that a number of new commodities will be produced in Finland for domestic use and for world markets (Fig. 6.11).

The world's deepest mines in South Africa extend to over 3 km in depth and the deepest mine in Europe, Finland's Pyhäsalmi mine, extends to more than 1.5 km in depth, but the vast majority of existing mines throughout the world exploit ores which crop out at the surface. Many potential ore-bearing geological formations extend to a depth of several kilometres, and there is no reason why ores should not

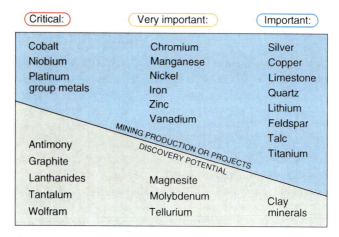

Fig. 6.9 Current and possible near-future production and discovery potential of the critical and economically important raw materials listed by the European Union (Anon 2010) (Data from Finland's minerals strategy 2010)

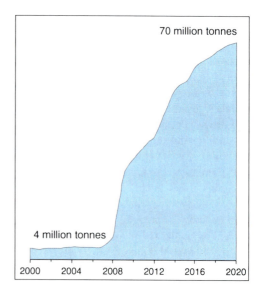

Fig. 6.10 A prognosis of ore output from metallic mines in Finland (Finland's minerals strategy 2010)

exist in similar quantities deep in the Finnish bedrock as on the surface. A particular challenge for the future will be to more effectively locate and develop ore deposits which have no surface outcrop. This will require more detailed geological data bases, better research-based understanding of ore formation, and efficient geophysical and geochemical exploration methods.

Fig. 6.11 A vision for metal mining in Finland in 2020 (Finland's minerals strategy, 2010)

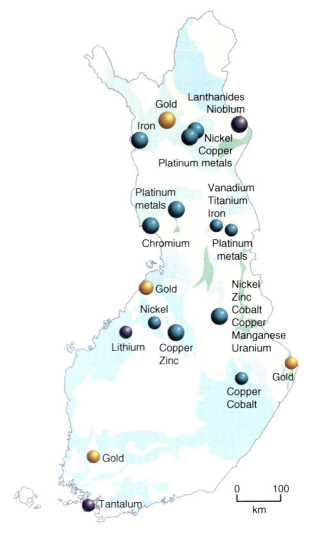

References

Alapieti T, Kujanpää J, Lahtinen JJ, Papunen H (1989) The Kemi stratiform chromitite deposit, northern Finland. Econ Geol 84:1057–1077

Amelin YV, Heaman LM, Semenov VS (1995) U-Pb geochronology of layered mafic intrusions in the eastern Baltic Shield; implications for the timing and duration of Paleoproterozoic continental rifting. Precambrian Res 75:31–46

Anon (2010) Critical raw materials for the EU. Report of the Ad-hoc Working Group on defining critical raw materials. European Commission, June 2010. http://ec.europa.eu/enterprise/policies/raw-materials/files/docs/report-b_en.pdf

Australasian Joint Ore Reserves Committee (2004) Australasian Code for Reporting of Identified Mineral Resources and Ore Reserves (The JORC Code), 2004 Edition Issued in December 2004. Australasian Institute of Mining and Metallurgy, Australian Institute of Geoscientists, and Minerals Council of Australia, 20pp. http://www.jorc.org/main.php

Buchert M, Schüler D, Bleher D (2009) Critical metals for future sustainable technologies and their recycling potential. Öko-Institut e.V. United Nations Environment Programme. http://www.unep.fr/shared/publications/pdf/DTIx1202xPA-Critical%20Metals%20and%20their%20Recycling%20Potential.pdf

Eckstrand OR, Hulbert L (2006) Magmatic nickel-copper-platinum group element deposit. Geol Assoc Can, Miner Depos Div Spec Publ 5:205–222

Eilu P (2011) Metallic mineral resources of Fennoscandia. In: Nenonen K, Nurmi PA (eds) Geoscience for society – 125th Anniversary volume. Geol Surv Finland Spec Paper 49:13–21

Eilu P, Sorjonen-Ward P, Nurmi P, Niiranen T (2003) A review of gold mineralization styles in Finland. Econ Geol 98:1329–1354

Eilu P, Hallberg A, Bergman T, Feoktistov V, Korsakova M, Krasotkin S, Lampio E, Litvinenko V, Nurmi PA, Often M, Philippov N, Sandstad JS, Stromov V, Tontti M (2007) Fennoscandian ore deposit database – explanatory remarks to the database. Geological Survey of Finland, Report of Investigation 168

Eilu P, Bergman T, Bjerkgård T, Feoktistov V, Hallberg A, Korsakova M, Krasotkin S, Muradymov G, Nurmi PA, Often M, Perdahl J-A, Philippov N, Sandstad JS, Stromov V, Tontti M (2009) Metallogenic map of the Fennoscandian Shield 1:2 000 000. Espoo: Trondheim: Uppsala: St. Petersburg: Geological Survey of Finland: Geological Survey of Norway: Geological Survey of Sweden: The Federal Agency of Use of Mineral Resources of the Ministry of Natural Resources of the Russian Federation

Ericsson M (2010) Global mining towards 2030. Geological Survey of Finland, Report of Investigation 187. http://arkisto.gtk.fi/tr/tr187.pdf

Fennoscandian Ore Deposit Database (2010) Geological Survey of Finland, Geological Survey of Norway, the Federal Agency of Use of Mineral Resources of the Ministry of Natural Resources of the Russian Federation, Geological Survey of Sweden http://en.gtk.fi/ExplorationFinland/fodd/

Finland's Minerals Strategy (2010) Geological Survey of Finland, www.mineraalistrategia.fi

Galley A, Hannington M, Jonasson I (2006) Volcanogenic-associated massive sulfide deposits (VMS). Geol Assoc Can, Miner Depos Div Spec Publ 5:141–162

Goldfarb RJ, Groves DI, Gardoll S (2001) Orogenic gold and geologic time: a global synthesis. Ore Geol Rev 18:1–75

Groves DI (1993) The crustal continuum model for late-Archaean lode-gold deposits of the Yilgarn Block, Western Australia. Mineral Depos 28:366–374

Groves DI, Condie KC, Goldfarb RJ, Hronsky JMA, Vielreicher RM (2005) Secular changes in global tectonic processes and their influence on the temporal distribution of gold-bearing mineral deposits. Econ Geol 100:203–224

Iljina M, Hanski E (2005) Layered mafic intrusions of the Tornio–Näränkävaara belt. In: Lehtinen M, Nurmi PA, Rämö OT (eds) Precambrian geology of Finland – key to the evolution of the Fennoscandian shield. Elsevier, Amsterdam, pp 101–138

International Copper Study Group (2010) The World Copper Fact Book 2009. http://www.icsg.org/index.php?option=com_docman&task=doc_download&gid=234&Itemid=61

Kontinen A, Peltonen P, Huhma H (2006) Description and genetic modelling of the Outokumpu-type rock assemblage and associated sulphide deposits. Final technical report for GEOMEX JV. Geological Survey of Finland, Archived Report M10.4/2006/1

Kukkonen IT, Heikkinen P, Heinonen S, Laitinen J, HIRE Working Group of the Geological Survey of Finland (2011) Reflection seismics in exploration for mineral deposits: initial results from the HIRE project. In: Nenonen K, Nurmi PA (eds) Geoscience for society – 125th Anniversary volume. Espoo. Geol Surv Finland Spec Paper 49:49–58

Loukola-Ruskeeniemi K, Heino T (1996) Geochemistry and genesis of the black shale-hosted Ni-Cu-Zn deposit at Talvivaara, Finland. Econ Geol 91:80–110

Lydon JW (1988) Volcanogenic massive sulphide deposits. Part 2: Genetic models. In: Roberts R, Sheahan PA (eds) Ore deposit models. Geological Association of Canada. Geoscience Canada Reprint Series 3:155–182

Moon CJ, Evans AM (2006) Ore, mineral economics, and mineral exploration. In: Moon CJ, Whateley KG, Evans AM (eds) Introduction to mineral exploration, 2nd edn. Blackwell Publishing, Malden, pp 3–18

Mutanen T (1997) Geology and ore petrology of the Akanvaara and Koitelainen mafic layered intrusions and the Keivitsa-Satovaara layered complex, northern Finland. Geol Surv Finland Bull 395

National Instrument 43-101 (2006) Standards for the disclosure of mineral projects. Canadian Institute of Mining and Metallurgy (CIM). http://www.ccpg.ca/guidelines/index.html

Peltonen P, Kontinen A, Huhma H, Kuronen U (2008) Outokumpu revisited: new mineral deposit model for the mantle peridotite-associated Cu-Co-Zn-Ni-Ag-Au sulphide deposits. Ore Geol Rev 33:559–617

Puustinen K (2003) Suomen kaivosteollisuus ja mineraalisten raaka-aineiden tuotanto vuosina 1530–2001, historiallinen katsaus erityisesti tuotantolukujen valossa (In Finnish). Geological Survey of Finland, Archived Report M 10.1/2003/3

Rasilainen K, Eilu P, Halkoaho T, Iljina M, Karinen T (2010) Quantitative mineral resource assessment of undiscovered PGE resources in Finland. Ore Geol Rev 38(3):270–287

Tiess G (2010) Minerals policy in Europe: some recent developments. Resour Policy 35:190–198

Tilton JE (2010) Is mineral depletion a threat to sustainable mining. SEG Newsletter 82, July 18–20

Weihed P, Arndt N, Billström C, Duchesne JC, Eilu P, Martinsson O, Papunen H, Lahtinen R (2005) Precambrian geodynamics and ore formation: the Fennoscandian shield. Ore Geol Rev 27:273–322

Weihed P, Eilu P, Larsen RB, Stendal H, Tontti M (2008) Metallic mineral deposits in the Nordic countries. Episodes 31:125–132

Whateley KG, Scott BC (2006) Evaluation techniques. In: Moon CJ, Whateley KG, Evans AM (eds) Introduction to mineral exploration, 2nd edn. Blackwell Publishing, Malden, pp 199–252

World Economic Forum (2010) Mining & metal Scenarios to 2030. http://www.mckinsey.com/clientservice/metalsmining/pdf/mining_metals_scenarios.pdf

Chapter 7
Isotopic Microanalysis: In Situ Constraints on the Origin and Evolution of the Finnish Precambrian

O. Tapani Rämö

7.1 Introduction

The Precambrian of Finland has been the target of systematic geochronological work since the adoption of modern mass spectrometric analytical methods in the 1950s. A radiogenic isotope laboratory utilizing thermal ionization mass spectrometry (TIMS) was founded by Olavi Kouvo at the Geological Survey of Finland in the early 1960s. Dr. Kouvo had participated in the early developments of modern isotope geochemical and geochronological methods in the United States, and he built Nier-type mass spectrometers and purification lines for pertinent isotopes (those of U and Pb in particular) in Finland. His doctoral thesis (Kouvo 1958) presented new radiogenic isotope data on the K-Ar and Rb-Sr isotopic systems in the mineral mica and the U-Pb isotopic system in the minerals zircon and monazite. These data gave the first glimpse of the Precambian geochronologic division of the Fennoscandian shield, as a >2.5 Ga age for zircons analyzed from a gneiss dome in eastern Finland was documented. This work was carried out further by Kouvo and his coworkers (Wetherill et al. 1962) who, on the basis of the isotopic systematics of zircon, feldspar, and mica, verified a Neoarchean (~2.7 Ga) age for the gneissic "Pre-Karelian" basement of eastern Finland, a Paleoproterozoic (~1.8 Ga) age for the Precambrian (Svecofennian) bedrock of southern and central Finland, and recognized ~1.8 Ga orogenic resetting of the Rb-Sr and K-Ar systems in mica from the Pre-Karelian basement. The current division of the Precambrian of Finland still builds on these early results (see Lahtinen this volume).

O.T. Rämö (✉)
Department of Geosciences and Geography, University of Helsinki, P.O. Box 64, FI-00014 Helsinki, Finland
e-mail: tapani.ramo@helsinki.fi

Since the 1960s, the isotope laboratory of the Geological Survey of Finland has provided invaluable boundary conditions and piercing points for the evolution of the Fennoscandian bedrock and is producing state-of-the-art radiogenic and stable isotope data on a wide variety of geological materials, also in collaboration with domestic and foreign universities and research institutes (see Huhma et al. 2011). At present, the Precambrian of Finland has, for example, more than 1100 U-Pb age determinations across the country (Huhma et al. 2011; see also http://geomaps2.gtk.fi/activemap/).

7.2 Materials

The mineral zircon ($ZrSiO_4$) has proven to be invaluable in U-Pb geochronology as a robust indicator of geological processes (e.g., Faure and Mensing 2005). Zircon crystallizes in high-temperature and high (and varying) pressure environments as silicate melts solidify and as solid rock material recrystallizes in lithospheric segments undergoing regional metamorphism. A fraction of the lattice-forming elements in zircon are substituted by elements such as uranium and hafnium that belong to radioactive decay series. After crystallization, the two radioactive isotopes of uranium (^{238}U and ^{235}U), for example, decay to lead (^{206}Pb and ^{207}Pb, respectively) at exceedingly slow rates, and thus provide a suitable isotopic clock that can be used in Precambrian (and Phanerozoic) geochronology to determine dates for past lithospheric processes (Fig. 7.1). The element hafnium contains the radiogenic isotope ^{176}Hf that is produced by radioactive decay of ^{176}Lu, also at a miniscule pace. Zircon incorporates little lutetium and thus serves as a good indicator of the hafnium isotope composition of the media from which it precipitated. The initial hafnium isotope composition of zircon (the $^{176}Hf/^{177}Hf$ ratio calculated back in time to the crystallization event) can be used as a geochemical fingerprint, e.g., in modeling magma sources and division of crustal terranes in Earth's lithosphere. The isotope composition of oxygen in zircon has also been recognized as an important fingerprint and is a valuable proxy for rock material that has been, at some point, exposed to exogenic processes at the surface of the Earth. These processes fractionate oxygen isotopes and increase the proportion of the heaviest oxygen isotope (^{18}O) relative to the lightest one (^{16}O) in supracrustal rock material. The isotope composition of oxygen is expressed as the $\delta^{18}O$ value relative to a global standard. The element boron has also been introduced as a fingerprint element and its isotope composition (expressed as the $\delta^{11}B$ value) has been applied, for example, in cosmochemical studies and in research involving mineralization processes in auriferous geological environments.

7 Isotopic Microanalysis: In Situ Constraints on the Origin and Evolution

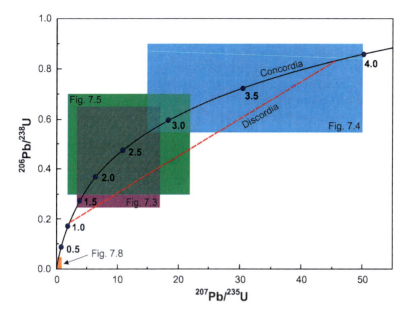

Fig. 7.1 The concordia diagram. This diagram is used in geochronology to determine ages of minerals that incorporate uranium (but little lead) upon crystallization. The two variables (Pb/U ratios) are related to the two radioactive isotopes of uranium (^{235}U and ^{238}U) that decay to lead (^{207}Pb and ^{206}Pb, respectively) at very low (and constant) rates. In a closed system, the position of an analytical point on the curve (concordia) denotes the time (expressed in billions of years, Ga) since crystallization of the analyzed solid material, as consanguineously defined by the two U-Pb decay systems. In a system that has not stayed closed, analytical points usually fall on a line (discordia) below the concordia; the discordia often intercepts the concordia at points of chronological significance. Shaded rectangles show the areas depicted in the respective figures of this chapter

7.3 Methods

Conventional TIMS analyses of zircon (e.g., the U-Pb and Lu-Hf isotope systems) usually involve bulk samples of several to hundreds (in the early days even much more) of zircon crystals and the U-Pb and Lu-Hf systematics revealed are always average values of the analysed crystal fraction. Zircon, however, is an utterly robust mineral and a single zircon grain usually preserves individual isotopic systems related to various crystallization and recrystallization events. Technical developments started in the late 1970s have provided geologists with in situ isotope analysis techniques that are capable of measuring isotopic systems in zircon and various other minerals on a μm-scale. This has revolutionized isotopic applications of geological materials and opened a completely new window to the origin, evolution, and correlation of Precambrian lithosphere.

In modern in situ analytical work, two basic techniques are used. In secondary ion mass spectrometry (SIMS), a polished mineral grain mounted in epoxy and coated with gold is bombarded with a primary ion beam (oxygen, cesium, or argon) in high vacuum, and material from the grain is ionized from a spot of 5 µm to 30 µm in diameter. In U-Pb chronology, the ionised material is analysed for its Pb isotope composition and the U/Pb ratio by a very high-resolution, double-focusing mass spectrometer in order to comply with substantial isobaric interferences caused by extra ions generated in the sputtering process. The SIMS technique was originally developed at the Australian National University (Clement et al. 1977). The prototype secondary ion mass spectrometer (SHRIMP I) was built in 1977–1981 and subsequently developed into an improved commercial version (SHRIMP II) that has been installed in more than ten laboratories worldwide. The SHRIMP is mainly used for U-Pb geochronology but it also has routines for stable isotope and ultra-high precision trace element analyses. Another commercially available secondary ion mass spectrometer is the French-made Cameca IMS1270/1280 that has a wide range of radiogenic and stable isotope and trace element applications. The Cameca IMS1270/1280 is also operational in more than ten locations worldwide, including the Swedish Museum of Natural History (the NORDSIM laboratory).

In situ isotopic analyses of geological materials are also carried out using laser ablation inductively coupled mass spectrometry (LAMS). The target is ablated at ambient pressure using a UV laser beam of circa 25–50 µm in diameter, which leaves behind a few tens of µm deep crater. The ablated material is carried in helium to an argon plasma induced by a radio frequency electric current, ionised in the plasma, and analyzed with a double-focusing multi-collector mass spectrometer. The LAMS method is suitable for a large variety of radiogenic and stable isotope systems (not oxygen, however) and is the state-of-the-art method for in situ Lu-Hf measurements on zircon.

The bulk of the in situ isotope analytical work thus far performed on samples from the Finnish Precambrian has involved the U-Pb in zircon and oxygen in zircon methods and has been carried out at the NORDSIM laboratory in Stockholm (http://www.nrm.se/en/menu/researchandcollections/departments/laboratoryforisotopegeology/nordsim.904_en.html). Some U-Pb zircon studies have been made at the SHRIMP facilities at the Research School of Earth Sciences, Australian National University in Canberra (http://shrimp.anu.edu.au/shrimp/instruments.htm) and at the A.P. Karpinsky Russian Geological Research Institute (VSEGEI) in St. Petersburg (http://www.vsegei.ru/way/247225/sx/obj/247276.html). The Cameca IMS1270 laboratory at the University of Edinburgh has been used for in situ boron isotope analysis (http://www.geos.ed.ac.uk/facilities/ionprobe/). During the last couple of years, in situ U-Pb and Lu-Hf zircon work has been conducted at the LAMS laboratory of the Department of Geosciences, University of Oslo (http://www.geo.uio.no/english/). In situ LAMS data on the Finnish Precambrian have also started to emerge from the newly established SIGL laboratory in Espoo, Finland. SIGL is a joint venture of the Geological Survey of Finland and Finnish universities with curricula in geology (http://en.gtk.fi/research2/infrastructure/isotopegeology/).

7.4 Twenty Years of In Situ Zircon Isotope Studies on the Finnish Precambrian

7.4.1 The First In Situ Analyses on Finnish Zircons

The first results of in situ isotopic work on the Finnish Precambrian were published by Huhma et al. (1991). They utilized the SHRIMP I in Canberra and measured the U-Pb isotope systematics of 13 zircon grains extracted from a metagreywacke (metamorphosed sandstone originally deposited in deep marine environment) from the Paleoproterozoic Tampere belt in southern Finland (Fig. 7.2). Conventional bulk-fraction zircon TIMS U-Pb analyses suggested approximate ages around 2.3 Ga for the detritus in the metagreywacke (Kouvo and Tilton 1966; Huhma et al. 1991). The in situ data (Fig. 7.3) showed positively that the conventional analyses represented mixtures of individual 2.0–1.9 Ga and 2.75–2.7 Ga grains and that the greywacke had two principal source areas (provenances) – Paleoproterozoic and Neoarchean continental crust. Huhma et al. (1991) found little evidence for the previously postulated 2.4–2.1 Ga crustal domains as the source area of the sedimentary detritus. A more extensive set of the in situ U-Pb zircon data acquired at the Canberra SHRIMP I laboratory for detrital zircons from the Finnish (and Swedish) bedrock (in total, 120 spots) was published by Claesson et al. (1993). These data revealed inheritance up to 3.44 Ga in the metagreywackes of the Tampere schist belt. Approximately one-third of the analyzed zircon crystals were Meso- to Neoarchean (2.97–2.6 Ga) and two-thirds Paleoproterozoic (2.12–1.88 Ga). These data further revealed an unrecognized ~2.1–1.9 Ga crust as an important provenance for the greywackes, and that there was no evidence for 2.5–2.1 Ga crustal domains as a paleodetritus source.

The pioneering studies by Huhma et al. (1991) and Claesson et al. (1993) illustrated the power of the in situ SIMS technique in acquiring completely new inferences about the evolution of Precambrian crust. The last two decades of in situ isotopic work have brought about a wealth of new data that have created completely new concepts (and considerably refined existing ones) regarding the origin and evolution of the Finnish Precambrian, as outlined below.

7.4.2 Archean Crustal Evolution

The bulk of the Archean crust in Finland belongs to the Karelian province of the Fennoscandia shield (Hölttä et al. 2008, Lahtinen this volume). The Archean crust is exposed in eastern and northern Finland and consists of multiply deformed sequences of tonalite-trondjemite-granodiorite (TTG) suite granitoids and associated amphibolites, sanukitoid rocks, paragneisses, and greenstone belts (Fig. 7.2). Most of the crust was formed at 2.9–2.7 Ga and the resulting crust is most likely an

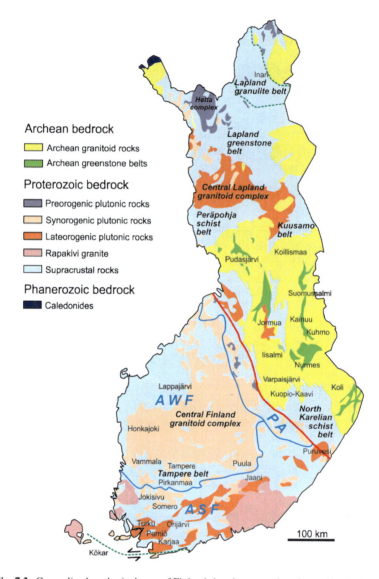

Fig. 7.2 Generalized geological map of Finland showing target locations of in situ isotope studies reviewed in this chapter. The diagonal red line is the boundary zone between Archean and Paleoproterozoic lithosphere, PA is the Primitive arc complex, AWF is the Arc complex of western Finland, and ASF is the Arc complex of southern Finland. The borders of the Lapland granulite complex in far north Finland and the sinistral South Finland shear zone in southwestern-most Finland are also shown (Modified from Nironen (2005) and Hölttä et al. (2008))

amalgamation of several crustal terranes brought together by Neoarchean (plate)tectonic processes (Hölttä et al. 2008). In situ isotopic work has been carried out in several parts of the Karelian province and has delivered new data regarding the age of the crust,

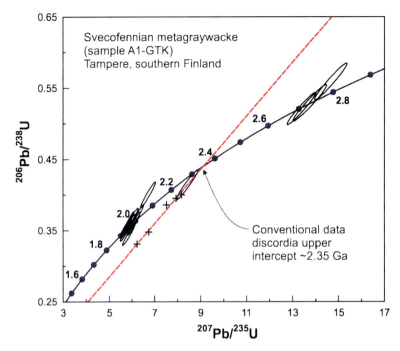

Fig. 7.3 U-Pb concordia diagram showing in situ SHRIMP data measured for 13 zircon grains from a metagreywacke (sample A1-GTK) from the Paleoproterozoic Tampere schist belt by Huhma et al. (1991) (individual data points are marked by error ellipses). Also shown are conventional multi-grain analyses of five zircon fractions (black crosses) from sample A1-GTK (Kouvo and Tilton 1966; Huhma et al. 1991). The red dashed line is a discordia line fitted through the conventional analyses; its intersection with the concordia implies an age of ~2.35 Ga (Modified from Huhma et al. (1991))

temporal evolution of granitoid magmatism, metamorphic events, magma sources, and terrane division.

7.4.2.1 The Earliest Crust in Finland

The westernmost segment (Pudasjärvi complex) of the Karelian province is characterized by high-grade mafic and felsic granulites, trondhjemitic gneisses, and local leucogranitoid rocks. Mutanen and Huhma (2003) analyzed 23 spots from zircons from a trondhjemite gneiss from the central part of the Pudasjärvi complex (Siurua). The analyzed grains indicated clearly older overall ages than previously measured from Finland. Most of the analytical spots (Fig. 7.4) cluster around 3.5 Ga and Mutanen and Huhma (2003) interpreted this as the crystallization age of the igneous precursor of the Siurua tonalite gneiss. Moreover, and quite importantly, the in situ data on the Siurua gneiss revealed a zircon core with an age of ~3.7 Ga, as well as a younger age of ~3.1 Ga. The former is a positive indication

Fig. 7.4 (a) Polished slab of the Siurua trondhjemite gneiss (the oldest rock in Finland and the European Union territory) from the Pudasjärvi complex in the western part of the Archean Karelian province in northern Finland. The rock slab is 10 cm wide at bottom part. Photo: O.T. Rämö. (b) U-Pb concordia diagram showing in situ SIMS data on 23 spots from the Siurua trondhjemite gneiss (sample A1602-GTK, Mutanen and Huhma 2003). Crosses show the results of five conventional multi-grain analyses from sample A1602-GTK and imply an upper intercept age of ~3.35 Ga. Inset in (b) shows a back-scattered electron image of a zircon crystal analysed from the Siurua trondhjemite gneiss. The crystal displays a rounded core (age ~3.73 Ga) and a rim with crystal faces around it (age ~3.45 Ga). The core dates the pre-existing source rock of the gneiss, the rim represents zircon precipitated during the crystallization of the throndhjemite. ((b) Modified from Mutanen and Huhma (2003))

of the existence of Eoarchean crustal material, and the latter shows that the gneiss registers a high-grade metamorphic event about 400 million years after the crystallization of the trondhjemite protolith. Conventional TIMS U-Pb zircon data on the Siurua gneiss yielded an upper intercept age of 3.36 ± 0.08 Ga (Mutanen and Huhma 2003, Fig. 7.4), and underlines the power of in situ techniques to unravel a complex geochemical history recorded in zircon grains from a single Precambrian sample.

7.4.2.2 Archean High-Grade Metamorphism, Crust-Building Magmatic Events, and Paragneisses

The first in situ U-Pb zircon studies on the Finnish Archean were published by Hölttä et al. (2000a) and Mänttäri and Hölttä (2002) on the Varpaisjärvi granulites (high-grade metamorphic rocks) in the southern part of the Karelian terrane (Fig. 7.2). They recognized a ubiquitous granulite-facies metamorphic and migmatizing event at 2.70–2.63 Ga in zircon from igneous enderbites and migmatite leucosomes. This high-grade event was, however, unable to reset the U-Pb systems of zircons in the crust-forming lithologic units. The latter revealed two distinct age groups, 3.2 Ga and 2.73 Ga, interpreted as the crystallization ages of the sources of the Varpaisjärvi granulites. The older group also revealed a metamorphic event at ~3.1 Ga.

The Kuhmo-Suomussalmi area in the central part of the Karelian province has been the target of active in situ analytical work (Käpyaho et al. 2006, 2007; Heilimo et al. 2011; Mikkola et al. 2011). The area comprises granitic rocks (TTGs, sanukitoid intrusions, and leucogranites) that characterize the Archean gneiss terranes on both sides of the greenstone belt (Fig. 7.2). With the ability to see through complexities related to inheritance and later metamorphic events, the aforementioned studies (in total, about 700 in situ U-Pb analytical spots) were able to distinguish three phases of TTG formation (2.95 Ga, 2.83–2.78 Ga, 2.76–2.74 Ga), a sharply defined episode of sanukitoid formation (2.74–2.70 Ga), and later emplacement of 2.70–2.68 Ga leucogranites reflecting crustal remelting. Käpyaho et al. (2007) further recognized two episodes of metamorphic zircon growth on pre-existing grains at 2.84–2.81 Ga and 2.73–2.70 Ga, the latter corresponding to the anatectic leucogranite event. Paleo- to Mesoarchean inheritance was also recorded in the leucogranites (<3.44 Ga, Mikkola et al. 2011) and sanukitoids (up to 3.2 Ga, Heilimo et al. 2011), again yielding a glimpse of ancient crust-forming processes in the sources of these Meso- to Neoarchean rocks. Mikkola et al. (2011) also published the first oxygen in zircon in situ results on the Finnish bedrock, with average $\delta^{18}O$ values of 6.1 ± 0.2‰ for the Kuhmo-Suomussalmi TTGs, 8.5 ± 0.5‰ for the sanukitoids, and 6.4 ± 0.1‰ for the Neoproterozoic leucogranites. These observations confirm the juvenile character of the TTGs and underscore the presence of supracrustal components in the mantle-derived sanukitoid rocks.

The Archean Nurmes paragneiss belt in the southern segment of the Finnish part of the Karelian province (Fig. 7.2) has also been studied utilizing new in situ U-Pb zircon data. Kontinen et al. (2007) delineated the deposition of greywacke source rocks of the Nurmes paragneisses at 2.71–2.69 Ga. They were also able to define 2.75–2.70 Ga TTGs/sanukitoids in the source of the paragneisses. These observations led Kontinen et al. (2007) to propose a Neoarchean juxtaposition of the Archean Karelian and Superior (present Canada) provinces.

7.4.2.3 Combined U-Pb and Lu-Hf Evidence for the Evolution of the Karelian Province

Lauri et al. (2011) applied both the U-Pb and Lu-Hf zircon in situ methods to assess the Archean crustal evolution of the Finnish part of the Karelian province in a study of the Archean Pudasjärvi, Koillismaa, and Iisalmi complexes (Fig. 7.2). Utilizing combined U-Pb (SIMS, LAMS) and Lu-Hf (LAMS) spots on individual zircon grains from the Siurua tonalite gneiss (Sect. 7.4.2.1) they were able to model the hafnium isotope composition of the zircons as revealing a $\geq$ 4.0 Ga precursor, thus bearing evidence from Hadean Earth. Initial hafnium isotope composition of zircons also showed that, in the pre-Neoarchean magma sources involved, there was no evidence for a mantle source with long-term depletion in incompatible elements (as is the case in the modern upper mantle of the Earth). For the

Neoarchean, however, such evidence exists and this probably reflects the onset of modern-type plate tectonics at the end of the Archean. Lauri et al. (2011) also showed that the three complexes examined may have evolved separately (as individual continental masses) prior to ~2.7 Ga.

7.4.3 Paleoproterozoic Crustal Evolution

A salient part of the Paleoproterozoic crust in Finland consists of the Svecofennian orogen that flanks the Karelian province southwest of the Archean-Paleoproterozoic boundary (Fig. 7.2). This crustal segment differentiated from the Earth's mantle in the 1.9–1.8 Ga Svecofennian orogeny and hence represents net crustal growth induced by orogenic processes (Lahtinen this issue). The Svecofennian orogen consists of at least three different crustal terranes – the Primitive arc complex, the Arc complex of western Finland, and the Arc complex of southern Finland (Fig. 7.2, Korsman et al. 1999). These segments have differing overall lithologic and isotopic compositions and probably represent different crustal entities that were amalgamated onto the Archean nucleus in the Svecofennian orogeny (Lahtinen and Huhma 1997; Lahtinen et al. 2005, Lahtinen this issue). Farther north, in the Finnish Lapland area (north of the Archean Karelian province) there are vast areas of Paleoproterozoic granitoid rocks (e.g., the Central Lapland granitoid complex and the Hetta complex) and supracrustal belts (e.g., the Lapland greenstone belt) that reflect, at least in part, reworking of the Archean craton during Paleoproterozoic orogeny. Still farther north, in the Inari area (Fig. 7.2), there is a high-grade granulitic crustal segment that represents a Paleoproterozoic thrust slice from the present northeast onto the Archean craton nucleus. All these lithologic units have been studied by in situ isotopic methods.

7.4.3.1 Northern Finland

Regarding the division of Archean and Paleoproterozoic continental crust in northernmost Finland, whole-rock Sm-Nd isotopes of paragneisses and mafic to intermediate plutonic rocks from the Lapland granulite belt (Tuisku and Huhma 2006, Huhma and Meriläinen 1991 as referred to in op.cit.) unequivocally showed that the granulite is Paleoproterozoic in age. The granulite complex hence separates the western parts of the Archean Belomorian and Kola provinces (see Lahtinen this volume). Tuisku and Huhma (2006) utilized the in situ U-Pb zircon method to study the evolution of the Lapland granulite belt and found that provenance ages of high-grade metapelites range from 2.9 Ga to 1.94 Ga, with a pronounced maximum at 2.2–1.97 Ga. They also recognized metamorphic zircon growth at 1.905–1.88 Ga, similar to ages of nearby intrusions. Combined with existing conventional geochronological data, these in situ U-Pb zircon results bracket the evolution of the Lapland granulite complex to 60 million years from the erosion of the detritus provenance to regional metamorphism and exhumation (Tuisku and Huhma 2006). The main (2.2–1.97 Ga) source of the detritus remains obscure, however.

Paleoproterozoic granitoid magmatism is widespread in northern Finland (Fig. 7.2). Early conventional whole-rock Sm-Nd isotope work (Huhma 1986) showed that the granites of northern Finland contain a significant (yet varying) Archean source component. In a comprehensive in situ U-Pb zircon study of the Central Lapland granitoid complex and the Hetta complex (in total, 164 zircon spots analyzed), Ahtonen et al. (2007) showed that the Paleoproterozoic granitoid magmatism of Lapland was episodic and took place over a remarkably long time, from 2.13 Ga to 1.76 Ga. The oldest granitoids were emplaced synchronously with Karelian cratonic metasediments, whereas the youngest register high-grade metamorphism and associated anatexis that occurred concurrently with the late Paleoproterozoic collisional event of the Svecofennian orogen in southern Finland (see Sect. 7.2.3.3). In line with the overall unradiogenic Nd isotope compositions of these granites, the zircons from them show inheritance of 2.9 Ga to 2.4 Ga crustal material (Ahtonen et al. 2007).

Mikkola et al. (2010) used the new SIGL laboratory to measure the in situ U-Pb isotope composition (in total, 61 grains) of zircon grains extracted from three small A-type granite intrusions (cf. Eby 1990) that cut the Archean bedrock in the Kainuu region in the central part of the Karelian province (Fig. 7.2). The data imply a crystallization age of 2.425 ± 0.006 Ga for the Rasinkylä and Pussisvaara granites west of Suomussalmi and 2.389 ± 0.005 Ga for the Rasimäki granite ~150 km farther to the south in the Nurmes area (Fig. 7.2). These ages confirmed earlier conventional U-Pb zircon ages on these granites and they also comply with results acquired utilizing the newly implemented chemical abrasion method (Mattinson 2005). The in situ data on these three A-type granites revealed no inheritance, which seems to be emerging as a typical feature of the A-type granite zircons, possibly owing to the relatively high temperature of this type of magma.

Hanski et al. (2005) published in situ U-Pb zircon age data on a Paleoproterozoic arkosite-amphibolite suite from the Peräpohja schist belt just south of the Central Lapland granitoid complex (Fig. 7.2). The felsic "arkositic" members of the suite were interpreted as volcaniclastic or epiclastic in origin and they display an A-type granite geochemical composition. The U-Pb in situ data delineate the formation of these rocks at 1.975 Ga, which implies they belong to the upper stratigraphic part of the Peräpohja schist belt. Hanski et al. (2005) found no inheritance in the examined zircons (17 from the arkositic samples analyzed). In a subsequent work, Hanski et al. (2010) presented in situ U-Pb zircon data on several mafic plutons of the gabbro-werhlite association from the Lapland greenstone belt and Peräpohja, Kuhmo, and Koli areas (Fig. 7.2). These intrusions typically comprise basal sills intruded at the Archean-Proterozoic unconformity over large areas in Lapland and the east-central part of the Karelian province. The in situ U-Pb data verified the crystallization age of the sills at 2.22–2.21 Ga; some of the analyzed crystals registered significant resetting at 1.9–1.8 Ga induced by the Svecofennian orogeny. Hanski et al. (2010) ascribed resetting of the U-Pb systems in these zircons to mobility of radiogenic Pb caused by hydrothermal $CaCl_2$-bearing fluids.

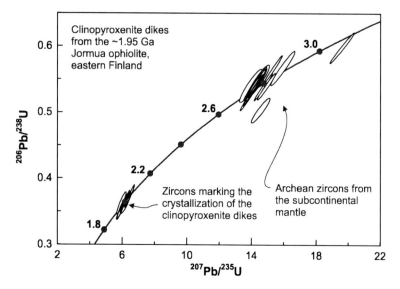

Fig. 7.5 U-Pb concordia diagram for zircons from two clinopyroxenite dikes from the Jormua ophiolite complex (fragment of Paleoproterozoic oceanic crust) in the western part of the Archean Karelian province. The in situ data show two populations that mark the crystallization of the clinopyroxenite dikes (at ~1.95 Ga) and Archean material (~3.1 Ga, ~2.8 Ga) in the dikes, derived from Paleoproterozoic subcontinental mantle (Modified from Peltonen et al. (2003))

Paleoproterozoic craton-margin supracrustal sequences and magmatic arc deposits within the western Karelian province margin host the Paleoproterozoic Jormua (Fig. 7.2) ophiolite that was identified as a fragment of oceanic lithosphere (aka ophiolite) in the 1980s (Kontinen 1987). Peltonen et al. (2003) studied zircon fractions extracted from clinopyroxenite dikes that crosscut the mantle peridotites of the Jormua ophiolite. Nineteen spots from 17 grains from two clinopyroxenite dikes revealed a ~1.950 Ga population registering the magmatic crystallization of the dikes and, quite remarkably, Meso- to Neoarchean (3.11–2.72 Ga) inherited grains (Fig. 7.5). The latter were interpreted by Peltonen et al. (2003) to have derived from deep subcontinental mantle sources in an ocean to continent transition zone with a major Archean subcontinental mantle component. The Archean zircons from the clinopyroxenite dikes of the Jormua ophiolite are the oldest zircons thus far found from upper mantle rocks on Earth.

7.4.3.2 Central Finland

The Paleoproterozoic (Svecofennian) bedrock of central Finland (the Primitive arc complex and the Arc complex of western Finland, Fig. 7.2) has also been a target of active in situ zircon isotopic work. Apart from studies that have dealt solely with sedimentary provenances (the sources of sedimentary detritus, see Sect. 7.4.4), the focus has been on the origin of the granitoids of the Central Finland granitoid complex (Rämö et al. 2001; Vaasjoki et al. 2003) and source rocks and metamorphism in the Vammala migmatite belt (Rutland et al. 2004). The ~44,000 km^2 Central Finland granitoid complex comprises two main Paleoproterozoic granitoid suites (Nironen 2003): Synorogenic, synkinematic granodiorites and granites (plus minor diorite and gabbro) dated at 1.89–1.88 Ga, and synorogenic, postkinematic granites and quartz monzonites (and coeval mafic intrusions) dated at 1.88–1.87 Ga. The latter are found as discordant, crosscutting plutons within the former. Rämö et al. (2001) studied the petrogenesis of the postkinematic magmatism of the Central Finland granitoid complex and used in situ U-Pb isotope systematics of zircon in search of source inheritance in the postkinematic granites. Forty-one spots analysed from zircons extracted from two postkinematic granites (Honkajoki and Puula, Fig. 7.2) revealed no substantial inheritance (one grain with an age of 1.95 Ga as opposed to crystallization ages of the plutons at 1.871 ± 0.009 Ga and 1.867 ± 0.006 Ga). This is despite the fact that the analyzed zircon grains exhibited many cores. In a further work, Vaasjoki et al. (2003) used the in situ U-Pb zircon method to study 1.93–1.91 Ga pre-orogenic granitoids from the Primitive arc complex area (Fig. 7.2). They also found minimal inheritance in these rocks (no zircons older than 1.95 Ga in the analyzed 1.92 Ga old granites). Therefore, no direct indication of an early Paleoproterozoic (≥2.0 Ga) crustal source was detected via these granitoids in the central Finnish part of the Svecofennian orogen.

Focusing on the Vammala schist belt on the southwestern margin of the Central Finland granitoid complex (Fig. 7.2), Rutland et al. (2004) studied the sources and metamorphic history of the local Svecofennian orogen. They observed Meso- to Neoarchean inheritance (2.9–2.5 Ga), no evidence for 2.4–2.1 Ga crustal terranes, and concluded, contrary to the prevailing perception (Sect. 7.2.4), that two orogenic phases (~1.92 Ga and ~1.88 Ga) separated by 40 million years of extension were instrumental in the assembly of the Svecofennian orogen.

7.4.3.3 Southern Finland

In situ zircon isotopic work on the Paleoproterozoic orogenic rocks of the Arc complex of southern Finland has thus far focused mainly on the collisional, late orogenic leucogranites of the Late Svecofennian granite-migmatite zone (LSGM, Ehlers et al. 1993), as well as some of their synorogenic country rocks. Pajunen et al. (2008) determined the crystallization ages of Svecofennian synorogenic tonalites, granodiorites, and charnockites at 1.88–1.86 Ga and noted minor inheritance in zircons from the tonalites (1.96–1.92 Ga). They also reported a 1.84 Ga crystallization age for a gabbro and a crystallization age of 1.83 Ga for a diabase dike in the central part of the Arc complex of southern Finland. In a meticulous survey of the geochronology of the late Svecofennian leucogranites that encompass the entire 500-km-long LSGM belt (Fig. 7.2b), Kurhila et al. (2005, 2010, 2011) utilized the in situ U-Pb zircon method for assessment of inheritance patterns of these anatectic granites that show a considerable range of emplacement ages (1.85–1.79 Ga). The main inherited zircon populations were found to be

2.1–2.0 Ga and 2.8–2.5 Ga. Kurhila et al. noted granites with either nil or strong inheritance and explained this by dominant igneous and sedimentary source components, respectively. Kurhila et al. (2010) further used the in situ Lu-Hf zircon isotope method in order to gain information on the age of the source material in these leucogranites. Samples from the western (Perniö), central (Jaani), and eastern (Puruvesi) parts of the LSGM zone (Fig. 7.2) have widely varying initial hafnium isotope compositions (Fig. 7.6). The Puruvesi granite, which is closest to the Archean-Paleoproterozoic crustal boundary, has the least radiogenic composition (the largest Archean source component; as verified by Nd whole-rock isotopes by Huhma already in 1986), and west of Puruvesi the samples show increasingly more juvenile compositions and thus have less material from Archean sources (Fig. 7.6).

Considerable effort has been invested in utilizing the in situ U-Pb zircon isotope method for examining the metamorphic and structural evolution of the bedrock that forms the Arc complex of southern Finland (Fig. 7.2). For the Orijärvi-Turku region in the western part of the LSGM zone, Väisänen et al. (2002) recognized the 1.9 Ga Orijärvi granodiorite as a pre-collisional intrusion in the Arc complex of southern Finland. A 1.85 ± 0.02 Ga tonalite was shown to have been emplaced at the syncollisional stage (during early Svecofennian D_2 deformation). Väisänen et al. were also able to date high-grade metamorphism (associated with D_3 folding) in the Turku area at 1.82–1.81 Ga, utilizing metamorphic growth rims on individual

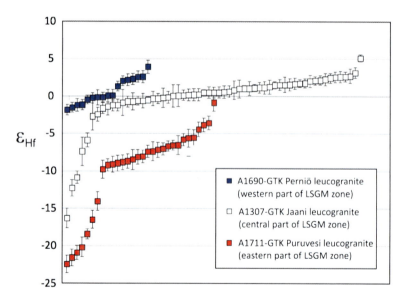

Fig. 7.6 Figure showing the initial hafnium isotope composition of individual zircon grains from three Paleoproterozoic lateorogenic leucogranites of the LSGM zone in the Arc complex of southern Finland. The ε_{Hf} value denotes the initial $^{176}Hf/^{177}Hf$ ratio of the analysed zircon crystals relative to the contemporaneous chondritic composition. The chondritic composition has been defined from meteorites and is thought to represent the primordial substrate of the solar system (e.g., Blichert-Toft and Albaréde 1997; Bouvier et al. 2008). The more negative the ε_{Hf} value, the larger the proportion of ancient crustal material in the zircons (Modified from Kurhila et al. (2010))

zircon crystals. Skyttä et al. (2006) provided a piercing point to the early Svecofennian D_2 deformation by dating syn-D_2 silicic and intermediate dikes in the Orijärvi area. Skyttä and Mänttäri (2008) focused on the Orijärvi-Karjaa migmatite terrain and dated, using in situ U-Pb zircon chronology, the D_3 extensional episode at 1.835–1.825 Ga, the D_4 convergence at 1.82 Ga, and the D_5 transpression at < 1.82 Ga. Substantial metamorphic recrystallization of zircon at 1.815 ± 0.003 Ga in the Orijärvi region, probably marking D_4 convergence, was also recognized by in situ analyses of Väisänen and Kirkland (2008). The studies by Skyttä and Mänttäri (2008) further revealed a typical pattern of inheritance with source ages of 2.7–2.6 Ga and 2.05–1.88 Ga.

Offshore the Finnish mainland, the archipelagic part of the LSGM zone (Kökar area, Fig. 7.2) was the target of a further study in pursuit of unravelling the deformation and reactivation history of a major, sinistral ductile shear zone (the South Finland shear zone) by Torvela et al. (2008). Building mainly on the in situ U-Pb zircon isotope method, Torvela et al. (2008) distinguished two ductile deformational phases at 1.85 Ga and 1.83 Ga; these were intense enough to partially recrystallize zircon in the gneisses deformed by the shearing.

7.4.4 Provenance of Paleoproterozoic Metasedimentary Rocks

The in situ U-Pb and Lu-Hf zircon isotope methods are most suitable for sedimentary provenance (source area) studies, as zircon is one of the most robust heavy accessory minerals that, mechanically and chemically, survive supracrustal processes (erosion, transport, deposition, diagenesis) and subsequent metamophism (recrystallization, even partial melting of the sedimentary rocks). Hence they convey information from the source rocks and the lithologic variation in the exposed bedrock at the time of deposition of their detritus (e.g., Andersen 2005). In situ U-Pb zircon provenance studies of the Finnish bedrock have thus far focused on the Paleoproterozoic sedimentary rocks of southern and central Finland (the Svecofennian domain) and eastern Finland (the Karelian domain).

7.4.4.1 Svecofennian Domain

Lahtinen et al. (2002) examined the source components of metasedimentary rocks from the Paleoproterozoic supracrustal belts surrounding the Central Finland granitoid complex (the Arc complex of western Finland) and south of it, in the Arc complex of southern Finland area (Fig. 7.2). On the basis of these data (in total, 193 analytical spots), they divided the central Svecofennian metasediments into stratigraphically lower and upper types with depositional ages of 1.92–1.91 Ga and 1.90–1.88 Ga, respectively. The lower type records major source at 2.02–1.93 Ga. The southern Svecofennian metasediments (which also include metamorphosed quartz arenites, originally quartz-rich sands) have younger depositional ages (1.90–1.86 Ga) and a dominant 2.1–2.0 Ga source population, and hence an older overall provenance that the metasediments from the Arc complex of western Finland. These data further reinforce the separate identities of the two arc complexes of the Svecofennian orogen (Lahtinen et al. 2002). In a follow-up study, Lahtinen et al. (2009) presented in situ U-Pb zircon data (in total, 251 spots on detrital grains and their metamorphic overgrowths) for Paleoproterozoic metasedimentary rocks from the Tampere and Pirkanmaa supracrustal belts on the southern flank of the Arc complex of western Finland. They delineated the depositional history of these supracrustal belts at 1.92–1.89 Ga (in contradiction to the model of Rutland et al. 2004, Sect. 7.4.3.2), and recognized pre-depositional overgrowths on pre-existing zircon grains of 2.0–1.91 Ga and postdepositional zircon growth at 1.885–1.875 Ga. In terms of bulk inheritance, the data show mainly Neoarchean (2.8–2.7 Ga) and Paleoproterozoic (2.1–1.9 Ga) populations, except in a sample from the western part of the study area (the Vammala Ni-zone), which has major 2.2–2.0 Ga and 2.5–2.4 Ga populations (Fig. 7.7).

Rare quartzites (metamorphosed quartz sandstones) in the Arc complex of southern Finland (as well as the Swedish Svecofennian) were studied by Bergman et al. (2008) utilizing the in situ U-Pb zircon method. The measured zircons (in

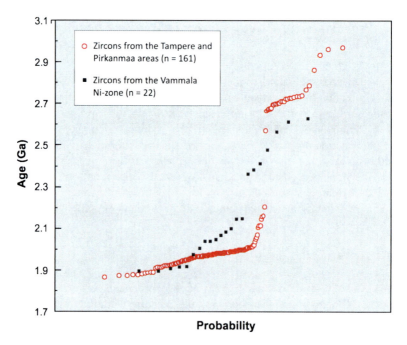

Fig. 7.7 A linear probability plot of zircon ages based on in situ U-Pb analyses of grains separated from metasedimentary rocks of the Tampere, Pirkanmaa, and Vammala areas on the southern flank of the Arc complex of western Finland. Overall, 1.95–1.90 Ga and ~2.7 Ga sources are dominant, but the sample from the Vammala Ni-belt shows substantial 2.1–2.0 Ga and 2.5–2.4 Ga populations, indicative of sources of anomalous age (see Lahtinen this volume) (Data from Lahtinen et al. (2009))

total, 145 spots from zircons from Finnish rocks) yielded main inherited populations at 2.95–2.60 Ga, 2.10–1.95 Ga, and 1.92–1.85 Ga, and a maximum value at 3.32 Ga. Using the youngest observed detrital ages, Bergman et al. (2008) determined the maximum age of deposition at three examined locations at 1.848 ± 0.010 Ga, 1.865 ± 0.011 Ga, and 1.848 ± 0.013 Ga. This implies the existence of intraorogenic sedimentary basins (marking tectonically quiet episodes) in the southern Svecofennian orogen at 1.86–1.83 Ga (see also Lahtinen et al. 2002).

7.4.4.2 Karelian Domain

The Paleoproterozoic Karelian supracrustal sequences (the North Karelian schist belt, Fig. 7.2) that flank the southern margin of the Archean Karelian domain in eastern Finland were assessed for their in situ U-Pb zircon isotopic systematics by Lahtinen et al. (2010). The studied samples (deep-water turbidites deposited offshore of the Archean crust) were divided into the autochthonous-parautochthonous Lower Kalevian and allochthonous Upper Kalevian strata. According to Lahtinen et al. (2010), zircons from the Lower Kaleva have a Neoarchean provenance, whereas the Upper Kaleva shows both Archean and Paleoproterozoic (dominant at 2.05–1.92 Ga) inheritance and a maximum deposition age of 1.95–1.92 Ga.

Farther to the north at the northern margin of the Karelian province, Laajoki and Huhma (2006) determined in situ U-Pb zircon ages for a Paleoproterozoic metaconglomerate-quartzite unit in the Paleoproterozoic Kuusamo belt (Fig. 7.2). Ten in situ analyses revealed Meso- to Neoarchean (2.91–2.63 Ga) and Paleoproterozoic (2.23–1.83 Ga) ages. The youngest in situ ages record the maximum depositional age for this unit at less than 1.9 Ga, which sets it apart from the main Karelian sequence of the Kuusamo belt.

7.4.5 Studies on Svecofennian Gold Mineralization

The in situ U-Pb isotopic method has also been applied to the determination of ages of orogenic-type gold mineralization in the Paleoproterozoic Svecofennian crust. These gold occurrences generally comprise auriferous veins hosted by Paleoproterozoic felsic metavolcanic rocks and quartz diorites and gabbros (Saalmann et al. 2009, 2010). In the Somero region in the west-northwestern segment of the Arc complex of southern Finland (Fig. 7.2), Saalmann et al. (2009) examined two gold prospects in Paleoproterozoic metavolcanic rocks and determined the in situ U-Pb isotope composition of 62 zircon and nine titanite ($CaTiSiO_5$) spots. Saalmann et al. (2009) were able to define the age of the host felsic volcanic rock (1.881 ± 0.003 Ga), inherited zircons in gold-bearing quartz veins (younger rims 1.832 ± 0.015 Ga), and a pegmatitic quartz-feldspar dike that crosscuts the gold-bearing veins (1.786 ± 0.007 Ga). The age of the gold mineralization was thus bracketed between 1.82 and 1.79 Ga, which complies with the age of ductile shearing in the aftermath of the Svecofennian orogeny. In a follow-up study, Saalmann et al. (2010) investigated a further gold occurrence at Jokisivu (Fig. 7.2). At Jokisivu, auriferous quartz veins are hosted by a synorogenic quartz diorite. Saalmann et al. (2010) determined the age of zircons from an unaltered quartz diorite and a quartz diorite recrystallized in the mineralization process. Zircon cores from these samples indicate magmatic crystallization at 1.884 ± 0.004 Ga and 1.881 ± 0.003 Ga, respectively, whereas zircon rims from the altered quartz diorite are 1.802 ± 0.015 Ga. This dates the gold mineralization event at Jokisivu and the event was correlated by Saalmann et al. (2010) to late Svecofennian shear tectonics related to WNW-ESE shortening.

In southwestern Finland, there is a clear spatial correlation between gold occurrences and tourmaline-bearing rocks. Talikka and Vuori (2010) studied tourmaline mineral chemistry and boron isotope geochemistry of tourmaline-bearing pegmatites, quartz-tourmaline veins, and tourmalinized intrusive contacts associated with gold occurrences in the Tampere (and Ronka) areas (Fig. 7.2). They utilized the SIMS laboratory in Edinburgh and measured 35 spots from ten tourmaline samples. These yielded mean $\delta^{11}B$ values between −13.2‰ and −8.7‰. Talikka and Vuori (2010) noted subtle, host rock-associated differences in the boron isotope composition of the analyzed tourmalines and attributed these to differences in the sources of boron and to fractionation related to tourmaline crystallization.

7.4.6 Origin of the Mid-Proterozoic locus classicus Rapakivi Granites and Related Rocks

In an in situ Lu-Hf zircon isotopic study, Heinonen et al. (2010) re-examined the petrogenesis of the mid-Proterozoic *locus classicus* rapakivi granites and related more mafic rocks (gabbroids, monzodiorite, massif-type anorthosite). Pre-existing (mainly whole-rock) isotope data (the Sm-Nd, Lu-Hf, Pb-Pb, Rb-Sr methods; Rämö and Haapala 2005 and references therein, Patchett et al. 1981) registered the existence of two principal source components (subcontinental mantle and continental crust), but were unable to separate these conclusively. New in situ Lu-Hf zircon data acquired for the 1.645–1.627 Ga rapakivi granites, gabbroids, anorthosite, and monzodiorite in southeastern Finland and 1.584–1.571 Ga rapakivi granites and a leucogabbronorite in southwestern Finland (Fig. 7.2) revealed homogeneous initial hafnium isotope compositions for the rapakivi granites (mean initial ε_{Hf} values for the older granites −0.3 to +1.0, for the younger granites −2.2 to +0.1) relative to the mafic and intermediate rocks (older group +0.7 to +4.3, younger group +2.7). Moreover, and most importantly, individual in situ analyzes of zircons from the mafic rocks revealed initial ε_{Hf} values that conform to the contemporaneous depleted mantle (ε_{Hf} up to +9). This is positive indication of the contribution from a subcontinental mantle source (with a long-term depletion in the REE elements), and strongly suggests that these Proterozoic massif-type anorthosites were derived from a mantle source.

7.4.7 Precambrian Sample in Neoproterozoic Kimberlite

The Archean Karelian province in eastern Finland hosts Neo- to Mesoproterozoic alkaline mafic-ultramafic rocks that comprise kimberlite diatremes, hypabyssal kimberlites, lamproites, and ultramafic lamprophyres (O'Brien and Tyni 1999, O'Brien and Lehtonen this volume). Some of these rocks have been explored for economic occurrences of diamond. Neoproterozoic to Cambrian (0.6–0.5 Ga) kimberlites in the Kuopio-Kaavi region (Fig. 7.2) carry mantle and lower crustal xenoliths, and the latter have been studied utilizing the in situ U-Pb zircon isotopic method. Hölttä et al. (2000b) determined zircon ages for two mafic granulites equilibrated (recrystallized) at 7.5–12.5 kbar and 800–900°C. Eighteen spot analyses showed a range of crystallization ages between 1.94 Ga and 1.78 Ga, 2.67 Ga and 2.64 Ga, and 2.39 Ga and 2.29 Ga. These lower crust samples from the Archean-Proterozoic crustal boundary thus reveal a mixture of Archean and Paleoproterozoic ages and show that both Archean and Paleoproterozoic crustal domains are present in the lower continental crust at this latitude. In a subsequent in situ U-Pb zircon isotope study of kimberlite-hosted lower crustal xenoliths equilibrated at 13–20 kbar, Peltonen et al. (2006) recorded Archean inheritance up to 3.5 Ga. They ascribed this to early crust that was formed in a major Paleoarchean plume event. Zircons with ages of ~1.90 Ga and ~1.80–1.74 Ga were also found and these were related by Peltonen et al. (2006) to zircon recrystallization caused by the Svecofennian orogeny and to post-orogenic heating of the lower crust by further thermal perturbations in the subcontinental mantle in these time periods.

7.4.8 Lappajärvi Impact Site

The in situ U-Pb zircon isotope method has also been successfully used to date an (Upper Cretaceous) impact crater at Lappajärvi in a Paleoproterozoic supracrustal belt in the western part of the Arc complex of western Finland (Fig. 7.2). Mänttäri and Koivisto (2001) analyzed spots from seven zircon crystals recovered from the suevite (impact breccia) of the impact structure (Lehtinen 1976). Four of these crystals were dull and thoroughly fractured (shock-affected), three were better preserved. Twelve spots analyzed by Mänttäri and Koivisto (2001) defined a discordia with intercepts at 0.0733 ± 0.0053 Ga and 1.85 ± 0.05 Ga (Fig. 7.8). The lower intercept age (73.3 ± 5.3 Ma) is essentially determined by spots from the shock-affected zircons. This finally resolved the controversy between an Ar-Ar whole-rock age (77.3 ± 0.4 Ma) determined from the impact melt (kärnäite) and a paleomagnetic age of ~195 Ma (Mänttäri and Koivisto 2001 and references therein).

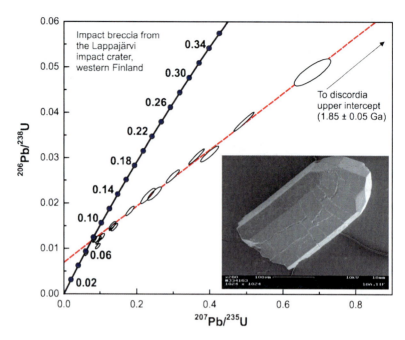

Fig. 7.8 U-Pb concordia diagram showing in situ SIMS data measured for zircon grains from the suevite (impact breccia) of the Lappajärvi impact crater in western Finland. The analytical spots (error ellipses) are centered at the lower end of the discordia. The lower intercept age (0.0733 ± 0.0053 Ga) dates the impact and the upper intercept (1.85 ± 0.05 Ga, beyond the panel area) indicates the overall age of the bedrock at the impact site. The inset shows a ~0.4 mm long zircon crystal recovered from the impact breccia, badly fractured by the impact event (scanning electron microscope image by courtesy of Irmeli Mänttäri)

7.5 Concluding Remarks

Application of the in situ isotopic microanalytical methods has profoundly enhanced our knowledge of the genesis and evolution of the bedrock of Finland. These new methods have recently become available for both established researchers and trainees and they currently form an essential part of many research projects that focus on the magmatic, sedimentary, metamorphic, and metallogenic evolution of the Finnish Precambrian. In situ isotopic research will continue to be a cornerstone in the modern study of solid geological materials and it will increase its relevance as in situ methods are developed and implemented for additional (radiogenic and stable) isotope systems.

References

Ahtonen N, Hölttä P, Huhma H (2007) Intracratonic Palaeoproterozoic granitoids in northern Finland: prolonged and episodic crustal melting events revealed by Nd isotopes and U-Pb ages on zircon. Bull Geol Soc Finland 79:143–174

Andersen T (2005) Detrital zircons as tracers of sedimentary provenance: limiting conditions from statistics and numerical simulation. Chem Geol 216:249–270

Bergman S, Högdahl K, Nironen M, Ogenhall E, Sjöström H, Lundqvist L, Lahtinen R (2008) Timing of Palaeoproterozoic intra-orogenic sedimentation in the central Fennoscandian Shield: evidence from detrital zircon in metasandstone. Precambrian Res 161:231–249

Blichert-Toft J, Albaréde F (1997) The Lu-Hf geochemistry of chondrites and the evolution of the mantle-crust system. Earth Planet Sci Lett 148:243–258

Bouvier A, Vervoort J, Patchett JP (2008) The Lu-Hf and Sm-Nd isotopic composition of CHUR: constraints from unequilibrated chondrites and implications for the bulk composition of terrestrial planets. Earth Planet Sci Lett 273:48–57

Claesson S, Huhma H, Kinny PD, Williams IS (1993) Svecofennian detrital zircon ages – implications for the Precambrian evolution of the Baltic Shield. Precambrian Res 64:109–130

Clement SWJ, Compston W, Newstead G (1977) Design of a large, high resolution ion microprobe. Proceedings of the international secondary ion mass spectrometry conference–, Springer-Verlag, Berlin, pp 12–17

Eby GN (1990) The A-type granitoids: a review of their occurrence and chemical characteristics and speculations on their petrogenesis. Lithos 26:155–134

Ehlers C, Lindroos A, Selonen O (1993) The late Svecofennian granite-migmatite zone of southern Finland – a belt of transpressive deformation and granite emplacement. Precambrian Res 64:295–309

Faure G, Mensing TM (2005) Isotopes – principles and applications, 3rd edn. Wiley, New Jersey

Hanski E, Huhma H, Perttunen V (2005) SIMS U-Pb, Sm-Nd isotope and geochemical study of an arkosite-amphibolite suite, Peräpohja Schist Belt: evidence for ca. 1.98 Ga A-type felsic magmatism in northern Finland. Bull Geol Soc Finland 77:5–29

Hanski E, Huhma H, Vuollo J (2010) SIMS zircon ages and Nd isotope systematics of the 2.2 Ga mafic intrusions in northern and eastern Finland. Bull Geol Soc Finland 82:31–62

Heilimo E, Halla J, Huhma H (2011) Single-grain zircon U-Pb age constraints of the western and eastern sanukitoid zones in the Finnish part of the Karelian Province. Lithos 121:87–99

Heinonen AP, Andersen T, Rämö OT (2010) Re-evaluation of rapakivi petrogenesis: source contrains from the Hf isotope composition of zircon in the rapakivi granites and associated mafic rocks of southern Finland. J Petrol 51:1687–1709

Hölttä P, Huhma H, Mänttäri I, Paavola J (2000a) P-T-t development of Archaean granulites in Varpaisjärvi, central Finland. II. Dating of high-grade metamorphism with the U-Pb and Sm-Nd methods. Lithos 50:121–136

Hölttä P, Huhma H, Mänttäri I, Peltonen P, Juhanoja J (2000b) Petrology and geochemistry of mafic granulite xenoliths from the Lahtojoki kimberlite pipe, eastern Finland. Lithos 51:109–133

Hölttä P, Balagansky V, Garde AA, Mertanen S, Peltonen P, Slabunov A, Sorjonen-Ward P, Whitehouse M (2008) Archean of Greenland and Fennoscandia. Episodes 31:13–19

Huhma H (1986) Sm-Nd, U-Pb and Pb-Pb isotopic evidence for the origin of the early Proterozoic Svecokarelian crust in Finland. Geol Surv Finland Bull 337:1–48

Huhma H, Claesson S, Kinny PD, Williams IS (1991) The growth of early Proterozoic crust: new evidence from Svecofennian detrital zircons. Terra Nova 3:175–178

Huhma H, Meriläinen K (1991) Provenance of paragneisses from Lapland granulite belt. In: Tuisku P, Laajoki K (eds) Metamorphism, deformation and structure of the crust. Joint Meeting of IGCP Projects 275, Deep Geology of the Baltic/Fennoscandian Shield, and 304, Lower Crustal Processes, Oulu, Finland, August 1991, Abstracts. Res Terrae, Serie A, Nr. 5, p.26

Huhma H, O'Brien H, Lahaye Y, Mänttäri I (2011) Isotope geology and Fennoscandian lithosphere evolution. In: Nenonen K, Nurmi P (eds) Geoscience for society 125th anniversary volume. Geol Surv Finland Spec Paper 49:35–48

Käpyaho A, Hölttä P, Whitehouse MJ (2007) U-Pb zircon geochronology of selected Archaean migmatites in eastern Finland. Bull Geol Soc Finland 79:95–115

Käpyaho A, Mänttäri I, Huhma H (2006) Growth of Archaean crust in the Kuhmo district, eastern Finland: U-Pb and Sm-Nd isotope constraints on plutonic rocks. Precambrian Res 146:95–119

Kontinen A (1987) An early Proterozoic ophiolite – the Jormua mafic–ultramafic complex, northeastern Finland. In: Gaál G, Gorbatschev R (eds) Precambrian geology and evolution of the central Baltic Shield. Precambrian Res 35:313–341

Kontinen A, Käpyaho A, Huhma H, Karhu J, Matukov DI, Larionov AS, Sergei A (2007) Nurmes paragneisses in eastern Finland, Karelian craton: provenance, tectonic setting and implications for Neoarchaean craton correlation. Precambrian Res 152:119–148

Korsman K, Korja T, Pajunen M, Virransalo P, GGT/SVEKA Working Group (1999) The GGT/SVEKA Transect: structure and evolution of the continental crust in the Paleoproterozoic Svecofennian orogen in Finland. Int Geol Rev 41:287–333

Kouvo O (1958) Radioactive age of some Finnish pre-cambrian minerals. Bull Comm Géol Finlande 182:1–70

Kouvo O, Tilton GR (1966) Mineral ages from the Finnish Precambrian. J Geol 74:421–442

Kurhila M, Vaasjoki M, Mänttäri I, Rämö OT, Nironen M (2005) U-Pb ages and Nd isotope characteristics of the lateorogenic, migmatizing microcline granites in southwestern Finland. Bull Geol Soc Finland 77:105–128

Kurhila M, Andersen T, Rämö OT (2010) Diverse sources of crustal granitic magma: Lu-Hf isotope data on zircon in three Paleoproterozoic leucogranites of southern Finland. Lithos 115:263–271

Kurhila M, Mänttäri I, Vaasjoki M, Rämö OT, Nironen M (2011) U-Pb geochrological constraints of the late Svecofennian leucogranites of southern Finland. Precambrian Res 190:1–24

Laajoki K, Huhma H (2006) Detrital zircon dating of the Palaeoproterozoic Himmerkinlahti Member, Posio, northern Finland; lithostratigraphic implications. Bull Geol Soc Finland 78:177–182

Lahtinen R, Huhma H (1997) Isotopic and geochemical constraints on the evolution of the 1.93–1.79 Ga Svecofennian crust and mantle. Precambrian Res 82:13–34

Lahtinen R, Huhma H, Kousa J (2002) Contrasting source components of the Paleoproterozoic Svecofennian metasediments: detrital zircon U-Pb, Sm-Nd and geochemical data. Precambrian Res 116:81–109

Lahtinen R, Korja A, Nironen M (2005) Palaeoproterozoic tectonic evolution of the Fennoscandian Shield. In: Lehtinen M, Nurmi PA, Rämö OT (eds) Precambrian geology of Finland: key to the evolution of the Fennoscandian Shield, vol 14, Developments in Precambrian Geology. Elsevier, Amsterdam, pp 418–532

Lahtinen R, Huhma H, Kähkönen Y, Mänttäri I (2009) Paleoproterozoic sediment recycling during multiphase orogenic evolution in Fennoscandia, the Tampere and Pirkanmaa belts, Finland. Precambrian Res 174:310–336

Lahtinen R, Huhma H, Kontinen A, Kohonen J, Sorjonen-Ward P (2010) New constraints for the source characteristics, deposition and age of the 2. 1–1. 9 Ga metasedimentary cover at the western margin of the Karelian Province. Precambrian Res 176:77–93

Lauri LS, Andersen T, Hölttä P, Huhma H, Graham S (2011) Evolution of the Archaean Karelian Province in the Fennoscandian Shield in the light of U-Pb zircon ages and Sm-Nd and Lu–Hf isotope systematics. J Geol Soc London 168:201–218

Lehtinen M (1976) Lake Lappajärvi, a meteorite impact site in western Finland. Geol Surv Finland Bull 282:1–2

Mänttäri I, Hölttä P (2002) U-Pb dating of zircons and monazites from Archean granulites in Varpaisjärvi, central Finland: evidence for multiple metamorphism and Neoarchean terrane accretion. Precambrian Res 118:101–131

Mänttäri I, Koivisto M (2001) Ion microprobe uranium-lead dating of zircons from the Lappajärvi impact crater, western Finland. Meteor Planet Sci 36:1087–1095

Mattinson JM (2005) Zircon U-Pb chemical abrasion ("CA-TIMS") method: combined annealing and multi-step partial dissolution analysis for improved precision and accuracy of zircon ages. Chem Geol 220:47–66

Mikkola P, Kontinen A, Huhma H, Lahaye Y (2010) Three Paleoproterozoic A-type granite intrusions and associated dykes from Kainuu, east Finland. Bull Geol Soc Finland 82:81–100

Mikkola P, Huhma H, Heilimo E, Whitehouse M (2011) Archean crustal evolution of the Suomussalmi district as part of the Kianta complex, Karelia; constraints from geochemistry and isotopes of granitoids. Lithos. doi:10.1016/j.lithos.2011.02.012

Mutanen T, Huhma H (2003) The 3.5 Ga Siurua trondhjemite gneiss in the Archaean Pudasjärvi Granulite Belt, northern Finland. Bull Geol Soc Finland 75:51–68

Nironen M (2003) Keski-Suomen granitoidikompleksi – kallioperäkartan selitys. Central Finland Granitoid Complex – Explanation to the bedrock map. Geol Surv Finland Rep Invest 157:1–45

Nironen M (2005) Proterozoic orogenic granitoid rocks. In: Lehtinen M, Nurmi PA, Rämö O (eds) Precambrian geology of Finland – key to the evolution of the Fennoscandian shield. Developments in Precambrian geology, vol 14. Elsevier, Amsterda, pp 443–480

O'Brien HE, Tyni M (1999) Mineralogy and geochemistry of kimberlites and related rocks from Finland. In: Gurney JJ et al. (eds) Proceedings of the 7th international kimberlite conference, University of Cape Town, South Africa, 2, 625–636

Pajunen M, Airo M-L, Elminen T, Mänttäri I, Niemelä R, Vaarma M, Wasenius P, Wennerström M (2008) Tectonic evolution of the Svecofennian crust in southern Finland. In: Pajunen M (ed) Tectonic evolution of the Svecofennian crust in southern Finland – a basis for characterizing bedrock technical properties. Geol Surv Finland Spec Pap 47:15–160

Patchett PJ, Kouvo O, Hedge CE, Tatsumoto M (1981) Evolution of continental crust and mantle heterogeneity: evidence from Hf isotopes. Contrib Mineral Petrol 78:279–297

Peltonen P, Mänttäri I, Huhma H, Kontinen A (2003) Archean zircons from the mantle: the Jormua ophiolite revisited. Geology 31:645–648

Peltonen P, Mänttäri I, Whitehouse MJ (2006) Multi-stage origin of the lower crust of the Karelian craton from 3.5 Ga to 1.7 Ga based on isotopic ages of kimberlite-derived mafic granulite xenoliths. Precambrian Res 147:107–123

Rämö OT, Haapala I (2005) Rapakivi granites. In: Lehtinen M, Nurmi PA, Rämö OT (eds) Precambrian geology of Finland – key to the evolution of the Fennoscandian shield. Developments in Precambrian geology, vol 14. Elsevier, Amsterdam, pp 553–562

Rämö OT, Vaasjoki M, Mänttäri I, Elliott BA, Nironen M (2001) Petrogenesis of the post-kinematic magmatism of the Central Finland Granitoid Complex I; radiogenic isotope constraints and implications for crustal evolution. J Petrol 42:1971–1993

Rutland RWR, Williams IS, Korsman K (2004) Pre-1.91 Ga deformation and metamorphism in the Palaeoproterozoic Vammala migmatite belt, southern Finland, and implications for Svecofennian tectonics. Bull Geol Soc Finland 76:93–140

Saalmann K, Mänttäri I, Ruffet G, Whitehouse MJ (2009) Age and tectonic framework of structurally controlled Palaeoproterozoic gold mineralization in the Häme belt of southern Finland. Precambrian Res 174:53–77

Saalmann K, Mänttäri I, Peltonen P, Whitehouse MJ, Grönholm P, Talikka M (2010) Geochronology and structural relationships of mesothermal gold mineralization in the Palaeoproterozoic Jokisivu prospect, southern Finland. Geol Mag 147:551–569

Skyttä P, Mänttäri I (2008) Structural setting of late Svecofennian granites and pegmatites in Uusimaa Belt, SW Finland: age constraints and implications for crustal evolution. Precambrian Res 164:86–109

Skyttä P, Väisänen M, Mänttäri I (2006) Preservation of Palaeoproterozoic early Svecofennian structures in the Orijärvi area, SW Finland – evidence for polyphase strain partitioning. Precambrian Res 150:153–172

Talikka M, Vuori S (2010) Geochemical and boron isotopic compositions of tourmalines from selected gold-mineralized and barren rocks in SW Finland. Bull Geol Soc Finland 82:113–120

Torvela T, Mänttäri I, Hermansson T (2008) Timing of deformation phases within the South Finland shear zone, SW Finland. Precambrian Res 160:277–298

Tuisku P, Huhma H (2006) Evolution of migmatitic granulite complexes: implications from Lapland Granulite Belt, part II: isotopic dating. Bull Geol Soc Finland 78:143–175

Vaasjoki M, Huhma H, Lahtinen R, Vestin J (2003) Sources of Svecofennian granitoids in the light of ion probe U-Pb measurements on their zircons. Precambrian Res 121:251–262

Väisänen M, Kirkland CL (2008) U-Th-Pb zircon geochronology on igneous rocks in the Toija and Salittu Formations, Orijärvi area, southwestern Finland: constraints on the age of volcanism and metamorphism. Bull Geol Soc Finland 80:73–87

Väisänen M, Mänttäri I, Hölttä P (2002) Svecofennian magmatic and metamorphic evolution in southwestern Finland as revealed by U-Pb zircon SIMS geochronology. Precambrian Res 116:111–127

Wetherill G, Kouvo O, Tilton G, Gast P (1962) Age measurements of rocks from the Finnish Precambrian. J Geol 70:74–88

Chapter 8
Fennoscandian Land Uplift: Past, Present and Future

Juhani Kakkuri

8.1 Introduction

A few years ago the author of this article was contacted from London by the Financial Times. They had heard that new land is rising up from the Gulf of Bothnia so rapidly that it can be seen by naked eye, and they asked whether or not this holds true. It is true, was the answer, 2 ha is being gained from the sea every day due to the regression of the shore line. In a year that makes 700 ha.

In the late nineteenth century the famous Finnish author Zachris Topelius (1818–1898; Fig. 8.1) depicted this phenomenon as follows (Topelius 1893): *Most noticeable are the effects of this still partly unexplained phenomenon. The land rises from the sea, the sea flees, shores are exposed, the slope moving forward. Where in the old days the ships were sailing, now hardly a ship can travel; where once the fisherman cast his net, now his cows go grazing on the coastal meadow. Banks and rocks appear out of the water, of which no sea chart has had knowledge before; banks expand into islets, these grow together and connect with the mainland. Beaches expand, harbours dry up, seaports must move after the fleeing sea. Every generation of men, new arable land rises from the sea, every century grants Finland a principality.*

8.2 The Oldest Document

The oldest document in which the subsidence of the sea level is mentioned goes back to the year 1491. In it the occupants of Östhammar complained to the Regent of Sweden that the shore line had retreated so far from the city that even fishing

---

J. Kakkuri (✉)
Finnish Geodetic Institute, P.O. Box 15, FI-02431 Masala, Finland
e-mail: juhani.kakkuri@fgi.fi

I. Haapala (ed.), *From the Earth's Core to Outer Space*,
Lecture Notes in Earth System Sciences 137, DOI 10.1007/978-3-642-25550-2
© Springer-Verlag Berlin Heidelberg 2012

Fig. 8.1 Zachris Topelius. Svenska Famil-Journalen 1866

boats were unable to arrive at the city. They were willing to establish a new town closer the shore and asked for permission. The permission was granted and the new town Öregrund founded to the better place.

Many a renowned Nordic scientist has studied the land uplift phenomenon, among them the Swede Gerard de Geer (1858–1943), who in the end of the nineteenth century proved that the land uplift is a residual rebound phenomenon from the Ice Age, as well as the Finnish geologist Wilhelm Ramsay (1865–1928), who separated the concepts of isostatic land uplift and eustatic sea level rise. De Geer's observational material consisted of elevated shorelines, which were later dated with stratigraphic methods. In addition to them, e.g. Joakim Donner (1995), Erkki Kääriäinen (1966), Erkki Niskanen (1939), Matti Sauramo (1918, 1923), and Tikkanen and Oksanen (2002; Fig. 8.2) have published essential studies on the land uplift. Erkki Niskanen was the first to study the phenomenon theoretically from the isostatic point of view.

8.3 Land Uplift Studies in the Nordic Countries

The land uplift maps shown in Fig. 8.3a (relative uplift) and Fig. 8.3b (absolute uplift) are compiled from the tide gauge and repeated leveling data from the Baltic States, Denmark, Finland, Germany, Norway, Russia, and Sweden.

The tide gauges along the shores of Finland are situated in Kemi, Oulu, Raahe, Pietarsaari, Vaasa, Kaskinen, Mäntyluoto, Rauma, Turku, Degerby, Hanko, Helsinki, and Hamina, of which Hanko is the oldest (1887) and Rauma the youngest (1933). Precise levelling extended over the Finnish territory has been performed

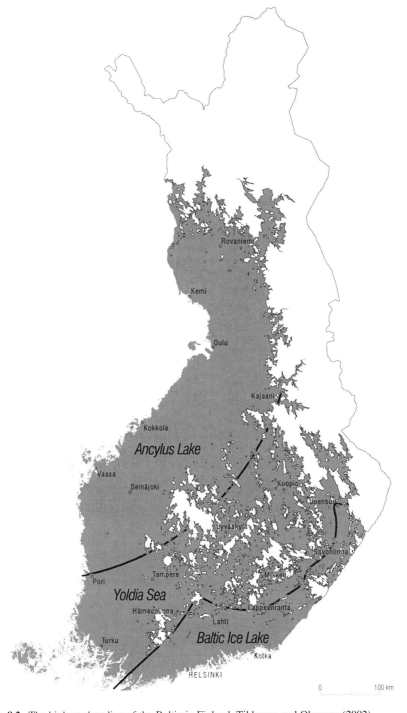

Fig. 8.2 The highest shoreline of the Baltic in Finland. Tikkanen and Oksanen (2002)

three times, of which the First Levelling of Finland took place in 1892–1910 (Blomqvist and Renqvist 1912), the Second Levelling of Finland in 1935–1975 (Kääriäinen 1966), and the Third Levelling of Finland in 1978–2004 (Lehmuskoski et al. 2008). Thus, the time span from the start of the First to the end of the Third Levelling of Finland is over hundred years. The absolute land uplift according to the gravity centre of the Earth is obtained by adding the eustatic rise and the uplift of geoid to the relative land uplift.

The present height system of Finland, N2000, is based on the data of the Third Levelling. It is connected through the Nordic levelling nets to the datum point of the European height system NAP (Normal Amsterdam Peil).

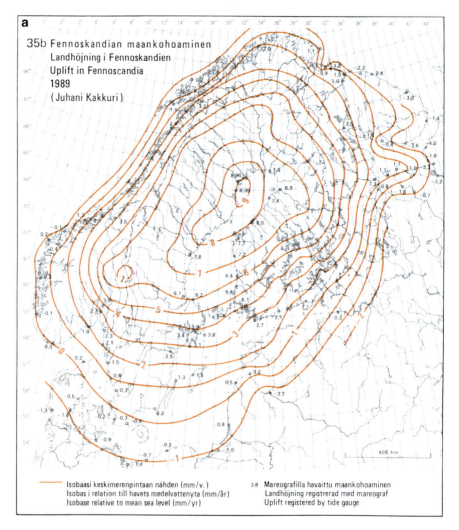

Fig. 8.3 (continued)

8 Fennoscandian Land Uplift: Past, Present and Future

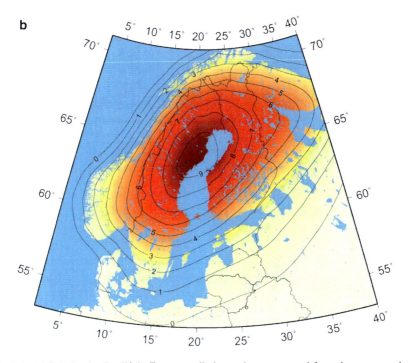

Fig. 8.3 (a) Relative land uplift in Fennoscandia in mm/year measured from the mean sea level (Alalammi 1992). (b) The absolute land uplift in Fennoscandia in mm/year. The Finnish Geodetic Institute

Numerous geophysical factors influence the height of the sea level. Eustatic rise is the most important factor of them. The other factors are isostatic changes in the Earth's crust, changes in the gravity field due to continental ice melt, thermal expansion of sea water, slow changes in the volume of the sea basin, and the influence of weather conditions.

8.4 Geoid

The geoid is the equipotential surface of the gravity potential which at the coasts coincides with the mean sea level of the world ocean. If the distribution of interior masses of the Earth changes, the shape of the geoid changes as well.

The gravimetric geoid of the Nordic Countries shown in Fig. 8.4 is the model computed by Rene Forsberg for the Nordic Geodetic Commission. Almost all the major tectonic features are visible in it, as well as in the Bouguer anomaly map shown in Fig. 8.5, including the Fennoscandian orogenies (Kola, Karelides, Svecofennides, Granulite Belt), the Caledonian orogeny, and the areas of rapakivi

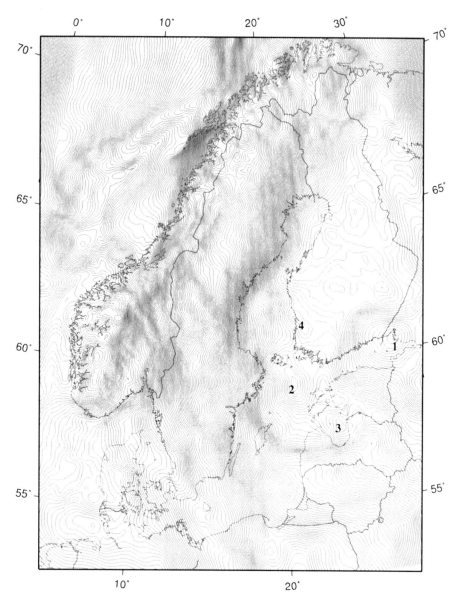

Fig. 8.4 Geoid model for the Nordic Countries. By courtesy of the Finnish Geodetic Institute. In the picture 1 = Viborg, 2 = Aland, 3 = Gulf of Riga and 4 = Laitila rapakivi batholites

granites. In the centre of the geoid is a large depression, which is partly due to the crustal structures, partly to the postglacial rebound. The depression, which is 10 m deep, is slowly recovering due to subcrustal mass flow from the mantle of the Earth. This reduces the gravity on the surface slowly, according to repeated measurements around 0.16 µGal for 1 mm uplift.

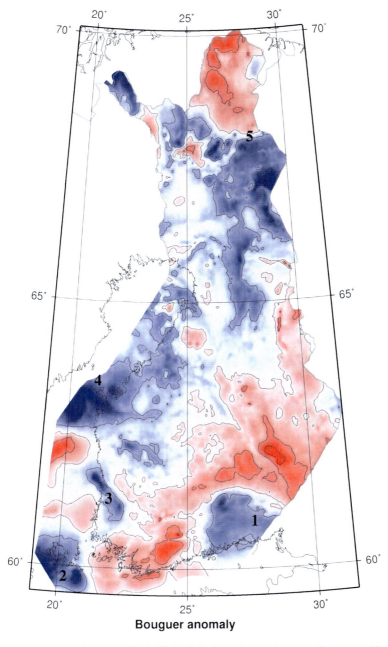

Fig. 8.5 Bouguer gravity anomalies in Finland. In the red areas the anomalies are positive and blue areas negative. In the figure, 1 = Viborg rapakivi, 2 = Aland rapakivi and 3 = Laitila rapakivi areas, 4 = maximal land uplift area, and 5 = the granulite belt. The Finnish Geodetic Institute

As well known, all pieces of mass attract each other as the Newton's gravitational law states. In accordance with this, the ice masses e.g. on Greenland attract the sea water of the surrounding ocean. On the contrary, the molten ice from Geenland will mostly flee to the southern hemisphere, while in its immediate vicinity, sea level will even subside (Vermeer 2010).

8.5 Influence of the Gravity on the Ocean Surface

According to theoretical investigations, if temperatures stay high over the next thousand years, the ice sheet on Greenland will melt totally. If this happens, for example, within 4,000 years, the eustatic rise of the sea level will be negative by 0.1–0.2 mm/year in the vicinity of Greenland (dark blue area in Fig. 8.6) but positive by 2.4–2.6 mm/year in the southern part of the Atlantic Ocean as well as in the Indian and Pacific Oceans. Thus, the total melt of Greenland does not essentially speed up the sea level rise in the North Atlantic and the Baltic Sea.

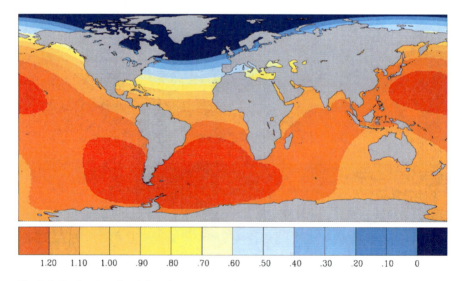

Fig. 8.6 Regional sea level rise after removal of ice from Greenland, including the effects of self gravitation, elastic rebound of lithosphere, and the Earth's rotation perturbations. Numbers of the scale are the coefficients by which the global average of the eustatic rise is to be multiplied to have the rise in the colored area. In the dark blue area the sea level is subsiding, in the area from light blue to yellow and red it is rising Mitrovica et al. (2001)

8.6 Future

The ice sheets and glaciers have started to melt in recent years due to global warming. As a consequence, the eustatic rise of the sea level has sped up from a typical annual value of around 1 mm/year being now about 2–3 mm/year.

If all the continental and mountain glaciers of the Earth would melt totally, the surface of the world ocean would rise approximately 90 m. About 80 m of this is due to the melting of Antarctica, 8 m to the melting of Greenland, and 0.6 m due to the melting of glaciers on the mountains.

Continents and sea bottoms are in fact pieces of the lithosphere, which float on the surface of the Earth's mantle following the Archimedean law. A rule of thumb is that a 100 m thick layer of ice presses the sea bottom down by 27 m, and a similar layer of water transfered from the sea onto the continents lifts the sea bottom 30 m up.

Post glacial land uplift on the areas which have been covered by Quaternary ice sheets (as now-a-days Fennoscandia and Canada), slow subsidence of the bottom of the world ocean, upheaval of the continents, and the changes in the Earth's gravity field are all resulting from the melting of these ice sheets.

The oceans cover around 70% of the Earth's surface. Subsidence of such a large sea bottom forces the continents to lift up. This phenomenon, called levering of the continents, was studied first time by R. A. Daly (1925). Later studies were performed by the north-americans R. I. Walcott (1972) and W. R. Peltier (1999), and by the Finn Lassi Kivioja (1967) from Purdue University. If, according to Kivioja, a 65 m thick ice sheet were transferred from the Antarctica to the oceans, the subsidence of the ocean bottom were 15 m but the levering of the continents at the same time as much as 35–40 m.

References

Alalammi P (1992) Geophysics of the solid Earth crust. Atlas of Finland Karttakeskus, Helsinki
Blomqvist E, Renqvist H (1912) Das Präcisionsnivellement Finlands 1892–1910. Fennia 31, No. 2
Daly RA (1925) Pleistocene changes of level. Am J Sci (5th Series) 10:281–313
Donner J (1995) The Quaternary history of Scandinavia, vol 7, World and regional geology. Cambridge University Press, Cambridge, UK
Kivioja LA (1967) Effects of mass transfers between land-supported ice caps and oceans on the shape of the earth and on the observed mean sea level. Bull Geod 1967:281–288
Kääriäinen E (1966) The second levelling of Finland. Publ Finn Geod Inst 61:1–313
Lehmuskoski P, Saaranen V, Takalo M, Rouhiainen P (2008) Bench mark list of the third levelling of Finland. Publ Finn Geod Inst 139:1–220
Mitrovica JX, Tamiciea ME, Davis JI, Milne GA (2001) Recent mass balance of polar ice sheets inferred from pattern of global sea level change. Nature 401:1026–1029
Niskanen E (1939) On the upheaval of land in Fennoscandia. Ann Acad Sci Fennicae AIII 10
Peltier WR (1999) Global sea level rise and glacial isostatic adjustment. Global Planet Change 20:93–122
Sauramo M (1918) Geochronologische Studien über die spätglaziale Zeit in Südfinnland. Fennia 41(1) and Bull Comm Géol Finlande 50

Sauramo M (1923) Studies on the Quaternary varve sediments in southern Finland, vol 60, Bull Comm Géol Finlande. Sällskapet för Finlands geografi, Helsingfors

Tikkanen M, Oksanen J (2002) Late Weichselian and Holocene shore displacement history of the Baltic Sea in Finland. Fennia 180(1–2):9–20

Topelius Z (1893) Maa suomalaistenkirjailijain ja taiteilijain esittämänä sanoin ja kuvin. Suomi 19:llä vuosisadalla. GW Edlund, Helsinki

Vermeer M (2010) Sea level rise in the Baltic Sea. Expert article 496, Baltic Rim Economics 28.4.2010

Walcott RI (1972) Past sea levels, eustacy and deformation of the Earth. Quat Res 2:1–14

Part II
Changing Baltic Sea

Cover page: Ice crusher Sisu in work Courtesy: Riku Lumiaro

Chapter 9
Ice Season in the Baltic Sea and Its Climatic Variability

Matti Lepparänta

9.1 Introduction

The Baltic Sea ice is located between temperate maritime and continental subarctic climate zones. Ice forms here annually, and in a severe winter the whole basin is ice-covered. The Baltic Sea lies on the edge of the seasonal sea ice zone of the northern hemisphere, and therefore even small climatic variations show up strongly in the characteristics of ice seasons. The length of ice season is 5–8 months, and the maximum annual ice extent is 12.5–100% of the area of the Baltic Sea (Fig. 9.1). Ice formation, ice cover, and ice melting are very important factors in the physics and ecology of the Baltic Sea and also have a major impact on the society. The future of Baltic Sea ice seasons in the next 100 years depends first of all on the atmospheric forcing, since in this period internal changes of the basin are not expected to have a notable influence on the heat budget.

9.2 Baltic Sea

The Baltic Sea is small and shallow marine basin, which is connected to the main North Atlantic via Danish Straits. For a detailed treatment of the physical oceanography of the Baltic Sea the reader is referred to a recent book by Lepparänta and Myrberg (2009). The mean depth of the Baltic Sea is 54 m and the maximum depth is 459 m. Water exchange through the Danish Straits is rather weak, and, consequently the Baltic Sea water is brackish, with mean salinity of 7.4 per mille or one-fifth of the salinity of normal ocean water. The Baltic Sea water body is stratified: at 40–70 m depth there is a permanent halocline, a thin layer of large increase of

---

M. Lepparänta (✉)
Department of Physics, University of Helsinki, P.O. Box 48, FI-00014 Helsinki, Finland
e-mail: matti.lepparanta@helsinki.fi

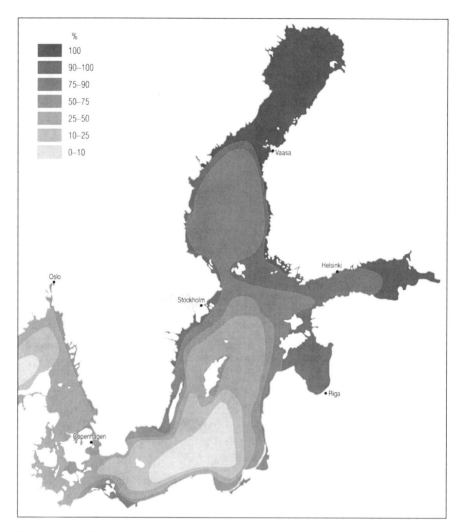

Fig. 9.1 Probability of annual ice occurrence in the Baltic Sea (Leppäranta and Myrberg (2009), based on SMHI and FIMR (1982))

salinity, which separates saline deep water and relatively fresh upper layer water from each other. The upper layer is mixed during the course of seasons all over the Baltic Sea, but the deeper water is renewed only by saline inflows from the North Sea. Thus the Baltic Sea is a marine basin, where the salinity stratification has a major role.

The Baltic Sea cools in fall as the water mass loses its summer heat to the atmosphere. Only the upper layer forms the active heat storage, which does not receive much heat from the deep water or the North Sea. Therefore, on average in the middle of November ice formation starts in the coastal areas of the Bay of Bothnia, at about the same time as large lakes in Finland in the corresponding latitudes. Interannual variability of the freezing date is about 1 month on both sides of the mean (Haapala and Leppäranta 1997). The time of freezing depends first of all on the evolution of the air temperature in the autumn. The thermal memory of the Baltic Sea is 2–3 months, and therefore the influence of the previous summer is not seen in the ice season. The salinity of seawater influences significantly on the water properties. In the Baltic Sea, the temperature of maximum density is 1.5–3ºC and the freezing point is 0.2–0.5ºC below zero centigrade, depending on the salinity of the water (in fresh water these temperatures are 4°C and 0°C). The implication of salinity is, however, quite small on the freezing date. In fact, if the Baltic Sea were a fresh water lake, autumn mixing would penetrate through the whole water body and freezing would be later than in the present brackish basin. North of Fennoscandia, there is relatively shallow Barents Sea. This basin is ice free for its southern half due to renewal of heat by the warm North Atlantic Current.

9.3 Ice Conditions in the Baltic Sea

Ice formation progresses from the northern Bay of Bothnia southward along the coast, while deeper central basins freeze slowly due to their large heat content. In the Vyborg Bay and Vaasa Archipelago freezing starts on average in mid-December, and thereafter ice cover expands, the Gulf of Finland and Sea of Bothnia freeze over in the same phase. In normal winters the Bay of Bothnia freezes over in mid-January and the Gulf of Finland, Gulf of Riga and Sea of Bothnia 1 month later. In mild winters only the Bay of Bothnia and the eastern Gulf of Finland freeze over. Ice extent is at largest in February 15 – March 15, and the maximum thickness of landfast ice is 50–120 cm. Melting of ice begins in south already in the beginning of March, at the time the ice is still growing in the north. The rate of melting is 1–2 cm per day and therefore the melting period is 1–2 months depending on the location. The date of ice break-up depends on the thickness of ice and net solar radiation, rain and air temperature. The Gulf of Finland and the Sea of Bothnia become ice free on average by the 1st of May, and the last ice floes are observed in the Bay of Bothnia at the end of May. In cold springs ice may persist until midsummer.

The Baltic Sea has marine ice cover (Fig. 9.2). This means two important characteristics: ice is saline, and in large basins a solid ice cover is unstable and breaks into floes, which drift under the forcing by winds, currents and sea level variations. The ice cover is similar in all freezing seas up to North Pole. Also in very large lakes, such as Lake Ladoga in Europe and North American Great Lakes, the ice cover breaks often into drift ice fields.

Brackish ice is in fine structure similar to normal sea ice. Ice sheet contains liquid brine pockets, which influence on the properties of ice and also act as life habitat for biota. Also there are active algae species in the brackish water ice of the Baltic Sea, and their growth is basically light-limited (Kuparinen et al. 2007). Brine pockets are small (diameter 0.1–1 mm, length several centimetres) limiting the amount to which biomass can grow. A simple food-web system develops in

Fig. 9.2 Envisat satellite picture of the central area of the Baltic Sea (Gotland island in the lower left corner and Gulf of Finland on the upper right corner). The satellite picture shows landfast ice zone in coastal and archipelago areas and broken drift ice fields with leads further out. Land areas are dark and shoreline is clearly shown as the boundary of land and snow-covered landfast ice. South of the Gulf of Riga clouds contaminate the picture. © ESA

the brine pockets, where algae produce organic matter to bacteria. In spring together with the warming of the ice, brine pockets expand, more sunlight penetrates into the ice, and in places patches of large algae population are found.

In coastal and archipelago areas *landafast ice* zone exists, with landscape similar to winter landscape in Finnish lakes. Landfast ice zone is solid and level, and the ice sheet is stable except for the very early and late seasons. In contrast, further out the ice cover is broken into floes and undergoes movement; this ice is called *drift ice* (Fig. 9.2). Due to ice motion *leads*, several kilometres wide, open into ice fields and can be utilized by winter shipping, while under pressure ice breaks into small blocks, which accumulate on top of each other to form *hummocked ice* and *pressure ridges* (Fig. 9.3). The largest ridge measured in the Bay of Bothnia has been 31.5 m in vertical extension, 28 m of it beneath the sea level (Palosuo 1975).

Drift ice landscape is thus variable with ice floes of different size and shape, hummocks and pressure ridges, and open water areas (Palosuo 1953; Leppäranta 2011). This landscape undergoes continuous changes, since forced by strong winds,

Fig. 9.3 Surface photograph of a pressure ridge

sea ice may drift 10–20 km/day. Ice reports, provided for shipping by operational ice services, are therefore needed daily to know the actual ice conditions. Drift of ice transports ice and latent heat from one sea area to another. Thus the evolution of ice conditions depends much on winds in addition to air temperature. Frequent strong winds keep the ice in motion and cause accumulation of hummocks and ridges, while in cold and less windy ice seasons ice grows thick and moves less. The thickness of landfast ice, as the thickness of lake ice, depends primarily on air temperature evolution. Thickness of half a meter results, when there are 100 cold days with mean air temperature 6–7°C below 0°C.

Drift ice is an interesting medium as a research object (Leppäranta 2011). There is still basic research done on that, and remarkable results have been obtained in the Baltic Sea. There management of observation campaigns is easier than in polar areas, and frozen basins of the Baltic Sea can be considered as 'physical laboratory models' for polar sea ice. The dynamics of drift ice contains strong nonlinearities. E.g., in the Gulf of Finland, under strong wind, thin ice (10–20 cm) can drift east and accumulate all in the eastern part of the basin, while thick compact ice (50 cm) does not move at all.

Ecology of the Baltic Sea has adapted into the annually freezing sea. The role of the ice season in the ecology has become better understood in the last 20 years. Baltic Sea ice acts as a habitat for biota and boundary condition for life. Ice has its own ecosystem, as discussed above, and water masses beneath or near ice feel the presence of ice. River waters flow just under *landfast* ice long distances, and the dispersion of nutrients brought by them is quite different between open water and ice seasons. Beneath the ice also algae blooms may occur in mid-winter in the river water layer; the stratification keeps the algae close to ice bottom where light is available. Optical thicknesses of ice and snow are 1–2 m and 10–20 cm, respectively, and especially when snow is absent or thin light conditions are favourable under the ice. In the melting season, organic matter and nutrients are released from ice to the ecosystem close to ice bottom and ice edge. Baltic Sea seals have as well adapted into the icy Baltic Sea. They give birth to their puppies on springtime thick ice floes, which is a safe environment for the puppies during their first weeks of life. In very mild winters the absence of ice may cause problems.

9.4 Ice Conditions and Human Life

As nature also human being has adapted its life into the annually freezing Baltic Sea. Before the time of trains and steamboats, traffic and transport was based on sailboats and ice cover isolated Finland from the rest of Europe (Palosuo 1953). It was said that the country was bounded by ice. With steamboat era coming, winter shipping started to expand, and in 1877s/s Express, a passenger ship, was kept in traffic all winter between Hanko, Finland and Stockholm, Sweden. By nearly 100 years later, in 1970, winter shipping had reached all the main harbours in the Baltic Sea, up to its northern ports in Kemi and Luleå. However, ice conditions still have a major impact on winter shipping, since Baltic Sea countries need to keep icebreaker fleets and ice often causes delays to cargo transport.

In marine constructions ice forces need to be considered in the Baltic Sea. Especially, pressure forces by drift ice on fixed structures and ships can be quite large. For example, Tammio lighthouse in the Gulf of Finland and Oulu outer lighthouse in the Bay of Bothnia have been damaged by ice pressure in 1960s and 1970s, and several ships have been sunk by the ice. On the other hand, on-ice traffic over landfast ice helps in the communication in archipelago areas, although in early and late winters neither small boats nor ice roads can be used. For example, in the Bay of Bothnia there is about 8 km long ice road from the mainland to the island of Hailuoto with 1,000 inhabitants. The bearing capacity of floating ice increases with thickness squared, so that 5 cm thickness is enough for one person and 20 cm for one car with mass of 1.5 t. In the northern part of the Baltic Sea ice grows to more than 20 cm in normal winters. Before the time of all-season winter shipping to the Bay of Bothnia, there was an ice road across the basin in cold winters, from Vaasa, Finland to Umeå, Sweden. In Helsinki, until 1985 city bus line #14 went across landfast ice to Suomenlinna castle island when the ice was thick enough.

Seal hunting used to be common from springtime drift ice in the past, and also specific techniques were developed for winter fishing. The importance of recreational hobbies has increased much, and winter sea ice fields have been found for fishing, skiing, long-distance skating and other outdoor activities as well. Sometimes landfast ice breaks off and drifts further out with several fishermen. Environmental accidents, especially oil spills, are difficult to manage in ice conditions. Detecting oil is particularly demanding in drift ice fields, and there is not yet good technology for collecting oil from among ice floes.

9.5 Baltic Sea During the Last 100 Years

Ice conditions in the Baltic Sea follow climatological variations in the atmosphere of northern Europe (BACC Author Group 2007). The basic picture of mild and severe winters tells that mild and moist Atlantic air is brought by the westerlies while cold and dry air is driven in by the continental high pressure over Russia. So-called NAO (*North Atlantic Oscillation*) index describes the strength of the westerly flow and it is correlated with characteristics of Baltic Sea ice seasons. This correlation is around 1/2, limited by the fact that warm and cold air come by other mechanisms also.

There are long records of the Baltic Sea ice season (Jevrejeva et al. 2004). Indirect information exists from medieval times in the form of opening and closing of sailboat traffic, and the longest phenological time series describes the annual maximum ice extent since 1720. This time series is dominated by large aperiodic variability, and the only systematic change has been a weak decreasing trend. The Baltic Sea has been totally covered by ice on average once in 30 years. Last time this took place in 1947, although in 1987 the ice coverage was 96% of the area of the Baltic Sea. Very mild winters occurred in 1988–1993, and in 2008 the maximum annual ice coverage was at its minimum so far, 12.5% of the area of the Baltic Sea. In the last 100 years the air temperature has increased by about 1°C. Because of land uplift the area of the Baltic Sea has slightly decreased and the salinity has gone *higher* by 0.5 per mille units. Salinity has increased across the whole vertical water column and therefore the strength of the stratification has not changed. In the state of the Baltic Sea, human influence has shown up strongly. Due to nutrient loads eutrophication has increased into a serious level, and poisonous algae blooms occur every summer. Eutrophication has increased the consumption of oxygen and consequently oxygen conditions in the lower layer have become worse.

It is anticipated that if climate warms there will be less ice. This has been the case so far, which is understandable since there have been no significant changes in the internal water mass structure and stratification in the Baltic Sea. Increasing air temperature has lead to corresponding increase in surface water temperature in the open water season, and the ice season has shortened by 1–2 weeks from both ends, depending on the site location. The area of ice cover relative to the area of the whole Baltic Sea has decreased about 10% units in the last 100 years, but the thickness of ice does not show any systematic changes. This is likely due to that ice thickness depends much on snow accumulation in addition to air temperature.

In addition to the slow trend toward milder ice conditions, there is quite large interannual variability in seasonal ice characteristics. When westerlies are dominating, Baltic Sea ice seasons are mild, moist and windy. In contrast, when westerlies cannot penetrate into the Baltic Sea, continental high-pressure situation prevails with cold air streams. In the last 100 years, on top of NAO driven variability, there has been a warming trend after the cold nineteenth century. Human activity has not influenced much on the ice conditions, apart from increasing the mobility of ice by the expanding winter shipping.

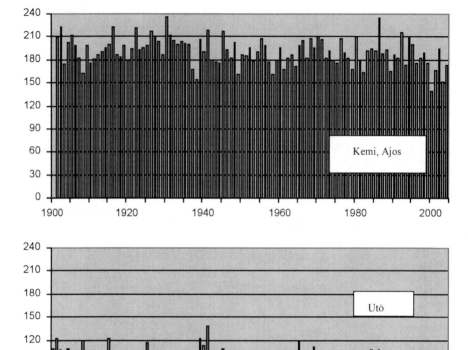

Fig. 9.4 The length of ice season in Kemi and Utö since 1900

The variability of ice conditions is well described by the time series of the length of ice season in Kemi and Utö (Fig. 9.4). In Kemi the sea freezes every winter, and the length of ice season is 5–8 months. Because of the large interannual variability, it is difficult to recognize the trend by eye. Ice season follow each other in a stable way. In contrast, Utö time series is qualitatively different. The length of ice season is 0–4 months; the location is at the edge of the climatological mean edge of the Baltic Sea ice cover, and therefore the variability is relatively speaking much larger than in Kemi. Winters are more difficult to predict, and in one winter out of five there has been no ice in Utö. In statistical analyses there is an interesting statistical problem: for example, how should mean freezing date be defined when freezing does not take place every year. Therefore, cumulative probability distribution is employed for medians and other fractiles.

A more detailed picture of ice seasons can be obtained with mathematical models. Construction of such models commenced in the 1970s, first for short-term (days) ice forecasting for winter shipping and since 1990s for investigations of ice climatology. These models provide future scenarios for ice seasons under a given new climate in northern Europe. Sea ice models contain four elements: (1) description of ice state (primarily the thickness field), (2) sea ice rheology, (3) equation of motion, and (4) ice conservation law. Sea ice is in intensive interaction with the underlying water body, and therefore ice models are essentially coupled ice–ocean models. Ice models are forced by atmosphere. Thermal processes produce ice and melt it, while winds and sea currents drive the drift of ice. Thermodynamics and dynamics interact, since drift ice strength depends on thickness and ice motion produces open water areas where ice production may be intensive. Thin ice grows faster than thick ice, and consequently dynamical processes usually increase the total ice production. The present ice models can simulate the evolution of ice seasons quite well, provided the atmospheric conditions are known.

9.6 Future of Baltic Sea Ice Seasons

The Baltic Sea is geologically rapidly changing basin. Since the Weichselian glaciation, during 13,000 years, it has undergone several fresh water and brackish water phases. Local land rise and eustatic changes of global sea level elevation still change the morphology of the Baltic Sea that will have further consequences into future ice seasons. But in the next 100 years no significant morphological changes are anticipated, and the main marine question is the future water exchange through the Danish straits (the Belts and Sound) and the consequences to the stratification of the Baltic Sea water masses. If these do not change, the response of the Baltic Sea to the regional weather conditions will be as now. This means that ice will stay, and the length and quality of ice seasons depend on the future global climate and the strength of the North Atlantic westerlies. The thermal memory of the Baltic Sea is just a few months, and therefore winter conditions depend only on the fall and winter weather.

Discussion around climate change and its implications is presently quite intensive. Recent climate changes show global warming trends during the last 100 years. The majority of climate scientists believe that to some degree this reflects human impact on nature, while there are others who consider that climate variations until present are only due to natural long-term behaviour of our planet. Such debate is anyway normal scientific communication, and the different views tell about large uncertainties in this matter. IPCC (Intergovernmental Panel on Climate Change) tries to find a consensus in its regular reports about the best forecast for the next 100 years. In these scenarios, which are based on simulations with global coupled atmosphere–ocean models, the anticipated future release of greenhouse gases to the atmosphere is the critical forcing factor. Consequences of rapid climate changes would be drastic to human living conditions, and therefore political strategies are sought to slow down anticipated rate of climate warming.

In the Baltic Sea region, based on IPCC scenarios air temperature is expected to rise by 2–4°C until 2100, more in winter than is summer (BACC Author Group 2007). Precipitation is expected to increase in winter and decrease in summer, so that there is no significant net change. Eustatic sea level rise is expected to 0.1–1.0 m, and it has likely not much influence on the water exchange between the Baltic Sea and the North Sea. With increasing air temperature, the surface temperature of the Baltic Sea will rise and temperature-dependent processes will consequently change. This means that the freezing date will be delayed and there will be less ice. Stratification of the Baltic Sea is not expected to change significantly. Model simulations have shown that changing the fresh water input by ±30% does not influence essentially on the stratification.

The long-term evolution of Baltic Sea ice seasons is closely connected to the future of the North Atlantic westerlies and continental high pressure in the east. If NAO index could be predicted for future, some predictability would follow for ice seasons but that is not the case. Consequently, we can say that large interannual variability of ice seasons will stay, connected to NAO, and decadal averages of ice seasons follow the evolution of global climate. The sensitivity of the ice seasons to air temperature can be examined from Table 9.1.

Assuming that in 2,100 average air temperature is 4°C more than now, we can conclude the following. Freezing date will be about 2 weeks later than now, the thickness of ice will be 30 cm less than now, and the ice area will be 40,000 km^2 less than now. This means that in a normal winter only the Bay of Bothnia and the eastern end of the Gulf of Finland, as presently take place in mild winters. Normal ice season therefore loses the ice from the Sea of Bothnia and most of the Gulf of Finland. Future severe winters would be as present normal winters, and in future mild winters the Bay of Bothnia would no more freeze over. However, ice would still form in the coastal zone and archipelago areas. Several scenario simulations with mathematical models have produced nearly the same result as shown in Table 9.1.

When the ice is thinner, instability of ice cover increases. Landfast ice breaks more easily and as well drift ice. The future of snow cover is the main open question. It influences on the insulation efficiency of the ice cover and on the albedo, and consequently start-up of melting season in spring. In Table 9.1 the change in ice breakup date is based on smaller thickness of ice, but if the amount of snow is less, melting season will begin earlier and consequently ice breakup comes earlier. The quality of ice conditions also would change.

Table 9.1 The sensitivity of the characteristics of Baltic Sea ice seasons to air temperature

Quantity	Change/1°C	Comments
Freezing date	Delay 5–10 days	Stratification of the sea does not change
Ice thickness	Thinner by 5–10 cm	Uncertainties due to snow
Quality	More unstable	Depends on ice thickness
Ice extent (area)	10,000 km^2	All above important
Ice breakup	2½–5 days earlier	Less ice

Sea ice is an active habitat of life. Sea ice algae grow inside ice, and after ice melting they continue to live in the surface water layer. Decrease of ice extent therefore results in change of species composition in spring bloom. Ice also influences on coastal ecosystems and cleans shore areas annually. Seal puppies are born on sea ice, and especially in the Gulf of Finland and Gulf of Riga the question is that how seal populations can adapt to possible climate warming with absence of ice in mild winters.

For human activities less ice sounds relieving, since winter shipping could be easier and less expensive. However, thinner ice is more dynamic, and shipping would meet more difficult ice during the shorter ice season. On-ice traffic – on foot or by car – will suffer since the time of good bearing capacity of ice becomes shorter. Also change toward milder ice seasons would have consequences in regional weather. More humid and rainy weather brings more fogs, lower visibility, and increase risks for ice accretion on ships. With less compact ice cover periods in Baltic Sea basins, there would be more variability in marine and coastal weather. If, as sometimes suggested, the frequency of storms increases, waves will be larger. Finally, an aesthetic consequence is that the landscape of the Baltic Sea becomes greyer.

The research of Baltic Sea ice has now many challenges. Future works needs integrate physics and ecology better together and result then in firmer understanding of future ice seasons and their further implications on the state and ecology of the Baltic Sea. By far the main problem in the Baltic Sea is presently not long-term climate change but the more immediate ongoing eutrophication and pollution.

References

BACC Author Group (2007) Assessment of climate change for the Baltic Sea Basin. Springer, Berlin/Heidelberg
Haapala J, Leppäranta M (1997) The Baltic Sea ice season in changing climate. Boreal Env Res 2:93–108
Jevrejeva S, Drabkin VV, Kostjukov J, Lebedev AA, Leppäranta M, Yeu M, Schmelzer N, Sztobryn M (2004) Baltic Sea ice seasons in the twentieth century. Climate Res 25:217–227
Kuparinen J et al. (2007) Role of sea-ice biota in nutrient and organic material cycles in the northern Baltic Sea. Ambio 36(2–3):149–154
Leppäranta M (2011) The drift of sea ice, 2nd edn. Springer, Berlin/Heidelberg
Leppäranta M, Myrberg K (2009) Physical oceanography of the Baltic Sea. Springer, Berlin/Heidelberg
Palosuo E (1953) A treatise on severe ice conditions in the Central Baltic Merentutkimuslaitoksen Julkaisu/Havsforskningsinstitutets Skrift 156
Palosuo E (1975) Formation and structure of ice ridges in the Baltic. Winter Navigation Research Board, Rep. No. 12. Board of Navigation, Helsinki
SMHI (Swedish Meteorologicl and Hydrological Institute) and FIMR (Finnish Institute of Marine Research) (1982) Climatological ice atlas for the Baltic Sea, Kattegat, and Lake Vänern. Sjöfartsverkets tryckeri, Norrköping, Sweden

Chapter 10
Baltic Sea Water Exchange and Oxygen Balance

Pentti Mälkki and Matti Perttilä

The Baltic Sea is surrounded by ten countries with 86 million people living in its drainage basin. Its environmental conditions depend very much on water exchange with the North Sea through the narrow Danish Straits. The strong vertical stratification decouples deep layers from atmospheric influence, and hence renewal of bottom layer water takes place only through inflow of saline water. During earlier decades renewal of the bottom layers has been an irregular phenomenon, but still it has occurred often enough. However, since the 1980s major inflows, able to renew deep bottom layers, have occurred only twice. As a consequence, even major inflows have been unable to remove the hydrogen sulfide from the deep layers. To improve the situation, major inflows would be needed almost annually. It is, therefore, possible that in the future the bottom layers will have a permanent oxygen deficit.

10.1 Introduction

The Baltic Sea is one of the largest brackish water basins in the world. It has evolved after the retreat of last continental ice sheet at he end of the Ice Age. Its total surface area is 370,000 km^2, and its average depth is only 55 m. In spite of that, the vertical stratification is very strong. The annual inflow of fresh river water is some 470 km^3. This volume is mixed with the inflowing North Sea water, which, due to higher salinity, is denser than the ambient water, and moves towards deeper

---

P. Mälkki (✉)
Department of Physics, University of Helsinki, P.O.Box 64, FI-00014, Helsinki, Finland
e-mail: pentti.malkki@helsinki.fi

M. Perttilä
Finnish Meteorological Institute, Erik Palmenin aukio 1, P.O. Box 503, FI-00101, Helsinki, Finland

layers. The annual water exchange, the strong stratification and the shallow bottom topography combined with meteorological variability produce complicated dynamics, which strongly influence the oxygen conditions in the Baltic Sea.

10.2 Water Balance

The total area of the drainage basin of the Baltic Sea is 1.7 million km^2. About one sixth of it, some 280,000 km^2 belongs to the Neva river basin, which is by far the largest river running into the Baltic Sea. Other large river basins, i.e. the Vistula (194,000 km^2), the Daugava (88,000 km^2) and the Neman (98,000 km^2) jointly add to the Baltic Sea a discharge equal to the Neva by itself.

About 40% of total river discharge runs to the Gulf of Bothnia, some 24% to the Gulf of Finland. As a consequence of this, the salinity of these gulfs is very low. Precipitation to, and evaporation from the sea compensate each other almost totally. Consequently, salinity of the surface layer of the sea depends basically on the mixing of incoming river water and incoming North Sea water under wind forcing. Vertical mixing in the sea is most intense in autumn and in winter, when the temperature of the surface layer is low, and thermal stratification is weak. Vertical mixing in the central Baltic Sea, which regulates water exchange with the North Sea, extends down to some 60–70 m depth. Surface layer salinity in the region between the south coast of Finland and Bornholm Isle is about 7 (on Practical Salinity Scale, grams salt in kilogram water). This reveals clearly the strong horizontal circulation due to atmospheric forcing. Below the surface layer, which in the winter is almost homothermal and homohaline, a strong halocline separates the deeper layers from the direct influence of the atmosphere.

10.3 Water Exchange

Because of the small size of the Baltic Sea, there are no significant tides. Nevertheless, the variations of the sea level are considerable. Mean sea level of the entire basin can be more than half a metre above or below the average. The duration of the variations extends from several days to weeks. The dominant factor for this is the in- or outflow through the Danish Straits. Sea level variations due to river runoff are minor compared with those due to water exchange between the Baltic and Kattegat/North Sea. East of the Danish Straits, surface salinity varies between 7.5 and 9, depending on weather conditions and circulation. On the Kattegat side salinity varies between 16 and 22. When the Baltic water level is higher than the North Sea, surface layer flow is directed to the west. The straits are very shallow, the sill depth of the Great Belt is only 22 m. Nevertheless flow in the deeper layers is directed towards the Baltic Sea. Change in weather conditions also causes a change in the direction of currents. The waters of the Kattegat and the Baltic Sea mix in the sounds and on both sides of the sounds.

Most of the time, even in the shallow sill region the water is vertically stratified. Surface layer salinity is close to that of the Baltic, deeper layer salinity close to that of Kattegat. Water exchange is considerable, but it is mainly a to-and-fro movement close to the sounds. The average direction of the movements of atmospheric disturbances in this region is from west to east. Stronger disturbances cause westerly winds, which in turn cause inflows to the Baltic Sea. The wind causes a tilt in the sea surface in the Kattegat, with the eastern side having a higher water level than the western side. Simultaneously on the Baltic side, water is transported east from the sill. This causes a water level gradient in the sill region, and enhances a flow of water towards the east. If weather conditions remain stable for several days, currents have time to transport Kattegat water down to the Arkona Basin east of the sounds. A strong current in the region is usually barotropic, i.e. has close to the same velocity from surface to bottom. By this mechanism, the Baltic Sea obtains water that has a higher salinity than the ambient water. It has been estimated that the inflow to the Baltic is some 1,200 km^3 annually. The outflow from the Baltic is about 1,700 km^3, corresponding to 4.5 m of water, if distributed evenly over the entire Baltic Sea. The outflowing current, the so-called Baltic current continues north along the Norwegian coast towards the Barents Sea. The Kattegat water which has reached the Baltic, due to higher salinity and hence higher density, continues its movement in the deeper layers. It mixes slowly with the ambient water and continues its way in the layer, which has same density as the mixed water. This water spreads all over the Baltic Proper and renews the water masses below the halocline. Most often the mixing in the southern Baltic is so strong that salinity decreases considerably and the current cannot penetrate the deepest layers, which typically have salinity 11–13. A renewal of these deep waters requires a very strong pulse of saline water, commonly called a Major Baltic Inflow (for recent analysis, see e.g. Matthäus and Franck 1992).

10.4 Major Baltic Inflow

Renewal of the deep layers of the Baltic Sea is totally dependent on occasional sufficiently strong inflows of saline sea water through the Danish Straits. That is why monitoring and research of these inflows has been through decades on the active agenda of both scientists and laymen concerned about the environment in the Baltic Sea. For the intensity of these inflows an index has been developed. It relates the strength of the inflow to the salinity of inflowing water and the duration of the inflow. Renewal of deep bottom waters is negligible if the salinity of the incoming water is less than 17, and if the duration of the inflow is less than 5 days. Strong inflows have salinities above 18, and duration often over 2 weeks. About half of cases are inflows that do not have a remarkable effect on the renewal of the deep layers of the central basin. A major inflow is possible if:

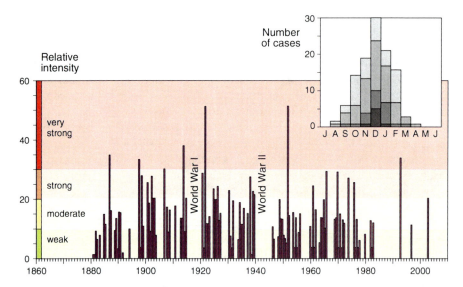

Fig. 10.1 Occurrence of major Baltic inflows during the past 100 years (Matthäus 2006)

- The sea level of the Baltic is considerably below the average. This is usually a consequence of an outflow caused by long lasting easterly winds.
- The stratification of deep layers is such that the inflow can penetrate below the ambient water.
- The water masses in both the Kattegat and the Skagerrak have been renewed from the North Sea so that their surface salinity is higher than 20.
- The duration of the westerly wind, which causes the inflow, is sufficient for driving the Kattegat water east of the Danish Straits.

The conditions described above can prevail during any season, but the highest probability for their occurrence is in autumn and winter. During those seasons winds are strongest, and the water level variations in the region reach their maxima. The winds causing outflow from the Baltic Sea are connected with a longstanding high pressure over the Baltic Sea. This so-called blocking high has a 25% probability during winter (Hurrell and Deser 2009), but its location and duration has considerable variability. Westerly winds causing an inflow are mostly connected with a positive phase of North Atlantic Oscillation (NAO). During this positive phase weather disturbances move eastwards over northern Europe. In wintertime the probability of a positive NAO is 29%. This type of weather has high variability. Long enough westerly winds are not frequent, hence conditions for strong inflows are seldom met. These two conditions alone can explain why major inflows are so rare.

Figure 10.1 shows the occurrence of major Baltic inflows during the past 100 years. During this time inflows which are classified as strong, and which have renewed water in the bottom layers of deep basins have occurred twenty-three times. Very strong inflows have occurred six times, the last one in 1993. As can be seen in the figure, after 1984 strong inflows have occurred only twice. The probability distribution in the upper right-hand corner of the figure reveals that the emphasis of occurrence is on the winter months.

Analysis of major inflows shows that the incoming water masses mix strongly with ambient water during their transport towards deeper basins. The masses first enter the Arkona Basin, which has a maximum depth less than 50 m. From there onwards, the flow continues along the strait between the island of Bornholm and southern Sweden to Bornholm Basin, some 100 m deep. This basin has on its eastern side a sill at the depth of 62 m, which the inflow has to pass in order to penetrate to the deep areas of the Eastern Gotland Basin or the Gdansk Basin. Figure 10.2 shows the route of major inflows from the Arkona Basin to the central basins, and Fig. 10.3 shows longitudinal profiles of the typical salinity (3a) and temperature (3b) from the Arkona Basin to the Gulf of Finland in 2002. Each deep basin has a dense, relatively high salinity layer, which has eddies of 10–70 km diameter. The inflowing water masses mix partly with the ambient water in the eddies, but major salinity decrease takes place on the route between the basins. The bottom water of the Bornholm Basin, which has salinity close to 15, does not cross the sill, and consequently does not proceed towards the central Baltic Sea. As the time elapsed from the previous major inflow is several years, practically the entire water layer below the halocline has an oxygen deficit.

The observations made during the exceptionally strong inflow in 1993, as well as measurements during the 2003 inflow, have enabled detailed studies of the inflow mechanisms (Piechura and Beszczynska-Möller 2004; Andrejev et al. 2002; Matthäus and Lass 1995; Zhurbas et.al. 2004). The basin north of Bornholm is

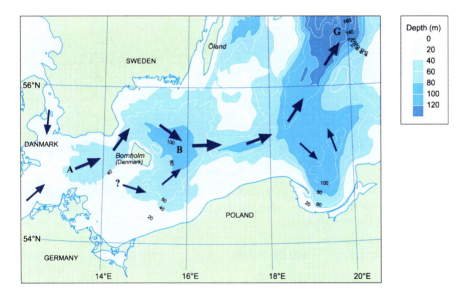

Fig. 10.2 Route of major inflows from Arkona Basin to central basins (Piechura and Peszczynska-Möller 2004). *A* Arkona Basin, *B* Bornholm Basin, *G* Gotland Basin

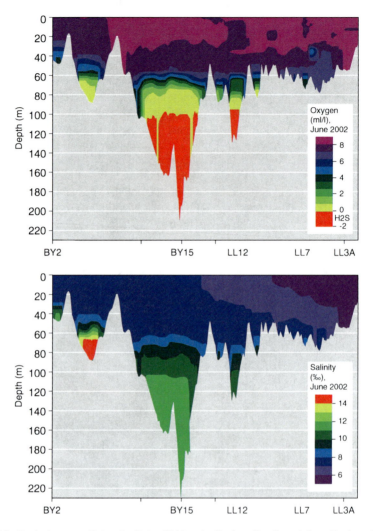

Fig. 10.3 Typical oxygen (3a) and salinity (3b) longitudinal profiles from Arkona Basin to Gulf of Finland in 2002 (Finnish Institute of Marine Research, unpublished). Marked points: *BY2* Arkona Basin, *BY15* Gotland Deep, *LL12* Entrance to Gulf of Finland, *LL7* Outside Helsinki, *LL3A* Outside Loviisa

the site where the mixing of inflowing saline and ambient less saline water mainly takes place. As a consequence of the mixing, the salinity of the bottom layer decreases and the pycnocline rises above the sill depth of 62 m at the eastern end of the basin. Overflow towards the east into Gdansk Bay and the Eastern Gotland Basin can start.

The inflow into the Gotland basin is in general slow. It is only during exceptionally strong inflows that the renewal of bottom layer waters is so rapid that the inflowing water pushes the ambient water mass forward to the northern parts of the

Baltic Proper and the Gulf of Finland. It has been estimated (Borenäs et al. 2007), that the flow over the sill can be at most 65,000–70,000 m^3/s. This means that filling a basin with 200 km^3 of inflowing water takes more than 1 month. Most often the inflow causes a rise of the pycnocline (a layer with rapid density increase in vertical direction). During the following winter this pycnocline is eroded and the surface layer salinity increases somewhat.

10.5 The Influence of Water Exchange on the Condition of the Bottom Layer

As a consequence of strong inflows the stratification in the bottom layers becomes stronger. Even winter storms mix water masses only to the depth of some 60–70 m. In the Arkona, Bornholm and Gdansk basins winds cause a tilt in the boundary layers, which induce bottom layer circulation. Material exchange between the layers is, however, limited. The biota produced by primary production in the surface layer has a limited life span, and after dying the organic material sinks to deeper layers. If the renewal of the deep water masses occurs seldom, the decay of organic matter rapidly consumes up the oxygen. After oxygen, the next efficient and abundant oxidizer in the marine environment is sulfate, one of the major anions constituting salinity. Organic carbon continues to be oxidized into carbon dioxide, but now sulfate is simultaneously reduced into hydrogen sulfide, which is a strong poison to the biota.

As a result, the redox potential becomes low enough to reduce iron(III) to iron (II), which breaks down the phosphate-iron complex holding phosphate in the precipitated form. Phosphate re-dissolves from the sediments into the deep water under the halocline. Because of the halocline, deep water and surface water mix only weakly. However, during upwelling occasions, especially close to coastal areas, deep water with high phosphate concentrations can be efficiently mixed with the surface layer. The importance of this transport of phosphate from the deep layer to the surface productive layer, "internal loading", is illustrated by the fact that a simple mass balance calculation indicates it being roughly of the same magnitude as the total external phosphorus loading from land-based sources into the Baltic Sea. Strong upwelling during the late summer cyanobacterial bloom may considerably increase the plankton production, leading to mass production of blue-green algae.

Lack of oxygen and the occurrence of hydrogen sulfide make the deep water under the halocline unsuitable for life. The area suitable for animals which require higher salinity for living or reproduction becomes smaller. This is highlighted by the diminishing of the cod populations. In order to develop, the roe of cod needs an environment with salinity over 11. Such salinities are found in the Baltic Sea mainly below the halocline. If salinity is less than that, the roe sinks down to the bottom and dies. The increase of the oxygen deficit, in particular in the

Bornholm and Gdansk Basins, has reduced the spawning grounds of the cod significantly. Jointly with illegal fishery this has led to serious reduction of cod stocks. As the abundance of the cod, a predator, has diminished, stocks of other fishes have also changed. The oxygen deficit induces also other changes in the entire ecosystem. Concentrations of nutrients, especially phosphate, increase in the bottom layer, and this enhances internal loading, in particular in the Northern Baltic Sea and the Gulf of Finland. Coastal upwelling, caused by winds blowing offshore and pushing away surface layer water from the coast, will bring these nutrients to the surface. One typical consequence of this is the mass production of blue-green algae in summer.

Underwater extensions of the large Salpausselkä esker between Hanko peninsula and Swedish coast form a shallow sill between the Baltic Proper and the Gulf of Bothnia. The sill prevents the higher salinity deep water from proceeding into the Gulf of Bothia. As a consequence, bottom salinities remain below 7, the stratification in the Gulf is weak, and surface water, rich in oxygen, can in winter mix with the deep layers. Thus in the Gulf of Bothnia, the oxygen conditions are generally good down to the bottom. The Bothnian Bay, which has an additional sill in the Quark, has even more favourable oxygen conditions.

In the Gulf of Finland, the state of oxygen is variable in the bottom layers, depending on the prevailing stratification and the depth of the halocline. The oxygen trends in the deep layer of the Gulf of Finland are often opposite to those in the Baltic Proper. During a prolonged stagnation period, the halocline tends to move deeper, and in consequence the vertical stratification in the Gulf of Finland, except for isolated deeps, weakens to the point that total mixing of the water column is allowed. This drastically improves the oxygen conditions in the Gulf of Finland, while those in the Baltic Proper turn worse. A strong inflow, on the other hand, improves the situation in the Baltic Proper, but may weaken the deep water oxygen conditions in the Gulf of Finland because of the strengthening halocline.

Primary production transforms the dissolved carbon dioxide into organic material, plankton. After dying, the plankton particles sink through the halocline, and the organic carbon begins to be oxidized back into carbon dioxide, first by oxygen. This reaction consumes up the oxygen in the deep layers because the halocline prevents the mixing of the surface water with the deep layers. After the oxygen has been consumed, sulfate continues to oxidize the organic carbon, resulting in hydrogen sulfide.

10.6 The Evolution of Stagnation During Recent Years

To prevent the continuous accumulation of hydrogen sulfide, oxygen is needed to oxidize the dead plankton, sinking down to the deep layers under the halocline. As the halocline forms an obstacle to the mixing of surface waters with the deep water layer below the halocline, a major inflow of saline water, rich in oxygen, remains the only natural way of renewing the deep water. Unfortunately, the available data indicates that the frequency and strength of inflows are decreasing.

There is, so far, no single explanation to this decrease. It has been proposed (e.g. Leppäranta and Myrberg 2009), that the increased river discharge into the Baltic is one of the major causes. This increase has lowered the salinity of Kattegat water, and hence reduced possibilities for strong saline inflows. There is a clear correlation between total river runoff and salinity in the bottom layers of the Gotland Basin (Matthäus and Schinke 1999). Necessary conditions for the formation of a major inflow are a long-lasting outflow and, after that, a long-lasting inflow. It is possible that change of meteorological disturbances towards faster passage and shorter duration could influence the probability of occurrence of major inflows. A considerable amount of climatologic analysis and modelling with coupled atmosphere–ocean models are needed for understanding these mechanisms.

The renewal of bottom layer water masses in the central Baltic Sea is a rare and slow process. This layer has no direct coupling to the atmosphere, which is the driving force of sea currents. That is why deep layers have quiescent conditions, and stagnations last for long times. As an example of development of oxygen conditions Fig. 10.4 shows bottom layer oxygen content in the years 2005–2008 (Finnish Institute of Marine Research). The conditions vary from year to year, but overall the long-term stagnation has consumed oxygen in wide areas of the central Baltic, partly also in the Bornholm and Gdansk Basins.

At worst, the anoxic deep water area in the Baltic Sea can extend 200–300 km in the north–south direction, and 50–100 km in the east–west direction, with the anoxic layer starting at a depth of 80–100 m, and extending down to the bottom. Assuming the amount of oxygen-deficient water to be roughly 1,000 km^3, with an average concentration of hydrogen sulfide of roughly 1 g/m^3, the total amount of hydrogen sulfide in the Baltic Sea basin is of the order of magnitude of one million tons. To oxidize this amount of hydrogen sulfide, some 1.3 million cubic metres of oxygen is needed. The only effective way of importing oxygen in the deep layers is a large inflow of highly saline sea water from the North Sea. Because this water has to pass the shallow Danish straits, it has been in contact with the atmosphere and is rich in

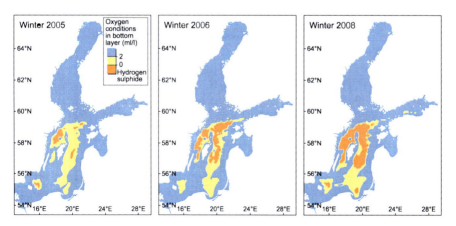

Fig. 10.4 Bottom layer oxygen content in the years 2005–2008 (Finnish Institute of Marine Research, unpublished)

oxygen when entering the Baltic Sea. For example, the major inflow during the winter of 1992–1993, was some 200 km^3 of sea water, containing as much oxygen as it could hold, some 1.6 million cubic metres. The inflow reached the anoxic central basin in the summer of 1993 (see Fig. 10.2 for the route), barely oxidizing the water mass below the halocline. Next winter the basin was again anoxic.

Similarly, the inflow in January 2003 removed hydrogen sulfide from the southern Baltic by the next autumn. Part of the removed water proceeded into the Gulf of Finland, which has no deep sills towards the Baltic Proper. All over the deep layers the amount of oxygen was barely able to support life, and by the autumn 2004 the conditions were close to the situation in 2001, when an oxygen deficit prevailed in the bottom layers. After 2004, oxygen conditions have deteriorated, in particular in the northern parts of the Baltic and in the Gulf of Finland (see e.g. Helcom 2007). The incessant lack of oxygen and the rare occurrence of major inflows has led to a discussion on artificially oxygenating the deep basins of the Baltic Sea.

10.7 Conclusions

The Major Baltic Inflow of saline Kattegat water, which is precondition for the renewal of the deep layers in the Baltic, is a very irregular process. During the last decades it has occurred only a few times. The necessary conditions for the occurrence of the inflow are well known. However, there is no clear explanation why it has occurred only twice after the year 1984. Climate change and change in the fresh water balance are possible candidates, but quantitative evidence is lacking. Long-term stagnations in the deep layers are a consequence of the more infrequent occurrence of these inflows. Even major inflows have been unable to remove hydrogen sulfide from the deep layers. Oxygen is also needed for the oxidation of organic carbon, produced annually in large amounts as a result of excess external and internal loading of nutrients.

To cover the need of oxygen in the deep water layers of the Baltic Sea, a major inflow would be needed almost annually. As, on the contrary, the frequency of the inflows seems to be weakening, it is probable that the oxygen deficit in the bottom layers has already become a permanent situation. Technical methods have to been developed to transport atmospheric oxygen or oxygen-containing surface water into the deep layers. These methods have given promising results when applied in restricted lake areas, but their suitability for use on the scale required by the size of the Baltic Sea is still an open question.

References

Andrejev O, Myrberg K, Mälkki P, Perttilä M (2002) Three-dimensional modeling of the main Baltic inflow in 1993. Environ Chem Phys 24:156–161

Borenäs K, Hietala R, Laanearu J, Lundberg P (2007) Some estimates of the Baltic deep – water transport through the Stope trench. Tellus 59A:238–248

HELCOM (2007) Climate change in the Baltic Sea area. Baltic Sea Environ Proc 111. Helsinki Commission, Helsinki

Hurrell JW, Deser C (2009) The North Atlantic climate variability: the role of the North Atlantic oscillation. J Mar Syst 78:28–41

Leppäranta M, Myrberg K (2009) Physical oceanography of the Baltic Sea. Springer, Berlin/Heidelberg/New York

Matthäus W (2006) The history of investigation of salt water inflows into the Baltic Sea – from the early beginning to recent results. Mar Sci Rep 65, Baltic Sea Res Inst, Warnemunde

Matthäus W, Franck H (1992) Characteristics of major Baltic inflows – a statistical analysis. Cont Shelf Res 12:1375–1400

Matthäus W, Lass HU (1995) The recent salt inflow into the Baltic Sea. J Phys Oceanogr 25:280–286

Matthäus W, Schinke H (1999) The influence of river runoff on deep water conditions of the Baltic Sea. Hydrobiologia 393:1–10

Piechura J, Beszczynska-Möller A (2004) Inflow waters in the deep regons of the Southern Baltic Sea – transport and transformations. Oceanologia 46(1):113–141

Zhurbas V, Stipa T, Mälkki P, Paka V, Golenko N, Hense I, Sklyarov V (2004) Generation of subsurface eddies in the southeast Baltic Sea: observations and numeric experiments. J Geophys Res 109:C05033

Chapter 11
Marine Carbon Dioxide

Matti Perttilä

11.1 Introduction

Industrialization has affected the global environment for about 200 years, but during that time, an immense amount has been used of the carbon accumulated on Earth during billions of years. Already in the 1800s, the Swedish scientist Svante Arrhenius indicated that the rise in the atmospheric carbon dioxide would lead into warming of the global climate (Arrhenius 1896). On the other hand, it was realized at that time also that reducing the carbon dioxide could lead into glaciations.

In 1958 Charles D. Keeling started an accurate time series of the atmospheric carbon dioxide concentration at Mauna Loa, Hawaii. The rise of the atmospheric CO_2 concentration was soon documented. The high precision CO_2 measurement series (the Keeling Curve) is still continued by the Scripps institute of Oceanography (http://scrippsco2.ucsd.edu/home/index.php).

After the beginning of the industrialization, in the beginning of the nineteenth century, around 250 Gt of carbon has been consumed. Half of this amount has been accumulated in the atmosphere, another half being dissolved in the oceans. Presently the use of fossil fuels liberates annually 7.4 Gt of carbon in the atmosphere, in the form of carbon dioxide. The documented atmospheric carbon rise, however, is only 3.3 Gt/year, resulting in a rise of about 2 ppm (2 µatm) in the carbon dioxide partial pressure. The increasing carbon dioxide accelerates the assimilation process of green plants, increasing the CO_2 consumption about 2 Gt each year. The rest, about 2 Gt/year, is transferred into the oceans. Thus the oceans constitute an important buffering factor partially controlling the atmospheric CO_2 increase. In the 1950s, the American oceanographer Roger Revelle showed that the oceans can absorb only about a half of the carbon dioxide increase, and that this buffering

---

M. Perttilä (✉)
Finnish Meteorological Institute, Erik Palménin aukio 1, P.O. Box 503, FI-00101 Helsinki, Finland
e-mail: matti.perttila@fmi.fi

capacity of the oceans would inevitably weaken with the increasing carbon dioxide (Revelle and Suess 1957).

11.2 The Global Carbon Dioxide Cycle

Most of the carbon of the planet Earth is in the form of carbonates in the bedrock and deep sediments, and does not take part in the global carbon cycle. Only about 1% of all carbon is in the atmosphere, fossil fuels, and biosphere. In the ocean surface layer, the amount of dissolved carbon dioxide in the form of inorganic components, is estimated to be 1,020 Gt. Some 3 Gt of carbon is in the form of living biomass, and 700 Gt in the form of dissolved organic matter (DOM). However, the deep water constitutes the marine carbon reservoir with the estimated 38,000 Gt of carbon.

The main carbon components taking part in the carbon cycle are thus the organic biomaterial and the inorganic forms of carbon (dissolved carbon dioxide, carbonic acid, bicarbonate and carbonate). In addition, small amounts of carbon exist in the form of carbon monoxide and methane. Carbon monoxide is formed by oxidation in low oxygen environment and methane mainly in fermentation processes where carbon disproportionates (is reduced and oxidized in the same reaction) into methane and carbon dioxide.

The basis for the formation of organic material is the assimilation reaction, requiring energy which is provided by the sun. Favourable conditions for the reduction reaction are found in the chlorophyll molecule, where electrons for the reduction reaction are readily available. The opposite reaction, breathing and decomposition, where organic carbon is oxidized, is spontaneous and liberates the sun's energy, accumulated in the organic material.

During the early ages of the Earth, in the prevailing anoxic, reducing environment, carbon existed mainly as methane and other hydrocarbons, but after the emergence of green plants capable of photosynthesis, the amount of oxygen started to increase in the atmosphere. Methane was first oxidized into carbon monoxide, then into carbon dioxide. The main carbon component in the atmosphere is now carbon dioxide with a partial pressure of about 390 μatm, totaling some 800 Gt of carbon. The methane and carbon monoxide partial pressures in the atmosphere are only 1.6 and 0.1 μatm, respectively.

In the preindustrial era, the global carbon cycle is assumed to have been roughly in balance, with some 74 Gt of carbon being assimilated and liberated annually. A small fraction of carbon was buried to deeper sediment and bedrock layers, transforming in time to the present reservoirs of fossil fuels.

Man makes now use, in a relatively short time, of the carbon reservoirs which have taken hundreds of millions of years to accumulate. It is easy to understand that this drastic liberation of enormous carbon quantities has severe consequences on the Earth's atmosphere, oceans and biosphere.

11.3 The Sea as Carbon Dioxide Sink/Source

Carbon dioxide constitutes the raw material for the biological primary production. In the chemical sense, primary production means the reduction of oxidized carbon (carbon dioxide) into organic carbon. Carbon dioxide controls the pH affecting chemical equilibria and biological processes. Finally, carbon dioxide is an important green house gas, affecting the global climate. The oceans constitute a vast reservoir capable of absorbing a large part of the carbon dioxide discharged into the atmosphere through human activities. Thus it is important to understand the basics of the CO_2 flow through the ocean – atmosphere interface.

The role of the oceans as carbon dioxide sink or source is controlled by the carbon dioxide partial pressures in the atmosphere and water;

$$\Delta pCO_2 = pCO_2^{atm} - pCO_2^{sw}$$

If the partial pressure in sea water (pCO_2^{sw}) is smaller than the partial pressure in atmosphere (pCO_2^{atm}), gas is absorbed into water (the sea acts as carbon dioxide sink). In case when $pCO_2^{sw} > pCO_2^{atm}$, the sea acts as carbon dioxide source. The net flow is controlled by this difference and the carbon dioxide solubility:

$$F = K_H{}^* k_g{}^* \Delta pCO_2$$

(K_H is the Henry's constant controlling the solubility of carbon dioxide in water and k_g is the factor characterizing the rate of the gas exchange).

The essential source/sink processes, called pumps, controlling ΔpCO_2 are the physical solubility pump and the biological primary production (biological pump). To some extent, the chemical process of calcium carbonate dissolution/precipitation affects the surface water carbon dioxide equilibrium (chemical pump).

11.3.1 Solubility Pump (Physical Pump)

Gases dissolve better in cold water than in warm water (K_H depends on temperature). Thus at high latitudes, more carbon dioxide dissolves in sea water. Because of the ice formation (sea ice contains only small amounts of salt) and the cooling of the ocean surface water, the density of the liquid ocean water increases, which starts the sinking (convection) process removing large amounts of dissolved carbon dioxide temporarily (1,000–1,500 years) from contact with the atmosphere.

11.3.2 Primary Production (Biological Pump)

Marine primary production uses annually some 92 Gt of carbon turning the oxidized form, carbon dioxide, into organic material. The process takes place in the surface layer, because the electrochemical reduction process is not spontaneous (organic carbon contains more energy than the starting material carbon dioxide), and the necessary energy is provided by the sun radiation. Phytoplankton growth is so intense that the aquatic partial pressure pCO_2^{sw} diminishes to less than 50 µatm, turning the sea water to an effective CO_2 sink. Marine scientists have discussed the possibility of slowing down or even reversing the climate change by increasing the oceanic primary production in high nutrient – low production areas. For example, it has turned out that in the Southern Ocean, iron deficiency is the cause of the low production, in spite of the high concentrations of the macronutrients, such as phosphorus, nitrogen and silica compounds (Powell 2007). By adding iron, the primary production could be boosted up to temporarily reverse the atmospheric CO_2 rise. The plan has, however, never been seriously attempted in a large scale, because of largely unknown consequences in the long run.

11.4 Carbon Dioxide Reactions in Sea Water

Carbon dioxide dissolves readily in water, reacting chemically with water to form carbonic acid H_2CO_3. Carbonic acid dissociates into bicarbonate ion HCO_3^- and carbonate ion CO_3^{2-}:

$$CO_2(aq) + H_2O \rightleftarrows H_2CO_3 \rightleftarrows HCO_3^- + H^+ \rightleftarrows CO_3^{2-} + 2H^+$$

According to the Henry's law, the dissolution of a gas in water is directly proportional to the partial pressure of the gas in the atmosphere (pCO_2):

$$[CO_2](aq) = K_H pCO_2 \, (K_H \text{ is the Henry's constant})$$

The dissolved aqueous carbon dioxide $CO_2(aq)$ and the non-dissociated carbonic acid H_2CO_3 constitute only about 1% of the total amount of the inorganic carbon in sea water. They are analytically undistinguishable, for which reason we can write

$$[CO_2]^* = [CO_2(aq)] + [H_2CO_3].$$

The present concentration of carbon dioxide in the atmosphere is about 390 ppm (390 µatm), increasing roughly 1.4 ppm annually. Assimilation/breathing reactions of green plants cause the well-known seasonality of around ±5 ppm in the otherwise constant increasing trend. Surface waters are quickly equilibrated with atmospheric changes, resulting in seasonality in the source/sink behaviour of the oceans. In closed systems (such as bottom water reservoirs without gas exchange with the atmosphere), carbon dioxide concentration can be much larger than in systems open to the atmosphere.

In sea water, also the dissolution reaction of calcium carbonate has to be taken in account:

$$CaCO_3 + CO_2 + H_2O \rightleftarrows Ca^{2+} + 2HCO_3^-$$

Accordingly, the increasing atmospheric concentration of carbon dioxide increases the dissolution of calcium carbonate. An example of this effect is given in the stalagmites and stalactites in many caves in areas with limestone bedrock. Water circulating in the bedrock contains large amounts of carbon dioxide originating from decomposing organic material. According to the above equation, the dissolved calcium carbonate concentration is high. Calcium carbonate is precipitated when the water is brought to contact with the air in the caves. Surplus carbon dioxide is evaporated to equilibrate the system with air, and excess calcium carbonate precipitates forming stalagmites and stalactites.

While the partial pressure of carbon dioxide in atmosphere increases, more gas will be dissolving into the oceans, according to Henry's law. However, also the relative amounts of the inorganic carbon components will change. Partial pressure of the aquatic CO_2 will rise, and so will also the concentrations of undissociated carbonic acid and bicarbonate, but that of carbonate will be diminished. Doubling of the atmospheric CO_2 leads to (almost) doubling of the aquatic CO_2, but the relative increase of the total inorganic carbon (sum of dissolved CO_2, non-dissociated carbonic acid, bicarbonate and carbonate) will only be about one tenth of the relative increase of carbon dioxide. This relation of the pCO_2 change to the C_{tot} change is called the Revelle factor, after the American oceanographer Roger Revelle. It explains why the oceans' capacity to absorb atmospheric CO_2 will decline with increasing CO_2 partial pressure.

11.5 Ocean Acidification

Adverse marine effects of the increasing CO_2 concentration in the atmosphere have been widely studied (see, e.g. Houghton et al. 2001; McCarthy et al. 2001; Metz et al. 2005; Solomon et al. 2007; Fernard and Brewer 2008). The chemical equilibrium of the inorganic carbon components, as defined by the above chemical reactions, controls the aquatic proton concentration, the ocean acidity. Increasing dissolved carbon dioxide leads to increasing carbonic acid, which in turn leads to increasing acidity. This process is called the ocean acidification (e.g. WBGU 2006). Since the last glaciations, until the start of industrialization era, the atmospheric carbon dioxide partial pressure remained roughly constant, around 280 μatm. Using thermodynamic equilibrium constants, available in the literature, one can calculate that sea water in equilibrium with such atmospheric CO_2 has a pH of ca. 8.28. With the present atmospheric CO_2 of about 390 μatm, the resulting pH of the surface layer (equilibrating quickly with the atmosphere), is around 8.12. The change in pH units seems small, but, because of the logarithmic pH scale, it corresponds to an increase of about 30% in the proton concentration (pH = $-\log[H+]$).

Ocean acidification, and especially its effects on calcification (production of shells and supporting structures out of calcium carbonate), are presently under intensive study (e.g. Raven et al. 2005). Research has already found that several species experience reduced calcification or enhanced dissolution when exposed to elevated CO_2. The modeling of future pH changes and their effects on ocean habitats demand high resolution measurement series, available in many monitoring systems around the world. In the Baltic Sea monitoring system (HELCOM), pH is measured against fresh water references, which in principle is wrong but this measurement system has been adapted because of practical reasons – salinity in the brackish water of the Baltic Sea varies from about 2 to 20 and thus several reference sets would be needed for the use of sea water scale. The errors thus introduced are larger than the pH changes that have taken place so far, and renders difficult the comparison of results collected from different sea areas and depths with different salinity. On the other hand, the use of fresh water scale does not affect the reliability of a trend analysis based on observations made at a single station, where salinity variations even in the long run are small. Figure 11.1 shows pH trend in the Baltic Sea (station BY15 in the Baltic Proper). Only winter observations have been used (in the absence of primary production, the aquatic PCO_2 is then close to the

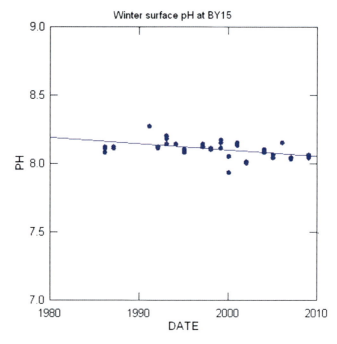

Fig. 11.1 Surface layer winter pH in the Baltic Sea (Baltic Proper) (Data source: Finnish Institute of Marine Research)

atmospheric PCO_2, and the sea – air system is in equilibrium). Downward pH trend is obvious, indicating an average change of 0.002–0.003 units per year.

The partial pressure of 560 µatm (double of that in the preindustrial era) will probably be attained somewhere in the middle of the present century, assuming the present rate of the pCO_2 increase continues. Such pCO_2 level corresponds to oceanic pH of 7.90. A pH so low would have an effect on the formation of the skeleton and other support structures of aquatic animals. These structures of many species are made of aragonite, which is a form of calcium carbonate. Increasing carbon dioxide will increases the solubility of calcium carbonate, hampering the formation of enough calcium carbonate for the bones and other support structures. In the oceans, the calcium concentration is high enough (because of high salinity; calcium is one of the major cations contributing to salinity) to still allow the precipitation of calcium carbonate (even though the carbonate concentration is diminishing). In the brackish water areas of the world, e.g. in the Baltic Sea, calcium concentration is more critical, which is demonstrated by the smaller size and thinner shell of the Baltic Sea aquatic species.

The inevitable pH changes will affect many other chemical equilibria and reactions affecting the solubility of compounds and controlling the biological processes. Acidification also weakens the solubility of carbon dioxide in water, which in the long run will inevitably weaken the oceans' ability to moderate the pCO_2 increase.

References

Arrhenius S (1896) On the influence of carbonic acid in the air upon the temperature of the ground. London, Edinburgh, and Dublin Philosophical Magazine and Journal of Science (5th Series) 41:237–275

Fernard L, Brewer P (eds) (2008) Changes in surface CO_2 and ocean pH in ICES shelf sea ecosystems. ICES (International Council for the Sea) Cooperative Research Report 290. ICES, Copenhagen

Houghton JT, Ding Y, Noguer M, van der Linden PJ, Dai X, Maskell K, Johanson CA (eds) (2001) Climate change 2001: the scientific basis. Contribution of Working Group I to the Third Assessment Report of the Intergovernmental Panel on Climate Change. Cambridge University Press, Cambridge/New York

McCarthy JJ, Canziani OF, Leary NA, Dokken DJ, White KS (eds) (2001) Climate change 2001: impacts, adaptation, and vulnerability. Contribution of Working Group II too the Third Assessment Report of the Intergovernmental Panel on Climate Change. Cambridge University Press, Cambridge/New York

Metz B, Davidson O, de Coninck H, Loos M, Meyer L (eds) (2005) IPCC special report on carbon dioxide capture and storage. Prepared by Working Group III of the Intergovernmental Panel on Climate Change. Cambridge University Press, Cambridge/New York

Solomon S, Qin D, Manning M, Chen Z, Marquis M, Averyt KB, Tignor M, Miller HL (eds) (2007) Climate change 2007: the physical science basis. Contribution of Working Group I to the Fourth Assessment Report of the Intergovernmental Panel on Climate Change. Cambridge University Press, Cambridge/New York

Powell H (2007) Fertilizing the ocean with iron. Oceanus 46(1). Online http://www.whoi.edu/oceanus/viewArticle.do?id=34167&sectionid=1000

Raven JA et al (2005) Ocean acidification due to increasing atmospheric carbon dioxide. Royal Society, London

Revelle R, Suess H (1957) Carbon dioxide exchange between atmosphere and ocean and the question of an increase of atmospheric CO_2 during the past decades. Tellus 9:18–27

WBGU (German Advisory Council on Global Change) (2006) The Future Oceans – Warming Up, Rising High, Turning Sour. Special Report. WBGU, Berlin

Chapter 12
Impact of Climate Change on Biology of the Baltic Sea

Markku Viitasalo

12.1 Introduction

The Baltic Sea is a semi-enclosed brackish-water sea where sea surface temperature varies seasonally from 0 to ca. 20°C and surface salinity varies from ca. 2‰ in the northern and eastern ends of the Gulfs of Bothnia and Finland to ca. 10‰ at the Danish Straits (Fig. 12.1). Its main basins – with the exception of Gulf of Bothnia – are strongly stratified, which creates a quasi-permanent anoxia in the deep water.

Due to the global warming, the air temperature in the Baltic Sea area has been predicted to increase by 3–5°C until year 2100 (Graham et al. 2008). This will increase sea surface temperature, with the most pronounced effects in the northernmost parts of the Baltic Sea during summertime. Temperature increase, together with other changes in atmospheric parameters, such as precipitation and wind conditions, will fundamentally influence the hydrography of the Baltic Sea. Because many ecosystem processes are driven by physical-chemical and physical-biological interactions, the biology of the Baltic Sea will also change. The basic chain of events from climatic forcing to community level effects can be summarised as follows (Fig. 12.2):

Large scale changes in *climatic factors* over the northern hemisphere are manifested in the Baltic Sea area as variations in air temperature, air pressure, wind conditions and precipitation. These parameters directly influence the *physics* of the Baltic Sea. With increasing heat flux to the sea, sea surface temperature will rise, and, with increasing amount of freshwater runoff, the surface water will become less saline. The average salinity of the Baltic Sea will decline, but, because of the continuing water exchange with the more saline North Sea, the Baltic Sea will remain brackish.

M. Viitasalo (✉)
Finnish Environment Institute, Marine Research Centre, P.O. Box 140, FI-00251 Helsinki, Finland
e-mail: markku.viitasalo@ymparisto.fi

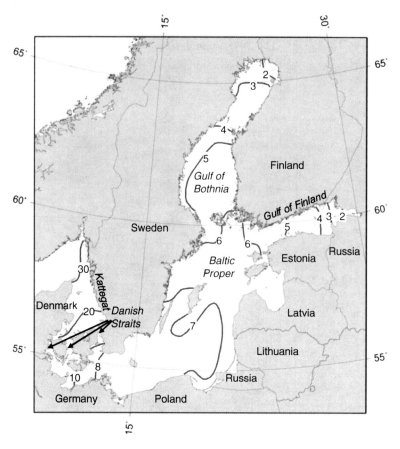

Fig. 12.1 The Baltic Sea with its sub-basins. Numbers and isolines denote surface salinity in June (in permilles; HELCOM 2010)

Variations in the physical properties of the sea, as well as changes in the amount of freshwater entering the Baltic Sea, will influence its *chemistry*. In the stratified Baltic Sea the sediments and deeper waters contain a large reserve of inorganic nutrients. Processes that affect horizontal and vertical movements of water masses influence the supply of nutrients to the surface layer. They are therefore important factors determining the growth rates and species succession of primary producers.

Marine physics and chemistry influence *marine biology* directly and indirectly. At the *individual level* physical and chemical characteristics of water affect the metabolism and biochemistry of organisms. Suboptimal temperature and salinity conditions cause physiological stress, affect energy allocation and impact organisms living at the edge of their temperature or salinity tolerance limits. Availability of nutrients, especially nitrogen, phosphorus and silicate, affect primary production. Increasing temperature also stimulates metabolism, feeding and *productivity* of secondary producers. These processes will affect cell division and birth and death rates of individuals, and hence influence the *population size*.

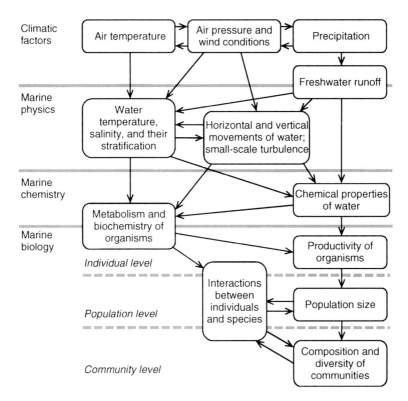

Fig. 12.2 A schematic presentation of the influence of climatic factors on physics, chemistry and biology of the Baltic Sea

Finally, changes in population sizes will alter the strength of *interactions between individuals and species*, such as grazing, predation and competition. These interactions will ultimately shape the *composition and diversity of communities*.

Although few biological processes in the Baltic Sea can be directly linked to climate change, they are however affected by marine physical and chemical parameters which most probably will change if the climate change proceeds as predicted. Such processes and interactions are reviewed and their effects on biota and ecosystem of the Baltic Sea are summarised below.

12.2 Direct Impacts of Climate-Related Parameters on Biota

12.2.1 *Temperature*

Both air temperature and sea surface temperature (SST) of the Baltic have already increased. According to Heino et al. (2008) air temperature warmed up by +0.08°C per decade during 1871–2004. Sherman et al. (2009), who studied SST trends in all

63 Large Marine Ecosystems (LME's) of the world during 1982–2006, reported that SST had increased in 61 LME's, and that the most pronounced increase, 1.35°C, had occurred in the Baltic Sea. The exceptionally rapid warming of the Baltic Sea can be explained by its northern location, small water volume and shallow mixed surface layer. Local long-term studies have confirmed that the summer SST has increased by at least 1°C during the past 50–60 years in the open sea of the northern (Rönkkönen et al. 2003) and southern Baltic (MacKenzie and Schiedek 2007).

According to regional climate projections air temperature over the Baltic Sea will increase by 3–5°C during the twenty-first century (Graham et al. 2008). Consequently the summer SST will increase, with the largest changes (up to 4°C) predicted for the Gulf of Bothnia (Döscher and Meier 2004). In the Baltic Proper the SST change will be ca. 2–3°C and in the Danish Straits and Kattegat ca. 2°C (Meier 2006).

Increasing water temperature increases metabolic rates of organisms. For instance, the metabolism and productivity of heterotrophic bacteria in the Baltic Sea are directly affected by temperature, provided that dissolved organic carbon is available in sufficient quantities. Temperature does not affect phytoplankton species as much and, if chemical factors associated with temperature change are not considered, the biomass of bacteria relative to phytoplankton may increase with warming up of water. On the other hand, increasing activity of bacteria will also increase remineralisation and release of nutrients which may be beneficial to algae (Dippner et al. 2008).

Population level responses to temperature are common in the Baltic Sea. Paleo-ecological evidence shows that abundances of certain phytoplankton taxa follow historic variations in temperature. During the relatively cold period from 1950 to 1960, diatom species composition in the Baltic Proper changed from a "thermophilic" community to one dominated by species adapted to cooler waters (Andrén et al. 2000). In more recent times, increases of dinoflagellates in spring (HELCOM 2007) and cyanobacteria and chlorophytes in summer (Suikkanen et al. 2007) have been reported with increasing temperature. Abundances of certain zooplankton taxa, such as cladocerans and rotifers, have also been shown to be positively correlated with long-term trends of SST (Viitasalo et al. 1995).

Warming up of water can also have negative effects on primary producers. Ehlers et al. (2008) showed that the shoot densities of eelgrass (*Zostera marina*) declined when they were subjected to simulated summertime heat waves in a mesocosm study. As this marine species is also disfavoured by declining salinity and increasing eutrophication, it will probably be declining along with the climate change in the Baltic Sea.

Fish also respond directly to temperature changes. The growth of larvae of many fish species, such as perch, roach (Karås and Neuman 1981) and herring (Hakala et al. 2003), is enhanced by higher temperatures. Sprat year classes have also been shown to be larger during mild winters compared to hard ones, which can be explained by the better survival of sprat eggs in warmer water (Nissling 2004).

Increasing temperature will also favour non-indigenous species originating from more southern sea areas. E.g. the ponto-Caspian fish-hook water flea (*Cercopagis pengoi*), that invaded the Baltic Sea in 1992, benefits from increasing temperatures in the Baltic Sea (Sopanen 2008).

12.2.2 Salinity

According to the classic concept of Adolf Remane the Baltic Sea is a hostile environment for both marine and freshwater organisms. Only truly estuarine species and species with a wide salinity tolerance range can cope with the salinity between 2‰ and 6‰. This makes many species sensitive to relatively small changes in salinity. If the general salinity level of the Baltic Sea will decline, the geographical limits of salinity-dependent species will shift accordingly. This applies to all groups – phytoplankton, zooplankton, benthic organisms and fish, as well as organisms living in the littoral zone.

But how large can the future salinity variations be? Meier (2006) estimated that surface layer salinity of the central Baltic Sea may decrease from 7‰ to as low as 4‰. This would mean that, e.g., the surface salinity isoline of 5‰, that now lies at the level of city of Vaasa in the Gulf of Bothnia and Helsinki in the Gulf of Finland (cf. Fig. 1), would move to the latitude of island of Bornholm (lat 55°), almost 1,000 km south. This prediction is based on a combination of a medium-high CO_2 emission scenario A2 by IPCC (2000) and a global general circulation model ECHAM4 by Max Planck Institute for Meteorology. Different combinations of emission scenarios and circulation models produce different salinity predictions and the estimate must therefore be taken with caution. It however exemplifies the possible range of salinity decline in the Baltic Sea.

Salinity of the Baltic Sea has fluctuated in the past. During the twentieth century the mean salinity of the Baltic Sea was at its lowest, ca. 7.2‰, in 1930–1935, increased steadily from 1936 to 1954, remained at 7.9–8.2‰ during 1955–1979 and declined again to ca. 7.3‰ during 1980–1993 (Heino et al. 2008: Fig. 2.52). This means that the current low salinity period is not yet exceptional. The predicted changes for the next century, in contrast, are larger than any changes during the recent history of the Baltic Sea.

Evidence for salinity-induced changes in species distribution and population sizes exist. During the "oceanization" of the Baltic Sea, that took place in 1936–1954, various marine taxa such as certain copepods, jellyfish *Cyanea capillata*, barnacle *Balanus improvisus*, as well as cod, garfish and mackerel spread hundreds of kilometres northwards, whereas species preferring low saline waters retreated (Segerstråle 1969). During the recent desalination period (from 1980) a reverse process has taken place.

Due to the restrictive effect of salinity, the number of macrozoobenthic species gradually diminishes from the Kattegat and southern Baltic towards the Gulf of Finland and Gulf of Bothnia. If the salinity of the northern Baltic will decline, the geographical limits of benthic species will shift, with associated changes in functional roles of benthic communities. This shift may have consequences for the benthic nutrient dynamics as well, since each functional group has a typical sediment reworking type and capacity. Another effect of the declining salinity would be a submergence of marine species – both benthic and planktonic. Furthermore, if salinity of the Baltic Sea decreased by several permilles, many originally marine species such as bladderwrack (*Fucus* spp.), eelgrass (*Zostera marina*) and blue mussel (*Mytilus trossulus*) might disappear from the sublittoral zones of the northernmost Baltic Sea. Marine fish species, such as cod and turbot would also diminish in abundance, while freshwater species would increase and spread southwards. In particular cyprinids (e.g. roach), many of which breed successfully only in the freshest parts of the Baltic archipelagos (Härmä et al. 2008), would be able to breed in significantly larger areas in the northern Baltic Sea.

12.2.3 pH

The pH of the world oceans has decreased by 0.1 pH units since 1750 (IPCC 2007), apparently due to the increased dissolving of atmospheric CO_2 into the water. For the Baltic Sea only few estimates exist. Swedish Environmental Protection Agency reports a decrease of 0.06–0.44 pH units for the period 1993–2007, depending on area (Naturvårdsverket 2008), and Perttilä (2012, this volume) reports a decrease of ca. 0.10–0.15 pH units since 1980 in the northern Baltic Sea. Thus acidification seems to proceed at a slightly faster rate in the Baltic Sea than in the ocean.

In freshwater, acidification is known to cause reproductive failure in fish, especially salmonids. Similar effects have not been observed in marine fish species. Instead, low pH of seawater is known to decrease calcification in species that use $CaCO_3$ in the formation of their shells and skeletons, such as bivalves and corals (Orr et al. 2005). In the Baltic Sea calcification of shells will probably slow down especially in juvenile stages of bivalves.

12.2.4 Effects on Uptake of Contaminants

Uptake of harmful substances increases with increasing metabolic rates. Simultaneously acting stress factors, such as declining salinity, may also make organisms more vulnerable to harmful substances. Invertebrates have been shown to take up contaminants, especially metals, faster at reduced salinities (Lee et al. 1998). Together with the temperature-induced speeding up of metabolic rates this could lead to more problems related to contaminants in biota.

12.3 Indirect Impacts of Climatic Forcing on Biology of the Baltic Sea

12.3.1 Sea Ice

The ice season has shortened by ca. 1 month (14–44 days) during the last century and, by the end of the present century, the ice cover has been predicted to diminish by 57–71%. Ice is however expected to regularly occur in the northern part of the Gulf of Bothnia, because winter temperature in the northernmost parts of the Baltic Sea is predicted to be affected the least. On average, ice season will be 1–2 months shorter than present in the northern basins of the Baltic Sea, and 2–3 months shorter in the central Baltic (Meier et al. 2004). Completely iceless winters will occur along longer and longer stretches of Baltic Sea coasts.

Disappearance of sea ice will influence living conditions of many species in the Baltic Sea. Lack of sea ice will affect littoral communities, because ice scraping influences the seasonal succession of littoral algae such as bladderwrack and vascular plants such as common reed. Sea ice also contains a maze of "brine channels" that harbour a diverse community of microbes, flagellates and other minute primary and secondary producers that store and recycle nutrients within the brine channels. During ice covered winters, a thin freshwater layer is formed right under the ice, especially close to the estuaries. Increasing freshwater runoff from rivers, predicted for the winter months, would make this layer thicker, which could prevent interaction between the sea ice community and the more saline water underneath. A total disappearance of the sea-ice and associated microbial community would of course drastically change these dynamics.

Earlier ice melt will make the phytoplankton spring bloom to start earlier, which may influence pelagic nutrient dynamics in spring and summer. If the bloom will also cease earlier, the "mismatch" between phytoplankton and zooplankton, emerging from resting eggs in spring and early summer, could be worsened. However, due to several confounding processes, the effects of changes in seasonality of the spring bloom on trophic dynamics of the Baltic Sea are difficult to predict.

Disappearance of sea ice will also affect some of the Baltic seal species. The Baltic ringed seal breeds almost exclusively on landfast ice, in a lair naturally formed by ice blocks, which offers the pup warmth and protection from predators. The disappearance of ice has obvious consequences for the survival of the pups (Meier et al. 2004). In contrast, the grey seal and harbour seal can also breed on land, and consequences of ice loss to their breeding success will probably be smaller.

12.3.2 Water Level

IPCC predicted in its Fourth Assessment Report (IPCC 2007) that the water level of the oceans will rise by 18–59 cm within the next 100 years. Despite the narrow connection between the Baltic Sea and the ocean, sea-level will also rise in the Baltic Sea. The more frequently occurring westerly winds may also contribute to this effect.

In the Gulf of Bothnia, where the crustal uplift is 50–90 cm in 100 years, the uplift will probably exceed the rise of water level induced by climatic warming. In the more southern parts of the Baltic, where the uplift is less than 20 cm per 100 years, the water is expected to rise. These estimates are based on the IPCC estimate from 2007. Certain recent studies however suggest that the global sea-level rise may be larger than predicted by IPCC. Different predictions vary between 40 and 215 cm (Rahmstorf 2010). If the ocean water level will rise more than 100 cm during the twenty-first century, the water level will probably rise even in the Gulf of Bothnia.

Effects of changes in the water level on Baltic biology are many. In the present situation the crustal uplift continuously exposes new shores, rocks and skerries in coastal areas of the northern Baltic Sea. It also contributes to the formation of "flads" (semi-enclosed bays growing to lakes), which harbour a specialized community of algae and water plants and which are important nursery areas for fish larvae. Rising of water would counteract flad formation and prevent the succession of these shallow-water habitats from aquatic to terrestrial ecosystems in the northern Baltic. In the more southern areas, rising of water level would naturally immerse large terrestrial areas, and influence coastal dynamics in many ways.

12.3.3 Stratification, Mixing and Oxygen Conditions

The present models do not allow reliable prediction of development of stratification in the Baltic Sea, and the consequences to the nutrient and oxygen dynamics in the different basins of the Baltic Sea are difficult to forecast. It is notable that many processes induced by changes in atmospheric parameters are counteractive in the Baltic Sea. The hydrographic processes may vary from area to area, and from season to season, and the effects may also be reversed after a certain threshold of temperature or salinity has been reached. A few possible scenarios are explained below and highlighted in Fig. 12.3.

In winter, rising air temperature will lead to diminishing ice cover in the northern Baltic Sea. Lack of sea ice will enhance wind induced mixing of water and improve benthic oxygen conditions, at least above the halocline. In summer, in contrast, warming up of surface layer will strengthen stratification and diminish mixing of water, thus deteriorating oxygen conditions of deeper waters. This has a dual effect for the planktonic primary producers: stronger stratification limits nutrient supply from deeper water, but strong thermal stratification also creates a shallower wind mixed layer, which retains phytoplankton in the well lit water layer, thus increasing capacity for photosynthesis (Dippner et al. 2008).

During winter, a very strong warming up of seawater could also decrease mixing in the northern Baltic. During late autumn and early winter, density difference

12 Impact of Climate Change on Biology of the Baltic Sea

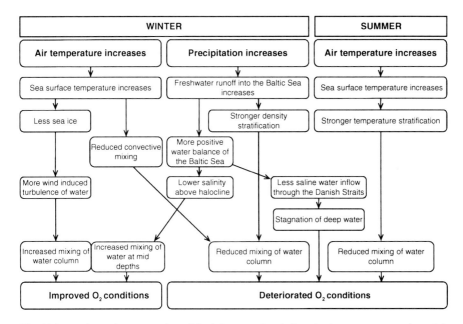

Fig. 12.3 A schematic presentation of the influence of variations in air temperature and precipitation on water mixing, stratification and oxygen conditions in the Baltic Sea

between water layers above the halocline is usually small and water is effectively mixed. In mid-winter the surface layer cools down to close to zero degrees, which creates an inverse temperature stratification. This stratification is again broken up in spring when surface water warms up to 2–3°C. At this temperature the surface water gets heavier than the water beneath it and starts to sink, creating "convective mixing". This mixing process is important for the replenishment of surface water layer with nutrients in spring. If winter air temperature increased by several degrees, temperature of the surface water might remain so high that convective mixing would not occur. This could diminish transport of nutrients from deep water to the surface layer and thus reduce the amount of nutrients available to the phytoplankton spring bloom.

Increasing freshwater runoff in winter may have both positive and negative effects on deep-water oxygen conditions. A high amount of freshwater runoff strengthens density stratification, which decreases mixing of waters below the surface layer and in the long run deteriorates oxygen conditions. Increasing freshwater runoff also makes the water balance of the Baltic Sea more positive which makes it more difficult for saline water to enter the Baltic Sea through the Danish Straits. If increased runoff leads to less frequent pulses of saline North Sea water, deep water will not be replenished by oxygen and will remain stagnant and anoxic.

Different effects can however be expected in different areas and at different depths of the Baltic Sea. A prolonged stagnation would keep the deep central parts of the Baltic Sea anoxic, but could in the long run improve oxygen conditions in shallower areas, such as the Gulf of Finland. This is highlighted by the past development of oxygen conditions and benthic communities in the Baltic Sea. During the stagnation period in 1977–1993 the macrozoobenthos communities deteriorated at sub-halocline (70–250 m) depths in the central Baltic. With continuing stagnation the stratification weakened and a gradual recovery of benthic communities took place at mid-depths (70–100 m) in the Gulf of Finland (Laine et al. 1997). Similar sequences of declines and recoveries of benthic communities may take place due to climate-induced variations in salinity and stratification of the Baltic Sea.

While a freshening of the Baltic Sea could benefit benthic communities in certain areas, a decline of salinity of the Baltic Sea associated with a prolonged stagnation would be clearly harmful to cod stocks. Cod reproduction is dependent on the amount of sufficiently saline oxic water because, after spawning, cod eggs sink to a depth level where they have neutral buoyancy. This water, which has a salinity of 11‰, exists at a depth of ca. 150 m and is nowadays usually anoxic. If the salinity of the Baltic Sea will decrease, cod eggs will sink even deeper, and cod reproduction may become impossible.

12.3.4 Nutrient Dynamics

One of the most severe consequences of climate change is expected to be the increasing runoff of nutrients into the Baltic Sea. This is thought to be a consequence of a 15% higher river runoff predicted for the twenty-first century (Graham et al. 2008). This, coupled with mild winters without deep ground frost, will enhance leaking of nutrients, as well as solids and harmful substances, into the watersheds and eventually into the sea.

Increasing amount of nutrients would speed up eutrophication. Planktonic production would increase, filamentous algae grow faster, and sediment loads carried by river water would turn many rocky bottoms to softer ones. Enhanced primary production in turn increases sedimentation of organic matter. This leads to increased oxygen consumption and consequent release of phosphorus from the anoxic bottoms. In shallower waters temperature increase will worsen the situation, because of enhancing microbial remineralisation of organic matter. Release of phosphorus will favour cyanobacteria, which in turn bind atmospheric molecular nitrogen into their cells. This nitrogen will be released into the water with the break-up of blooms, which again contributes to increasing planktonic production. Climate change may thus speed up the "vicious circle" of internal loading of the Baltic Sea.

On the other hand, nutrient dynamics vary in different Baltic Sea basins. For instance in the Gulf of Bothnia severe anoxia does not occur because the sills in the Archipelago Sea and the Åland Sea prevent penetration of saline water into the Gulf of Bothnia. Lack of density stratification secures oxygenation of the deep water and limits internal loading of phosphorus. This makes the Gulf of Bothnia more dependent on external loading of nutrients than, e.g. Gulf of Finland, where a large fraction of phosphorus derives from the sediment. Increasing runoff of river-borne nutrients should therefore speed up eutrophication in the Gulf of Bothnia. Recent studies have however shown the opposite: phytoplankton production was *lower* during the years when the amount of freshwater runoff into the gulf was high (Wikner and Andersson 2009). This can be explained by the fact that the rivers running to the Gulf of Bothnia carry a large load of dissolved organic carbon. This DOC serves as a substrate for bacteria which in this relatively oligotrophic environment compete for nutrients with phytoplankton. Thus if river runoff to the Gulf of Bothnia increases the ratio of bacterial versus phytoplankton biomass increases, which may be seen as lowered primary production and, hence, decreasing eutrophication.

12.3.5 Cascading Trophic Effects

Climate change may also induce cascading food web effects. Increases and decreases of populations will give rise to new or modified interspecific interactions which may lead to "top-down controlled" trophic cascades. An example is the case of cod, sprat, herring and zooplankton in the Baltic.

The unfavourable environmental conditions have, together with overfishing, eradicated cod from the fauna of the northern Baltic Sea. Cod is the most important fish predator of Baltic clupeids, and alleviation of cod predation has led to a significant increase of sprat populations. Meanwhile, herring populations have not increased as drastically, but weight-at age of herring has diminished by 50% between years 1982 and 1992 (e.g. Flinkman et al. 1998).

While the increase of sprat stocks can be attributed to a release of top-down control by cod, the thinning of herring can be explained by a bottom-up process. The food supply for herring probably decreased because of changes in zooplankton community composition. Along with the salinity decline, that started in 1980, populations of larger marine copepods diminished which probably decreased energy supply to herring. This "herring growth anomaly" was further worsened by interspecific competition with sprat (Möllmann et al. 2005). In contrast, sprat feed relatively more on zooplankton species that had not decreased in abundance and therefore changes of zooplankton community were initially less harmful to sprat than for herring. However, evidence for thinning of sprat has also accumulated, suggesting food competition between clupeids (Casini et al. 2006).

Hänninen et al. (2000) suggested that this chain of events was ultimately driven by climatic processes. The influence of climatic variations on herring and sprat stocks has also been demonstrated by MacKenzie and Köster (2004). If the climate change further decreases the salinity of the Baltic Sea, it is unlikely that cod will return to the northern Baltic Sea – unless the oxygen conditions of the deep basins will be improved and overfishing of cod is stopped. The system will most probably remain sprat dominated.

Changes in fish stocks can influence other predator groups as well. Österblom (2006) noted that guillemot chicks at the island of Stora Karlsö, west of Gotland, had lost weight in the late 1990s. Guillemots feed their chicks almost exclusively with sprat, and, taking into account the good availability of sprat, thinning of chicks was surprising. It could however be explained by the fact that guillemot adults fly to hunt for prey at a rate that cannot be significantly increased. The amount of energy gained by the chick is therefore more dependent on the size and energy content of each fish rather than their availability in the sea. As sprat had become less fatty due to intense intraspecific competition for food, the chicks started to lose weight.

Interestingly, recent studies show similar changes in Baltic grey seals as well. The females today have a thinner fat layer under their skin than in the late 1990s (Karlsson and Bäcklin 2009), which could be due to food web effects ultimately caused by climatic change.

12.4 Conclusions

Increasing air temperature and precipitation over the Baltic Sea area will influence the Baltic Sea ecosystem through their influence on water temperature, salinity, stratification and nutrient dynamics. If sea surface temperature increases and salinity declines, geographical ranges of many species living close to their temperature and salinity tolerance levels will shift accordingly. Due to the desalination of the Baltic Sea, freshwater species will increase in abundance and spread horizontally as well as vertically, while marine species will retreat. Due to warming up, metabolism of many organisms and remineralisation of organic matter will be enhanced. Non-indigenous species originating from warmer areas may increase in abundance. Other important effects include changes in water level and decrease of sea ice. Runoff of nutrients from land during wintertime will increase which may increase primary production and sedimentation of organic matter, which again favours internal loading of phosphorus. Several confounding processes may however influence the nutrient dynamics. Loss of sea ice and decreased salinity above halocline may also increase mixing of water and thus improve oxygen conditions in shallower areas. Also, in the Gulf of Bothnia the high runoff may enhance transport of dissolved organic carbon into the sea which would increase bacterial production but slow down eutrophication.

In conclusion, it is currently difficult to predict how climate change will influence the state, productivity and biodiversity of the Baltic Sea ecosystem. Because of the high influence of physical factors on the Baltic Sea ecosystem, it is likely that climatic influences will become increasingly visible – and scientifically proofed. Complex system-level effects can take place both in pelagic and benthic ecosystems. Timing and consequences of the climatic forcing may however vary from basin to basin and from season to season, and some of the processes may counteract or alleviate each other. A thorough analysis of effects of climate change on the biogeochemistry and biology of the Baltic Sea remains to be done.

References

Andrén E, Andrén T, Kunzendorff H (2000) Holocene history of the Baltic Sea as a background for assessing records of human impact in the sediments of the Gotland Basin. Holocene 10:687–702

Casini M, Cardinale M, Hjelm J (2006) Inter-annual variation in herring, *Clupea harengus*, and sprat, *Sprattus sprattus*, condition in the central Baltic Sea: what gives the tune? Oikos 112:638–650

Dippner JW, Vuorinen I, Daunys D et al (2008) In: BACC Author Team (ed) Assessment of climate change for the Baltic Sea Basin. Springer, Berlin/Heidelberg, pp 309–377

Döscher R, Meier HEM (2004) Simulated sea surface temperature and heat fluxes in different climates of the Baltic Sea. Ambio 33:242–248

Ehlers A, Worm B, Reusch TBH (2008) Importance of genetic diversity in eelgrass *Zostera marina* for its resilience to global warming. Mar Ecol Prog Ser 355:1–7

Flinkman J, Aro E, Vuorinen I, Viitasalo M (1998) Changes in the northern Baltic zooplankton and herring nutrition from 1980s to 1990s: top-down and bottom-up processes at work. Mar Ecol Prog Ser 165:127–136

Graham LP, Chen D, Christensen OB et al (2008) Projections of future anthropogenic climate change. In: BACC Author Team (ed) Assessment of climate change for the Baltic Sea Basin. Springer, Berlin/Heidelberg, pp 133–219

Hakala T, Viitasalo M, Rita H, Aro E, Flinkman J, Vuorinen I (2003) Temporal and spatial variability in the growth rates of Baltic herring (*Clupea harengus membras* L.) larvae during summer. Mar Biol 142:25–33

Hänninen J, Vuorinen I, Hjelt P (2000) Climatic factors in the Atlantic control the oceanographic and ecological changes in the Baltic Sea. Limnol Oceanogr 45:703–710

Härmä M, Lappalainen A, Urho L (2008) Reproduction areas of roach (*Rutilus rutilus*) in the Northern Baltic Sea: potential effects of climate change. Can J Fish Aquat Sci 65:2678–2688

Heino R, Tuomenvirta H, Vuglinsky VS et al (2008) Past and current climate change. In: BACC Author Team (ed) Assessment of climate change for the Baltic Sea Basin. Springer, Berlin/Heidelberg, pp 35–131

HELCOM (2007) Climate change in the Baltic Sea Area. HELCOM Thematic Assessment in 2007. Baltic Sea Environment Proceedings 111. Helsinki Commission, Helsinki

HELCOM (2010) Atlas of the Baltic Sea. Helsinki Commission, Helsinki

IPCC (2000) Emission scenarios special report of the intergovernmental panel on climate change. Cambridge University Press, Cambridge/New York

IPCC (2007) Climate change 2007: the physical science basis. Cambridge University Press, Cambridge

Karås P, Neuman E (1981) First year growth of perch (*Perca fluviatilis* L.) and roach (*Rutilus rutilus* (L.)) in a heated Baltic Bay. Rep Inst Freshw Res Drottningholm 59:48–63

Karlsson O, Bäcklin B-M (2009) Magra sälar i Östersjön. Havet 2009:86–90

Laine AO, Sandler H, Andersin A-B, Stigzelius J (1997) Long-term changes of macrozoobenthos in the eastern Gotland Basin and the Gulf of Finland (Baltic Sea) in relation to the hydrographical regime. J Sea Res 38:135–159

Lee BG, Wallace WG, Luoma SN (1998) Uptake and loss kinetics of Cd, Cr and Zn in the bivalves *Poamocorbula amurensis* and *Macoma balthica*: effects of size and salinity. Mar Ecol Prog Ser 175:177–189

MacKenzie BR, Köster FW (2004) Fish production and climate: sprat in the Baltic Sea. Ecology 85:784–794

MacKenzie BR, Schiedek D (2007) Daily ocean monitoring since the 1860s shows record warming of northern European seas. Global Change Biol 13:1335–1347

Meier HEM (2006) Baltic Sea climate in the late twenty-first century: a dynamical downscaling approach using two global models and two emission scenarios. Climate Dyn 27:39–68

Meier HEM, Döscher R, Halkka A (2004) Simulated distributions of Baltic sea-ice in warming climate and consequences for the winter habitat of the Baltic ringed seal. Ambio 33:249–256

Möllmann C, Kornilovs G, Fetter M, Köster FW (2005) Climate, zooplankton and pelagic fish growth in the Central Baltic Sea. ICES J Mar Sci 62:1270–1280

Naturvårdsverket (2008) Trends and scenarios exemplifying the future of the Baltic Sea and Skagerrak. Ecological impacts of not taking action. Swedish Environmental Protection Agency Report 5875

Nissling A (2004) Effects of temperature on egg and larval survival of cod (*Gauds morgue*) and sprat (*Sprattus sprattus*) in the Baltic Sea – implications for stock development. Hydrobiologia 514:115–123

Orr JC, Fabry VJ, Aumont O et al (2005) Anthropogenic ocean acidification over the twenty-first century and its impact on calcifying organisms. Nature 437:681–686

Österblom H (2006) Complexity and change in a simple food web – studies from the Baltic Sea (FAO area 27.IIId). PhD thesis, Stockholm University, Stockholm

Perttilä M (2012) Marine carbon dioxide. In: Haapala I (ed) From the Earth's core to outer space Lecture notes in Earth system sciences 137. Springer, Berlin/Heidelberg, pp 163–170

Rahmstorf S (2010) A new view on sea level rise. Nat Rep Climate Change 4:44–45

Rönkkönen S, Ojaveer E, Raid T, Viitasalo M (2003) Long-term changes in the Baltic herring growth. Can J Fish Aquat Sci 61:219–229

Segerstråle SG (1969) Biological fluctuations in the Baltic Sea. Prog Oceanogr 5:169–184

Sherman K, Belkin IM, Friedland KD, O'Reilly J, Hyde K (2009) Accelerated warming and emergent trends in fisheries biomass yields of the World's large marine ecosystems. Ambio 38:215–224

Sopanen S (2008) The effect of temperature on the development and hatching of resting eggs of non-indigenous predatory cladoceran *Cercopagis pengoi* in the Gulf of Finland, Baltic Sea. Mar Biol 154:99–108

Suikkanen S, Laamanen M, Huttunen M (2007) Long-term changes in summer phytoplankton communities of the open northern Baltic Sea. Estuar Coast Shelf Sci 71:580–592

Viitasalo M, Vuorinen I, Saesmaa S (1995) Mesozooplankton dynamics in the northern Baltic Sea: implications of variations in hydrography and climate. J Plankton Res 17:1857–1878

Wikner J, Andersson A (2009) Increased freshwater discharge shift the carbon balance in the coastal zone. 7th Baltic Sea Science Congress 2009. Abstract Book. Tallinn University of Tecnology, Tallinn

Part III
Climate Change

Cover page: Heating plant pollutes air locally Courtesy: Heikki Ketola/Kuvaliiteri

Chapter 13
Evolution of Earth's Atmosphere

Juha A. Karhu

13.1 Introduction

Earth's atmosphere plays an important role in maintaining habitable conditions on the surface of the Earth. The atmosphere helps to block out harmful ultraviolet radiation, and surface temperatures have an intimate connection to atmospheric greenhouse gas concentrations. Several inorganic as well as biologically mediated reactions and processes are responsible for the concentrations of different components in the atmosphere. Clearly, these same processes have also been active in the past. It is easy to imagine that the intensity of these processes may have varied over time, and, therefore, we have no reason to assume that the atmospheric composition had remained constant over the planetary history of the Earth.

The surface temperature of Earth is directly related to the intensity of the energy flux from the sun. Considering the present-day radiation flux, a surface temperature of $-18°C$ would be expected. This is, however, contrary to what is observed. The average surface temperature on Earth is 15°C, which is more than 30°C higher than that expected from radiation balance calculations. This significant difference is related to the effects of the greenhouse gases in the atmosphere.

Nitrogen is the most abundant gas species in the atmosphere, with a 78% proportion among the atmospheric components. It is, however, relatively inert, and it is probable that the nitrogen content of the atmosphere has not varied greatly. The most fundamental topics in discussions of the past atmospheric composition are related to the levels of greenhouse gases and the oxidation state of the atmosphere. The oxidation state is directly related to the concentration of free oxygen molecules (O_2), which now amount to 21% of the total composition of the atmosphere. Geological evidence and arguments derived from the stellar evolution

J.A. Karhu (✉)
Department of Geosciences and Geography, University of Helsinki, P.O. Box 64, FI-00014 Helsinki, Finland
e-mail: juha.karhu@helsinki.fi

model for the Sun indicate that major changes in these parameters have occurred during the geological history of the Earth. In the past, greenhouse gas levels have been higher by several orders of magnitude. Atmospheric oxygen contents, in turn, have experienced a transition from an early atmosphere with very low oxygen contents to one with oxygen in the range of the present atmospheric level.

13.2 Greenhouse Gases

13.2.1 General

The dominant greenhouse gases in Earth's atmosphere are carbon dioxide (CO_2), methane (CH_4) and water vapor (H_2O). They all have the capacity to absorb infrared radiation emitted by the surface of the planet and thereby increase the average surface temperature. Water vapor, however, only provides feedback on climate, and it is not a climate forcer. This is because the contents of water vapor in the atmosphere are always near its condensation temperature (Kasting and Ono 2006). In contrast, carbon dioxide and methane do not have similar constraints and, accordingly, the levels may have varied greatly.

During the Phanerozoic Eon (the last 542 Ma), Earth's surface has been inhabited by macroscopic life, and the atmosphere has contained significant quantities of oxygen gas. In the presence of oxygen, the life time of methane is short, preventing its buildup in the atmosphere. No such limits exist for carbon dioxide, which, therefore, is a potential driver of Phanerozoic climates.

13.2.2 Last Million Years

Direct analyses of the contents of carbon dioxide and methane in ancient air exist for the last 800,000 years. Ice sheets in polar regions grow slowly by accumulating snow. As the snow pack turns into ice, it includes tiny gas bubbles of ambient air that can be extracted for analyses. The results indicate distinct variations in the atmospheric concentrations of carbon dioxide and methane during the last 800,000 years (Siegenthaler et al. 2005; Loulergue et al. 2008). The concentrations of both greenhouse gases have been higher during warm interglacials and lower during the cold glacial periods. These records provide strong evidence that carbon dioxide and methane have acted as either a climate driver or an important forcer during the last 800,000 years. Methane is rapidly oxidized and destroyed in an oxygen bearing atmosphere, and, therefore, its level in the present-day atmosphere is relatively low at 1.8 ppm (parts per million by volume). The Pre-anthropogenic methane levels are even lower than this and have varied between 0.8 and 0.4 ppm during interglacial and glacial periods, respectively (Loulergue et al. 2008).

13.2.3 Phanerozoic Eon

In the Phanerozoic (542-0 Ma) direct analyses of atmospheric gases are missing, with an exception of the last 800,000 years. Therefore, information on the levels of carbon dioxide has been derived from various proxy variables. The values of these proxy variables are in a predictable way dependent on the content of carbon dioxide in air. In addition, levels of carbon dioxide have been estimated from carbon cycle models. Royer et al. (2004) compared the records from four different proxies with data from a carbon cycle model. The proxies included the isotopic composition carbon in carbonate and goethite minerals that have been precipitated in ancient weathering horizons, the isotopic composition of carbon in phytoplankton, the stomatal distribution in the leaves of C3 plants and the isotopic composition of boron in planctonic foraminifera. The records from these proxies and those from a carbon cycle model suggest generally higher carbon dioxide levels for the earlier Phanerozoic atmosphere compared to those in the present-day atmosphere (Fig. 13.1). Two well known cold periods are an exception. The data suggest generally low carbon dioxide concentrations during the Permo-Carboniferous and late Cenozoic glaciations (Fig. 13.1). Accordingly, the coupling between the levels of carbon dioxide in air and the climate appears to be strong during the last 542 million years of Earth's history.

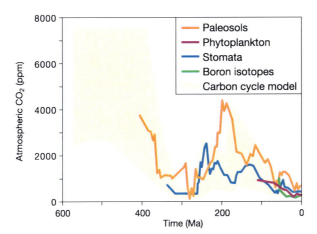

Fig. 13.1 Atmospheric concentration of carbon dioxide in the Phanerozoic Eon based on evidence from various proxies and carbon cycle modeling after Royer et al. (2004). The paleosol proxy is based on carbon isotope data from fossil weathering horizons. The phytoplankton proxy is derived from the $\delta^{13}C$ values of marine phytoplankton. The number of stomatal pores on the leaf surface is inversely related to atmospheric carbon dioxide content, shown as the stomatal proxy. The boron isotope proxy refers to isotopic composition of boron in planctonic foraminifera, which is strongly dependent on the pH of seawater and thereby on the partial pressure of carbon dioxide. Time on the horizontal axis is presented as millions of years (Ma)

13.2.4 Precambrian Supereon

The Precambrian Supereon covers the largest amount of geologic time from the beginning of Earth's formation about 4,560 million years ago to the beginning of the Phanerozoic Eon 542 million years ago. In total, the Precambrian covers 88% of the existence of the planet. The geologic record of the Precambrian is more limited in comparison to the wealth of detailed geologic data from the Phanerozoic Eon. Life was present in the oceans, but metazoan life was missing, with an exception of the latest Precambrian time. Sedimentary proxies providing information on carbon dioxide levels have generally been metamorphosed with a concomitant loss of primary information. Furthermore, the isotopic records for marine dissolved carbon are not known with a sufficient accuracy to allow well constrained carbon cycle modeling. Therefore, a completely different approach is used to gain information on the levels of carbon dioxide in the Precambrian atmosphere.

According to the standard solar model, the energy output from the Sun is slowly increasing with time. Immediately after the formation of the planetary system, 4,500 million years ago, the luminosity of the Sun was 30% lower in comparison to the modern value (Kasting 1987). Using this level of energy output and the present-day greenhouse gas concentrations in air, any water exposed on the surface of the Earth would have been frozen (Fig. 13.2). The geological record of sedimentary rocks, however, extends back 3,800 million years, providing evidence for liquid water on Earth's surface. New evidence from the oxygen isotope records in zircon minerals suggest that the hydrosphere might have been present already 4,400 million years ago (Wilde et al. 2001). These contradictory lines of evidence form the apparent "Faint Young Sun Paradox", first pointed out by Sagan and Mullen (1972).

As a solution to the paradox, Sagan and Mullen (1972) suggested that the early atmosphere provided a considerably stronger greenhouse effect in comparison to the present-day situation. Originally, they suggested that higher concentrations of ammonia (NH_3) in the atmosphere might be responsible for the stronger greenhouse effect. Later it was realized that ammonia is rapidly destroyed by photolysis in the

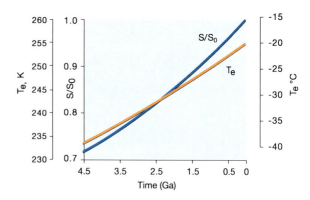

Fig. 13.2 Diagram illustrating the evolution of the luminosity of the Sun relative to today (S/S_0), and the resulting Earth's effective radiating temperature (°C) in the absence of any greenhouse gas heating according to Kasting (1987). Time on the horizontal axis is presented as billions of years (Ga)

atmosphere, and high concentrations of ammonia could only exist under an efficient ultraviolet shield. On these grounds, the focus has been directed to other greenhouse gases that in higher concentrations could have been able to eliminate the impact of the early faint Sun.

Kasting (1987) calculated that sufficiently high levels of carbon dioxide would possess a greenhouse effect powerful enough to compensate for the missing radiation energy. The estimated concentrations of carbon dioxide, are, however, very high, exceeding 0.4 bar. Rye et al. (1995) noticed that such high carbon dioxide contents would have an effect on the mineralogical composition of weathering horizons. Under a high partial pressure of carbon dioxide, iron carbonate (siderite) would be precipitated in soil horizons. In detailed studies, the estimated mineralogical effects were not observed, which places an upper limit to the carbon dioxide content of the atmosphere. This is lower by a factor of 20 or more compared to the level that is needed to keep Earth's surface from freezing (Rye et al. 1995). These results pointed out that another greenhouse gas, in addition to carbon dioxide, is needed to provide a sufficiently efficient greenhouse effect.

Methane has been identified as a potential greenhouse gas providing a strong heating effect for the early atmosphere (Walker 1987). Pavlov et al. (2000) estimated that methane concentrations of 100 ppm or more would provide enough heating to supplement the greenhouse effect from carbon dioxide at a level in agreement with the weathering horizon data of Rye et al. (1995).

Methane at high concentrations has an interesting side effect. If the ratio of methane to that of carbon dioxide in air is >1, methane tends to polymerize and form organic haze (Pavlov et al. 2000). Instead of heating, the haze would have the tendency to cool the surface of the planet.

13.3 Oxygen

13.3.1 Connection to Carbon Cycle

The evolution of atmospheric oxygen is closely related to the evolution of the Earth system. Oxygen is created almost exclusively by photosynthesis, which can be presented by the following generalized chemical reaction formula from left to right:

$$CO_2 + H_2O \leftrightarrow CH_2O + O_2. \qquad (13.1)$$

According to Eq. 13.1, carbon dioxide and water molecules are combined to form organic matter, presented in the reaction by the general formula CH_2O. This formula closely represents the average composition of organic matter considering its main constituents, carbon, hydrogen and oxygen. It is important to notice, that as a side product of photosynthesis oxygen gas is created. Photosynthesis does not occur spontaneously but requires an input of energy from the Sun.

Reaction 13.1 can also proceed from right to left. In the biological carbon cycle, nearly all of the organic matter is recycled back to carbon dioxide and water via decomposition and decay. There is, however, a tiny leak in the recycling process. A small fraction of the organic matter produced is buried in sediments and thereby avoids decomposition. This corresponds to a net production of oxygen, available for the build-up of the atmospheric reservoir. The fraction of organic matter avoiding decomposition has been estimated to be about 0.2% (Holland 1978).

Equation 13.1 indicates that for each atom of carbon in sedimentary organic matter, one molecule of free oxygen is created. Therefore, an estimate the sedimentation rate of organic carbon directly leads to an approximation of the net quantity of liberated oxygen. Geological records are incomplete, and generally it is not possible to estimate past sedimentation rates of organic carbon. However, the isotopic composition of carbon in sedimentary carbonates offers insight into the geological history of carbon burial.

Carbon has two stable isotopes, ^{12}C and ^{13}C, whose relative proportions in Nature are affected by chemical and physical processes. The isotopic composition of carbon is measured as a ratio of the two isotopes ($^{13}C/^{12}C$) and reported as a $\delta^{13}C$ value (13.2), which gives a per mil difference of the ratio from that of the international VPDB standard:

$$\delta^{13}C = (R_{sample}/R_{VPDB} - 1) \times 1000, \qquad (13.2)$$

where R gives the $^{13}C/^{12}C$ ratio of the sample or the standard.

One part of carbon is buried in sediments as carbonates and another part as organic matter. These two forms of carbon have distinct isotopic compositions. The isotopic composition of carbon in sedimentary carbonates is inherited from the isotopic composition of marine dissolved carbonate, whereas the isotopic composition of carbon in sedimentary organic matter is similar to that of organic carbon in the biosphere. In the present-day system, the $\delta^{13}C$ values of sedimentary carbonates are close to 0‰. The isotopic composition of sedimentary organic matter is more depleted in ^{13}C, with $\delta^{13}C$ values at -27 ± 7‰ (Schidlowski et al. 1983). The depletion in ^{13}C is related to fractionation effects in photosynthesis, where the light carbon isotope, ^{12}C is strongly favored. Considering that the average value of all surficial carbon is -5‰, it can be calculated on the basis of isotope mass balance that about 20% of all carbon in sedimentation is buried as organic matter. Organic burial here refers to that part of organic matter, which has avoided decomposition reactions back to water and carbon dioxide (Reaction 13.1), and instead has been buried in sedimentary formations.

For the last 3,500 Ma, the carbon isotope values of sedimentary carbonates have generally remained very close to the present-day $\delta^{13}C$ value of about 0‰. (Schidlowski et al. 1983). However, more recent studies have shown that certain time periods are characterized by unusually high $\delta^{13}C$ values between 5‰ and 10‰ (Fig. 13.3). Mass balance considerations indicate that these periods represent times of large burial rates of organic carbon relatively to carbonate carbon. If the total

13 Evolution of Earth's Atmosphere

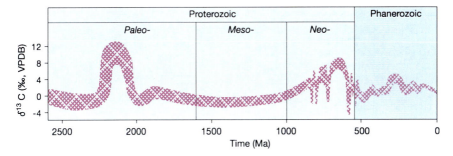

Fig. 13.3 Range for the carbon isotope composition of sedimentary carbonates vs. time after Karhu (1999). For sources of data see Karhu (1999). Sedimentary carbonates record accurately the isotopic composition of dissolved inorganic carbon in ancient seawaters

carbon sedimentation rate has remained constant, large absolute burial rates of organic carbon are implied and, accordingly, large volumes of excess oxygen would have been liberated to the environment (e.g. Karhu and Holland 1996). The most notable positive carbon isotope excursions are recorded for the Permo-Carboniferous, the Neoproterozoic and the Paleoproterozoic (Fig. 13.3). A major release of oxidative power to the environment is suggested for these time periods.

13.3.2 Phanerozoic Eon

Earth's atmosphere in the Phanerozoic is known to have contained significant amounts of free oxygen. This general notion is in agreement with the existence of multicellular, complex life forms from the beginning of the Phanerozoic. Considerations of possible variations in oxygen levels have been largely ignored (Berner et al. 2003). Nevertheless, detailed theoretical calculations indicate major fluctuations in the oxygen content of the atmosphere in the Phanerozoic. The proportion of oxygen in Phanerozoic atmospheres has been estimated using tabulations of reduced carbon and sulfur abundances in sedimentary rock formations (Fig. 13.4, Berner 1999). Although both carbon and sulfur cycling was considered, for estimating atmospheric oxygen levels the cycle of carbon is definitively more important compared to that of sulfur. Many other redox sensitive elements also participate in oxidation reactions and affect atmospheric oxygen, but their impacts are insignificant in comparison to carbon and sulfur (Holland 1978).

Also carbon and sulfur isotope data can be used as starting points for carbon and sulfur cycle models (Berner et al. 2003), and these results seem to confirm the data obtained from simple mass balance models. It appears that the contents of atmospheric oxygen have varied widely from about 15% to 35% during the Phanerozoic (Fig. 13.4). The most notable positive excursion in the oxygen levels occurred in the Permo-Carboniferous, about 300 million years ago, with concentrations reaching

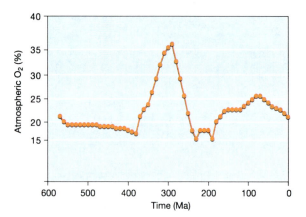

Fig. 13.4 Atmospheric level of oxygen vs. time for the Phanerozoic Eon calculated using a sediment abundance model with rapid recycling after to Berner (1999). Time on the horizontal axis is presented as millions of years (Ma)

35% in air. The high oxygen levels have been interpreted to have resulted from the rise of vascular land plants in the Devonian (416–359 Ma) and the resulting globally increased burial rate of organic matter in coal beds and in sedimentary formations in general (Berner et al. 2003). It has been speculated that the high oxygen levels in the Permo-Carboniferous are related to the presence of giant insects, such as huge dragonflies with a wingspan of 75 cm.

13.3.3 Precambrian Supereon

The Precambrian time witnessed a most dramatic rise of atmospheric oxygen from a virtually oxygen-free environment to an atmosphere with detectable amounts of oxygen. Many geologic indicators have suggested a major increase in the levels of atmospheric oxygen at about 2,000 million years ago (Holland 1984). However, the most convincing evidence for a major shift in the oxygen content of the atmosphere is preserved by sulfur isotope ratios in sedimentary sulfur minerals, such as pyrite (Farquhar and Wing 2003).

Sulfur has four non-radioactive isotopes, which participate in chemical, physical and biological processes and fractionate in a predictable way, depending on the mass of the isotope. In atmospheric reactions catalyzed by ultraviolet radiation, these isotopes, however, show mass independent fractionation effects (Farquhar et al. 2000; Farquhar and Wing 2003), which are different for the oxidized and reduced sulfur species. In the presence of atmospheric oxygen, the reduced sulfur species are oxidized and the mass independent signal is erased. In addition, molecular oxygen is the source for ozone that shields the lower atmosphere from ultraviolet radiation and thereby prevents mass independent reactions.

The sharp disappearance of the mass independent signal in sedimentary sulfur isotopes has been dated to about 2,400 million years (Bekker et al. 2004). This shift has been interpreted to mark a shift from a practically oxygen-free atmosphere to one with an oxygen content $>10^{-5}$ times the present atmospheric level (Kasting

2001). The limiting oxygen concentration is low in comparison to the modern atmosphere, but still significant considering the oxidation potential of the atmosphere.

The rise in atmospheric oxygen is closely associated with the appearance of oxygen producing photosynthesizing cyanobacteria. It is, nevertheless, surprising that the earliest chemical evidence for photosynthesizing cyanobacteria has been found in sedimentary successions deposited already 2,700 million years ago (Brocks et al. 1999), which precedes the rise of atmospheric oxygen by 300 million years. The delay may be understandable, if vast reducing sinks for oxygen existed that were able to remove oxygen at the same rate as it was created. Several possibilities for the reducing buffer have been presented, including a shift in the oxidation state of volcanic gases to a higher level as subaerial volcanism became abundant after 2,500 million years ago (Kump and Barley 2007).

It is also unexpected that the initial rise in atmospheric oxygen is not associated with a positive carbon isotope excursion, which would mark massive burial of organic carbon (Fig. 13.3). A major positive carbon isotope excursion occurred between about 2,200 and 2,100 million years ago (Karhu 1993; Karhu and Holland 1996), but it clearly postdates the first appearance of atmospheric oxygen by ca. 200 Ma. Evidence for the excursion was first reported from the Zimbabwe Craton (Schidlowski et al. 1975) and from the Fennoscandian Shield (Karhu 1993), where the excursion is widely present in sedimentary carbonate formations. The excursion was by Karhu and Holland (1996) interpreted to mark a significant global carbon burial event, providing enough oxygen to produce the transition from an anoxic to an oxic atmosphere. Clearly, the event leading the rise of atmospheric oxygen is more complicated than has been thought in the pioneering studies. One possibility is that the oxygenation of Earth's atmosphere progressed stepwise, reaching near-modern levels only at the end of the Precambrian time, approximately at the same time as the first metazoan animals appeared in the fossil record (Kah et al. 2004).

The rise of atmospheric oxygen produced a drastic environmental change, with a wide range of consequences on weathering reactions, atmospheric chemistry, oceanic chemistry and the biosphere. Weathering reactions shifted from anoxic to oxidative weathering, which had a vast impact on the geochemical cycles of several redox sensitive elements, such as iron. Also the concentrations of some other atmospheric gases were strongly affected by the appearance of oxygen. Before the appearance of oxygen, methane probably acted together with carbon dioxide to produce a strong greenhouse effect to compensate the low luminosity of the faint young Sun. The rise of oxygen removed methane from the atmosphere, and thereby decreased the heating effect of greenhouse gases. It has been suggested that rapid elimination of methane possibly triggered the first glaciations on Earth at about 2,400 million years ago (Pavlov et al. 2000; Kasting and Ono 2006).

As a result of oxidative weathering, sulfide minerals in continental weathering horizons became unstable and were oxidized producing sulfate. Sulfate and oxygen were delivered to the ocean, transforming its chemical composition. On the basis of the deposition of sedimentary banded iron formations (BIFs), it is known that iron was mobile in the oceans till 1,800 million years ago. To be mobile, iron needs to be present in its reduced Fe^{2+} form. In an oxidative environment Fe^{2+} is oxidized to Fe^{3+}, which is very sparingly soluble in water. The deposition of BIFs indicate that iron in the oceans at that time was present as Fe^{2+}. It is commonly believed that after 1,800 Ma, the mobility of iron was suppressed by oxidizing conditions in marine waters.

As a result of the rise of atmospheric oxygen, the biosphere probably faced the largest catastrophic event in the history of Earth. Before the rise of oxygen, marine waters were dominated by anaerobic, single-celled archaea and bacteria, but when oxygen appeared anaerobic microbes were forced to restricted environments separated from the oxidized atmosphere and surface waters. As a result, Earth's surface was open to new aerobic life forms, such as eukaryotes. Importantly, the oldest eukaryotic body fossils have been recovered from formations deposited around 2,000 million years ago.

13.4 Conclusions

Earth's ancient atmospheres have been characterized by higher contents of greenhouse gases in comparison to the present-day atmospheric composition, which has compensated for the lower energy flow from the faint young Sun. The concentrations of carbon dioxide show a generally increasing trend, going back in time. Before the rise of atmospheric oxygen at about 2,400 million years ago, methane may have been another greenhouse gas compensating for the lower luminosity of the Sun. The rise of atmospheric oxygen has been dated to have occurred 2,400 million years ago. After that, oxygen levels in the atmosphere may have increased stepwise, until high concentrations, comparable to the modern atmospheric levels were reached in the beginning of the Phanerozoic Eon, around 540 million years ago.

References

Bekker A, Holland HD, Wang P-L, Rumble D III, Stein HJ, Hannah JL, Coetzee LL, Beukes NJ (2004) Dating the rise of atmospheric oxygen. Nature 427:117–120

Berner RA (1999) Atmospheric oxygen over Phanerozoic time. Proc Nat Acad Sci 96: 10955–10957

Berner RA, Beerling DJ, Dudley R, Robinson JM, Wildman RA Jr (2003) Phanerozoic atmospheric oxygen. Annu Rev Earth Planet Sci 31:105–134

Brocks JJ, Logan GA, Buick R, Summons R (1999) Archean molecular fossils and the early rise of eukaryotes. Science 285:1033–1036

Farquhar J, Wing BA (2003) Multiple sulfur isotopes and the evolution of the atmosphere. Earth Planet Sci Lett 213:1–13

Farquhar J, Bao H, Thiemens M (2000) Atmospheric influence of Earth's earliest sulfur cycle. Science 289:756–758

Holland HD (1978) The chemistry of the atmosphere and oceans. Wiley Interscience, New York

Holland HD (1984) The chemical evolution of the atmosphere and oceans. Princeton University Press, Princeton

Kah LC, Lyons TW, Frank TD (2004) Low marine sulphate and protracted oxygenation of the Proterozoic biosphere. Nature 431:834–838

Karhu J (1993) Paleoproterozoic evolution of the carbon isotope ratios of sedimentary carbonates in the Fennoscandian Shield. Bull Geol Surv Finland 371:1–87

Karhu J (1999) Carbon isotopes. In: Marshall CP, Fairbridge RW (eds) Encyclopedia of Geochemistry. Kluwer Academic Publishers, Dordrecht, pp 67–73

Karhu JA, Holland HD (1996) Carbon isotopes and the rise of atmospheric oxygen. Geology 24:867–870

Kasting JF (1987) Theoretical constraints on oxygen and carbon dioxide concentrations in the Precambrian atmosphere. Precambrian Res 34:205–229

Kasting JF (2001) Earth history – the rise of atmospheric oxygen. Science 293:819–820

Kasting JF, Ono S (2006) Palaeoclimates: the first two billion years. Phil Trans R Soc B 361:917–929

Kump LR, Barley ME (2007) Increased subaerial volcanism and the rise of atmospheric oxygen 2.5 billion years ago. Nature 448:1033–1036

Loulergue L, Schilt A, Spahni R, Masson-Delmotte V, Blunier T, Lemieux B, Barnola J-M, Raynaud D, Stocker TF, Chappellaz J (2008) Orbital and millennial-scale features of atmospheric CH_4 over the past 800,000 years. Nature 453:383–386

Pavlov AA, Kasting JF, Brown LL, Rages KA, Freedman R (2000) Greenhouse warming by CH_4 in the atmosphere of early Earth. J Geophys Res 105:11981–11990

Royer DL, Berner RA, Montañez IP, Tabor NJ, Beerling DJ (2004) CO_2 as a primary driver of Phanerozoic climate. GSA Today 14:4–10

Rye R, Kuo PH, Holland HD (1995) Atmospheric carbon dioxide concentrations before 2.2 billion years ago. Nature 378:603–605

Sagan C, Mullen G (1972) Earth and Mars: evolution of atmospheres and surface temperatures. Science 177:52–56

Schidlowski M, Eichmann R, Junge CE (1975) Precambrian sedimentary carbonates: carbon and oxygen isotope geochemistry and implications for the terrestrial oxygen budget. Precambrian Res 2:1–69

Schidlowski M, Hayes JM, Kaplan IR (1983) Isotopic inference of ancient biochemistries: Carbon, sulfur, hydrogen and nitrogen. In: Schopf JW (ed) Earth's earliest biosphere, its origin and evolution. Princeton Universtity Press, Princeton, pp 149–186

Siegenthaler U, Stocker TF, Monnin E, Lüthi D, Schwander J, Stauffer B, Raynaud D, Barnola J-M, Fischer H, Masson-Delmotte V, Jouzel J (2005) Stable carbon cycle-climate relationship during the late Pleistocene. Science 310:1313–1317

Walker JCG (1987) Was the Archaean biosphere upside down? Nature 329:710–712

Wilde SA, Valley JW, Peck WH, Graham CM (2001) Evidence from detrital zircons for the existence of continental crust and oceans on Earth 4.4 Ga ago. Nature 409:175–178

Chapter 14
Late Quaternary Climate History of Northern Europe

Antti E.K. Ojala

14.1 Introduction

Recent decades have brought growing interest in past climate variability as a result of ongoing global change. As stated by Raymond S. Bradley (2008): *Whatever anthropogenic climate changes occur in the future, they will be superimposed on, and interact with, underlying natural variability.* The same also applies to the recent past in climate history. Climate change has been acknowledged as one of the main factors behind the expansion and collapse of human cultures in the past (e.g. Berglund 2003), but anthropogenic activities are also likely to have punctuated climate variables via land use, deforestation and animal agriculture for thousands of years (e.g. Ruddiman 2003).

Earth's climate is a complex system that includes five main components: the geosphere (solid earth), hydrosphere (oceans, lake, rivers, and groundwater), cryosphere (ice sheets, glaciers and sea ice), atmosphere and biosphere. It varies on many different time scales and has experienced large changes throughout the planet's history. The generally accepted theory by Milutin Milankovitch proposes that three separate cyclical variations in Earth's orbital characteristics – eccentricity, precession, and obliquity – are responsible for the past glacial-interglacial cycles due to variations in the amount of solar radiation that is received on different parts

---

Due to the diversity of time scales discussed in the present review, the following abbreviations have been used: (1) "ka ago" means thousands of years ago, and is used for the glacial period, (2) "yrs BP" means calibrated calendar years before AD 1950 and is used for the post-glacial period, and (3) "AD/BC" are designations according to the Julian and Gregorian calendars and are used for the period of historically documented climate excursions (i.e. the last ca. 2,000 years).

A.E.K. Ojala (✉)
Geological Survey of Finland, Betonimiehenkuja 4, FI-02150 Espoo, Finland
e-mail: antti.ojala@gtk.fi

of Earth's surface (Milankovitch 1941; Rapp 2009). Orbital models demonstrate that the Milankovitch cycles have the greatest impact on Earth's climate when the peaks of all three cyclical variations coincide (Milankovitch 1941; Bradley et al. 2003; Rapp 2009). Glacial-interglacial climate fluctuations are global or hemispheric phenomena, whereas rapid, annual to centennial oscillations are minor in amplitude and typically appear within a limited region. In trying to understand the background to natural climate variability, it is important to understand the extreme climate fluctuations of glacial-interglacial periods, but also to comprehensively recognize the climate change of the more recent past. After all, natural changes in global insolation due to Earth's orbital variations and changes in the Sun's energy output have evidently also played an important role in the climate variability of the Holocene interglacial (Mayewski et al. 2004).

The longest instrumental temperature observations barely cover the last 300 years of climate history, and only ca. 100 years or less for most parts of the world (Bradley 2008). These global series are sufficiently long to document climate phenomena at inter-annual to decadal time scales. Beyond that period, however, there is a need for historical sources of climate-related documents as well as accurately-dated palaeoproxy records that can be used to investigate the past climate fluctuation and factors that have been forcing the climate change at different time scales (Bradley et al. 2003).

Current understanding of past climate change beyond the instrumental observations is based on a variety of palaeoproxy records, calculations of climate forcing factors and palaeoclimate modelling. By definition, a palaeoproxy record of the climate contains a climatic signal, which can be used to infer qualitative and quantitative information about the past climate on inter-annual to multi-millennial time scales. It is noteworthy that the interpretation of palaeoproxy records is not straightforward, because their climatic signal may be weak, they may contain a great deal of random background noise, or their signal may reflect a combination several climatic parameters that are interacting. For example, alpine and maritime glaciers in the Northern Hemisphere have constantly changed their altitude and extensions, plant and animal species have spread and retreated according to temperature and precipitation patterns, and periglacial processes have occurred in different places at different times during glacial and interglacial periods. Sedimentary records in oceans, lakes and peat bogs provide a variety of physical, geochemical and biological parameters that can be used as palaeoproxies. Ice core records from Antarctica and Greenland are perhaps the most exquisite archives for detailed studies of late Quaternary temperature and precipitation changes, sea-level oscillation and ocean volume, the chemical composition of the lower atmosphere, and the occurrence of volcanic eruptions. For example, the EPICA Dome C ice core record from Antarctica spans the last eight glacial-interglacial cycles, going back ca. 800 ka (Jouzel et al. 2007). Several other annually resolved palaeoproxy records, such as tree rings, speleothemes, varved sediments and corals, are also important sources of inter-annual to millennial-scale climate fluctuations (Fig. 14.1).

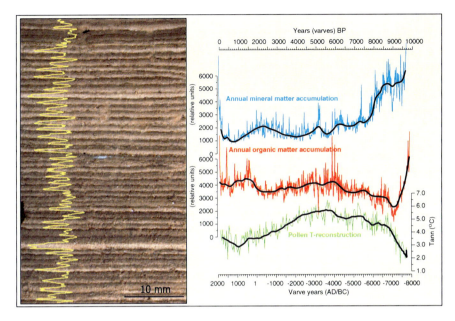

Fig. 14.1 An example of an annually resolved palaeoproxy record that is used to infer information on past climate variability. Lake Nautajärvi (Längelmäki, Finland) varves consist of two layers, a *pale* layer of mineral material that reflects winter and spring temperatures and precipitation, and a *darker* organic layer that is a proxy for the temperature during the growing season. The Holocene varve- and pollen-based temperature reconstructions are based on data from Ojala and Alenius (2005) and Ojala et al. (2008b)

14.2 The Ice Age World

According to current understanding of past long-term climate fluctuations, periods with glacial ice have occurred with variable intervals more or less throughout the entire past of Earth. This understanding is based on scattered findings of ancient glacigenic deposits around the world, in South Africa, Canada and Australia, for example, that have been dated to 2–3 billion years ago (Hannah et al. 2004). In Fennoscandia, the oldest glacigenic deposits formed during ice ages appear as conglomerates, diamictites and varve-like sedimentary rocks, and are associated with the Early Proterozoic Sarioli formations that have been dated to 2.2–2.4 billion years ago (Marmo and Ojakangas 1984). In the Neoproterozoic eon, ca. 850–630 million years ago, Earth's climate was characterized by severe glaciations that also extended to lower latitudes. Even a global-scale glaciation, the "snowball Earth" hypothesis, has been suggested for this eon in several studies on climate history and modelling (e.g. Kirschvink 1992). This hypothesis is based on geological findings that indicate concurrent glacial activity worldwide, and that no other geological deposits of the same age are found anywhere. It is not exactly known what caused and ended this massive glaciation, but scientists hypothesize that elevated greenhouse forcing due to the emission of CO_2 and methane from volcanic activity finally made the surface ice melt.

Earth's climate has been slowly cooling since the Late Paleocene thermal maximum at around 55 million years ago (Zachos et al. 2001). According to an array of marine and terrestrial climate proxies, ice sheets have been present in Antarctica for more than 30 million years, whereas the first evidence of the occurrence of Northern Hemispheric ice sheets has been dated to ca. 14 million years ago (Zachos et al. 2001). More extensive glaciations, periods when continental ice sheets have covered larger areas in North America and Eurasia, have occurred for the last ca. 2.6 million years, a period we know as the Quaternary (e.g. Shacleton et al. 1984). During the Quaternary, the climate of the Earth has become unstable and begun to follow the well-defined orbital variation cycles described by Milankovitch (Milankovitch 1941; Rapp 2009). Initially, glacial-interglacial changes were dominated by a 41-kyr astronomical cycle and climate changes were moderate in amplitude in relation to the more recent glacial-interglacial cycles (Imbrie et al. 1984; Lisiecki and Raymo 2005) (Fig. 14.2).

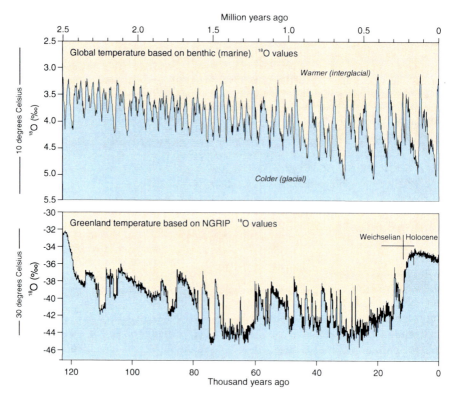

Fig. 14.2 Past temperature fluctuations on two different time scales. Global temperature variability during the past 2.5 million years based on a benthic oxygen isotope stack by Lisiecki and Raymo (2005) (above), and Greenland temperature variability during the last glacial-interglacial cycle based on the $\delta^{18}O$ temperature proxy from the NGRIP ice core record (NGRIP Members 2004) (below)

About 800 ka ago, the amplitude increased and since then Northern Hemisphere has undergone eight major glaciations that have been roughly spaced at 100 ka intervals (Rapp 2009). It is likely that the orbital eccentricity cycle has now become more dominant, instead of the obliquity amplitude that dominated earlier. Between the colder phases (glacials) there have been ca. 10 ka long warmer periods known as interglacials, during which continental ice sheets have melted almost entirely. According to Jouzel et al. (2007) and Lisiecki and Raymo (2005), it is likely that due to orbital factors the temperature differences between consecutive glacial and interglacial stages have increased during the past 500 ka by 1–2°C, which is about 10% of the overall magnitude of variation between ice ages and interglacials during the late Quaternary. Furthermore, the last four glacial stages have been the coldest and the recent interglacials have been characterized by more pronounced warmth than previously during the Quaternary Period.

14.3 The Last Interglacial-Glacial Cycle in Northern Europe

Extensive continental glaciations are typical characteristics of the late Quaternary climate cycle in the Northern Hemisphere, for example in Fennoscandia, NW Russia and the British Isles, and Finland lies in the central part of the often glaciated terrain. The recent EIS (Eurasian Ice Sheets) and QUEEN (Quaternary Environment of the Eurasian North) projects were international and interdisciplinary research programmes that were established to study the glacial-interglacial history of the Eurasian region during the last 200 ka and to understand the processes involved in environmental changes in the Arctic regions (see Svendsen et al. 2004 for a synthesis of the main conclusions).

The Saalian cold stage (ca. 130–190 ka ago) was characterized by very extensive glaciation, and continental ice sheets probably advanced further in continental Europe and in NW Siberia than during any other Quaternary glaciation (e.g. Svendsen et al. 2004). It was followed by a warmer Eemian interglacial stage that occurred between the Saalian and Weichselian glaciations, ca. 115–130 ka ago. During the Early Eemian, the global sea level was approximately 5–7 m above the present sea level, and Fennoscandia was an island of the saline Eemian Baltic Sea. An array of marine and terrestrial records from central and northern Europe suggest that the annual mean temperature in Europe was about 2–3°C warmer during the Eemian stage than during the present interglacial, the Holocene (e.g. Field et al. 1994). The climate development of the Eemian was broadly similar to the Holocene development in Fennoscandia and NW Russia (e.g. Saarnisto and Lunkka 2004). According to studies in eastern Finland by Saarnisto et al. (1999), the Early Eemian climate warmed very rapidly and there was a continental and dry climate during the early part of the Eemian, which later switched towards to a pronounced oceanic climate towards the end of the interglacial. Simulations of the Arctic climate in Greenland by Huybrechts (2002) indicate that at around 123 ka ago, temperatures peaked 5–7°C higher than at present, which resulted in considerable shrinkage of the ice sheet. The ice sheet never disappeared from Greenland during the Eemian, but it covered only less than half of its present area and had only a third of its present volume (Huybrechts 2002). Palaeorecords from Arctic parts of North America also indicate that the extension of the Arctic Sea ice was considerably reduced during the Eemian interglacial, and climate models suggest up to 5°C higher temperatures than presently (Ehlers and Gibbard 2004).

The climate became cooler towards the end of the Eemian interglacial, concurrently with the insolation minimum in the Northern Hemisphere. As a consequence, several ice sheet nucleation areas started to build up in various parts of North America and Eurasia (Svendsen et al. 2004). This was the beginning of the latest glacial period in Eurasia known as the Weichselian (ca. 115–11.7 ka ago). The geological findings indicate that the Weichselian was characterized by sharp but short-lived glaciation-deglaciation cycles, which in Scandinavia and northern Russian had a different geographical focus at different times (e.g. Svendsen et al. 2004; Rapp 2009), and in between there existed warmer periods that are termed interstadials (Helmens et al. 2000) (Fig. 14.2). The significance of geological interstadials is that they mark a temporary retreat of the continental ice, and it is thereby known that certain parts of glaciated terrains in Northern Europe evidently remained free of continental ice during the earliest and middle stages of the Weichselian. The prevailing climate and environmental conditions in such areas must have been either warm interstadial or cold periglacial. According to a review by Saarnisto and Lunkka (2004), the mean July temperature in NW Finland could have been +10°C or more during the warmer interstadial periods of the early Weichselian. Some studies even suggest that summer temperatures were warmer that at present in the Sokli area in NE Finland, northern Russia and Greenland at around 80 and 100 ka ago (e.g. Väliranta et al. 2009). On the other hand, stratigraphical studies indicate that a colder ice-free environment and tundra-type vegetation prevailed in eastern Finland during the Middle Weichselian ca. 50 ka ago (Helmens et al. 2000; Saarnisto and Lunkka 2004).

In general, the Barents-Kara Ice Sheet became progressively smaller and the Scandinavian Ice Sheet generally increased in size in continental Eurasia throughout the Weichselian glaciation (Svendsen et al. 2004). The Scandinavian Ice Sheet reached its maximum extension (the Last Glacial Maximum, LGM) during the Late Weichselian about 20 ka ago. Based on oxygen isotope data from mammoth bones in central and southern Finland, Arppe and Karhu (2006) suggested that already between 32 and 25 ka ago the annual mean temperature was up to 6°C colder than today. Tarasov et al. (1999) estimated that the mean summer and winter temperatures in the Moscow area were, respectively, 5–11°C and ca. 20°C lower during the LGM than at present. The Greenland ice sheet was thicker than presently and expanded in most locations offshore onto the continental shelf during the LGM (Funder 1989). The Greenland temperatures are estimated to have been about 20°C lower during the LGM than at present (Johnsen et al. 2001) (Fig. 14.2).

Following the LGM, the deglaciation initiated along the ice margin of the entire Eurasian Ice Sheet around 15–17 ka ago at the latest (Svendsen et al. 2004). By the period known as the Younger Dryas (YD, 12.9–11.7 ka ago), the ice margin had already retreated to the Salpausselkä region in southern and eastern Finland. The YD represented a rapid return to glacial environmental and climatic conditions for a period of about 1,000 years. During the YD, the retreating continental glacier halted or even re-advanced to some extent, forming the Salpausselkä ice-marginal formations (end moraines) and their correlatives in Russian Karelia. Forest vegetation in northern Europe was replaced by glacial tundra vegetation and periglacial processes were common (e.g. Walker et al. 1994). Isarin (1998) has suggested that during the coldest part of the YD there was continuous permafrost in Europe north of 54°N, which includes the whole of Fennoscandia, among many other localities. Mean annual air temperatures could have been near or below −8°C at sea level and mean temperatures during the coldest month well below −20°C for the entire zone (Isarin 1998). Based on palaeobotanical records, Isarin and Bohncke (1999) interpreted the mean July temperatures at sea level in the northern part of the British Isles and in ice-free Scandinavia to be about +10°C during the YD. It has been estimated from Greenland ice core records that the mean annual temperature decreased by as much as 10–15°C during the YD (Johnsen et al. 2001).

The most likely explanation for the YD cooling was a disturbance in the North Atlantic thermohaline circulation (THC) in response to a sudden influx of a massive amount of fresh water from a North American glacial lake or lakes (e.g. Alley et al. 1993). Bradley and England (2008) have also suggested that the Younger Dryas cold episode was at least partially due to thick sea ice being formed and driven from the Arctic Ocean, dampening or shutting off the THC. Geological and palynological evidence indicates that even though the Younger Dryas caused considerable cooling in the North Atlantic region, there were large local and regional differences in the temperature change around Eurasia (e.g. Väliranta et al. 2006). Results from the Greenland ice-core records suggest that the termination of the Younger Dryas was as rapid as, if not even more rapid than the beginning of this climate anomaly (Alley et al. 1993). A considerable warming of up to 10°C occurred in just a few decades in Greenland.

14.4 The Post-Glacial Climate History of Northern Europe

The end of the Younger Dryas at 11,650 years BP is considered to mark the beginning of the most recent period in the geological timetable, the Holocene interglacial. A typical feature of the Holocene interglacial in the Northern Hemisphere is that Greenland is the only place that is occupied by a massive ice sheet, while there are smaller glaciers and valley-type glaciers in Svalbard, in the mountains of Norway, Sweden and the Russian Urals in the Eurasian Arctic, in the Alps, Alaska, the northern Canadian Rockies, and in the Canadian Arctic (Fig. 14.3). From a climatic perspective, the Holocene epoch is broadly similar to previous interglacials such as the Eemian, but is marked by a few distinctive characteristics. It has lasted, and probably will continue to last somewhat longer than previous interglacials, and its temperature has been slightly lower and more stable than during the previous

Fig. 14.3 The melting ice sheet in Kangerlussuaq in SW Greenland (67° 09′ 25″; 50° 04′ 56″) (Photo by T. Ruskeeniemi, 2010). The Greenland summit ice cores show past climate fluctuations over 100,000 years (Johnsen et al. 2001; NGRIP Members 2004)

interglacials. Nevertheless, the climate has constantly and regionally fluctuated in the Northern Hemisphere throughout the Holocene, with a typical amplitude of variation in annual mean temperatures of roughly 1–3°C. Proxy-based palaeotemperature reconstructions indicate that both the magnitude and timing of temperature fluctuation has varied substantially between regions (Mayewski et al. 2004; Wanner et al. 2008; Renssen et al. 2009). Palaeorecords also indicate that there have been changes in the inter-annual variability of the climate as well as in the mean annual climate during the Holocene (e.g. Giesecke et al. 2010).

Climate changes during the Holocene have been driven by the same natural factors that have caused ice ages. They are: (1) *orbital forcing*, which has an effect on the geographical and seasonal distribution of insolation, (2) *solar forcing*, which depends on the activity of the Sun, (3) *volcanic activity*, which potentially blocks the energy transmission to and from Earth's surface, (4) *the concentration of CO_2 and CH_4* and other greenhouse gases in atmosphere, and (5) *atmospheric and oceanic circulation*, which distributes heat and moisture globally (see Bradley et al. 2003 for an overview of climate forcing factors and related large-scale environmental changes during the Holocene). In Fennoscandia, for example, the winter temperature is heavily influenced by the North Atlantic Oscillation (NAO) (Chen and Hellstrom 1999). The NAO is a large-scale fluctuation in atmospheric sea-level pressure between the Azores high and the Icelandic low in the Atlantic. Based on instrumental observations, variations in the NAO are known to affect precipitation and temperature anomalies in the North Atlantic region. In winters dominated by the positive phase of the NAO, warm air masses moving from the Atlantic over the Baltic Sea region and Fennoscandia produce milder and wetter conditions. Winters with a negative NAO, on the other hand, are more severe and dryer. It is believed that the NAO has varied considerably during the Holocene (e.g. Trouet et al. 2009).

Based on findings of ice-rafting material in deep ocean sediment cores, Bond et al. (1997) proposed a comprehensive theory of climate cyclicity of 1,470 ± 500 years in the North Atlantic region during the Holocene (Fig. 14.4). The "Bond cycles" are probably caused by variations in solar output during the Holocene. The first Bond cycle coincides with the transition from the Younger Dryas to the Holocene, and some others with additional established climate shifts, but not all of them are linked to well-known climate shifts of the Holocene interglacial in the Northern Hemisphere. Bond et al. (2001) concluded that the Arctic-Nordic Seas may be a key region where solar-induced atmospheric changes are amplified and distributed globally because of their impact on sea ice and the North Atlantic THC. Likewise, a recent examination of about 50 globally distributed palaeoclimate records by Mayewski et al. (2004) revealed six periods of significant rapid climate fluctuation during the Holocene. These took place 9,000–8,000, 6,000–5,000,

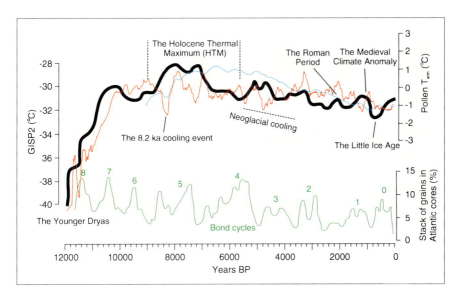

Fig. 14.4 A generalized sketch of Holocene climate oscillation (black thick line) superimposed upon the GISP2 ice-core temperature reconstruction from central Greenland (*red thin line*) (Alley 2000) and the North-European pollen-based stacked annual mean temperature (T_{ann}) reconstruction (*thin blue line*) (Seppä et al. 2009). Bond cycles (0–8) are indicated with a *green thin line* (Bond et al. 2001)

4,200–3,800, 3,500–2,500, 1,200–1,000, and 600–150 years BP. They argued that most of these periods are globally identified and appear as contemporaneous polar cooling, tropical aridity and changes in the atmospheric circulation. However, the individual records studied by Mayewski et al. (2004) are fairly heterogeneous or even possess discordant data, and the results therefore only provide a generalized picture of Holocene climate oscillations. More recent studies have actually argued against their simplified interpretation (e.g. Wanner et al. 2008).

Perhaps the most pronounced longer-term climate feature of the Holocene period is an early to mid-Holocene warm event known as the Holocene Thermal Maximum (HTM), which evidently occurred at slightly different times in different parts of the Northern Hemisphere (e.g. Davis et al. 2003; Kaufman et al. 2004; Renssen et al. 2009; Seppä et al. 2009) (Fig. 14.4).

14.4.1 The Early Holocene

Due to orbital forcing, summer insolation reached maximum values in the Northern Hemisphere during the Early Holocene. Since then, the summer insolation receipts have steadily declined, whereas winter insolation receipts have increased through the Holocene (e.g. Bradley et al. 2003). Orbital forcing was responsible for the early Holocene warming in the Northern Hemisphere, causing melting of the remains of continental ice sheets and a rise in the sea level (Bradley et al. 2003). Palynological evidence suggests that Northern Europe experienced a more oceanic climate soon after Late Weichselian deglaciation, but a general rise in summer temperatures was due to higher insolation during the Early Holocene that was accompanied by a trend towards a more continental climate (Snowball et al. 2004). The temperature increase was much slower in southern Scandinavia than in Greenland during the Early Holocene.

Evidence from palaeoproxy records indicates that many parts of the Northern Hemisphere experienced an unstable and variable climate during the Early Holocene between 8,000 and 11,600 years BP (e.g. Björck et al. 1997; Alley et al. 1997). It has been suspected that one of the main reasons for these oscillations was that the gradual intensification of the North Atlantic THC was frequently interrupted by meltwater discharges from retreating continental ice sheets (e.g. Alley et al. 1997; Mayewski et al. 2004; Svendsen et al. 2004). At least three cooling episodes during the early Holocene are commonly recognized in palaeoproxy records in the NH. The first is called the Preboreal oscillation (ca. 11,150–11,300 years BP), which has been documented by vegetation changes, increased soil erosion, re-advances and/or standstills of the ice sheet and ingression of brackish water into the Baltic Sea Basin. It is believed that climate conditions were cool and humid throughout northwestern and central Europe (Björck et al. 1997). Preboreal cooling was a short climate anomaly that is also visible in ice core records in Greenland (Johnsen et al. 2001). The second cooling episode occurred around 10,300 years BP, which according to Björck et al. (2001) may have been partly due to a temporary decline in solar activity.

The third cooling episode has been referred to as the "8.2 ka climate cooling event", and it has been reported from numerous palaeoproxy records globally, but particularly in the Northern Hemisphere (e.g. Alley et al. 1997; Veski et al. 2004; Ojala et al. 2008a). According to pollen-based quantitative temperature reconstructions and several other climate palaeoproxies, cooling was more pronounced at most sites south of 61°N in Europe, but there is less significant evidence of climate fluctuation in the North-European tree-line region (Seppä et al. 2007). One explanation for this behaviour may be that the 8.2 ka climate cooling event was mainly a winter and spring cooling (Seppä et al. 2007). Based on palaeoproxy data, it seems possible that the 8.2 ka climate cooling episode was actually not a single event but rather a longer-lasting colder period that terminated at one or several cooling events around 8,000 years BP (Alley et al. 1997; Snowball et al. 2002; Ojala et al. 2008a).

Today, the Greenland ice sheet is the only source for such a massive amount of ice-cold fresh water in the North Atlantic region, but it is unlikely to melt so rapidly that it would cause a sudden and pronounced interruption of the THC. However, the issue of a possible slowdown of the THC and weakening of the Gulf Stream due to climate change is under continuous investigation, because most coupled climate model integrations indicate weakening of the THC as CO_2 increases in the atmosphere.

14.4.2 Mid-Holocene Warmth

The early to mid-Holocene warm period known as the Holocene Thermal Maximum (HTM) is clearly documented in a wide range of terrestrial and marine palaeoproxy records. The HTM is most clearly recorded at mid- and high latitudes in the Northern Hemisphere (Renssen et al. 2009). It is probable that the HTM more strongly reflects summer warming rather than year-round warmth, and was probably associated with significant changes in precipitation patterns globally (Davis et al. 2003; Renssen et al. 2009).

The HTM occurred at different times and to varying degrees in different places at high latitudes (Kaufman et al. 2004). For example, in Alaska and northwest Canada the HTM occurred between ca. 9,000 and ca. 11,000 years BP, but about 4,000 years later in northeast Canada, probably due to the Laurentide Ice Sheet residual, which locally impacted on the surface energy balance and ocean circulation (Kaufman et al. 2004). Fennoscandia experienced the HTM roughly between 5,500 and 7,500 years BP (e.g. Snowball et al. 2004). Given the widespread evidence for low glacial activity, high organic productivity in lakes, and more northern distribution of many tree species than at present in Fennoscandia, there is no doubt that the HTM was warmer than the twentieth century AD (e.g. Nesje et al. 2001; Snowball et al. 2002, 2004; Ojala et al. 2008b). In central southern Finland, for example, a combined lacustrine varve record and a pollen temperature reconstruction suggests low but rising summer temperatures for the Early Holocene (9,500–8,500 years BP), with a clear peaking of the HTM at about 7,500–4,500 years BP, and post-HTM cooling at about 4,500 years BP (Fig. 14.1) (Ojala et al. 2008b). According to the review by Snowball et al. (2004), the climate in Fennoscandia was about 2°C warmer than at present during the HTM, but also dryer than during most of the Holocene. This led to a retreat of valley glaciers in Scandinavian mountains and low lake levels in northern Finland (e.g. Hyvärinen and Alhonen 1994; Nesje et al. 2001).

14.4.3 Neoglacial Cooling

By the beginning of the second half of the present interglacial, the summer insolation receipt had decreased considerably while the winter isolation receipt had increased (Bradley et al. 2003). Many palaeorecords in Fennoscandia indicate abrupt mid-Holocene cooling and rather instant termination of the HTM at around 5,000–6,000 years BP (e.g. Karlén 1988). This applies to northern Fennoscandia in particular. Glaciers in northern Sweden started to re-advance (Karlén 1988), interannual variability increased in the Finnish tree-ring records (Eronen et al. 1999), and maritime glaciers in western Norway reacted to temperature and precipitation instability (Nesje et al. 2001). Davis et al. (2003) suggested that the cooling was more a summer temperature decline in NW Europe, whereas winter temperatures continued to rise in western and NW Europe. In NE Europe, both winter and summer temperatures declined during neoglacial cooling, while southern Europe has experienced a steady seasonal temperature increased since the early Holocene.

Evidence of climate instability and a transition towards a more variable climatic regime is well preserved in varved lake sediment records in Finland and Sweden (Ojala and Alenius 2005; Snowball et al. 2002). These records reflect seasonal-scale fluctuation in winter severity and spring meltwater intensity, as well as the duration and intensity of the growing season (Fig. 14.1). Varve data indicate that periods of variable and intense erosion due to severe winters coupled with a higher than normal accumulation of snow occurred in Scandinavia at ca. 2,000–3,700 years BP. Ojala and Alenius (2005) also observed that the annually resolved palaeoproxy record of Lake Nautajärvi was characterized by periodic features of 87 and 126 years during the time interval from 2,000 to 3,700 years BP. The same periodicities have been detected in the cosmogenic Δ^{14}C isotopes and interpreted to indicate changes in solar variability (Stuiver and Braziunas 1993). It is likely that in the Lake Nautajärvi varved sediment sequence these periodicities reflect a complicated response to both spring and winter temperature and precipitation patterns, which are linked to sedimentation via spring surface runoff and catchment erosion. Furthermore, peat bog investigations in southern Finland indicate a simultaneous increase in carbon accumulation, suggesting a wet and cool climate about 3,000 years BP (Mäkilä and Saarnisto 2008).

14.4.4 The Climate Change of the Last 2,000 Years

A typical characteristic for the climate in the Northern Hemisphere during the last 2,000 years, and in particular during the last millennium, is that there have been significant regional and seasonal variations in temperature and precipitation fluctuations. Humans and the environment do not experience hemispheric mean conditions; they live in a limited area that is subjected to a certain regional temperature and precipitation (Bradley 2008). An areal appearance of climate change, whether it is due to precipitation or temperature patterns, poses a great challenge to past climate research, particularly for recent decades when anthropogenic activities have probably become an ever increasing component of environmental and climate change. From instrumental measurements, historical sources and palaeoproxy records we know that there have been several historical climate anomalies such as cold spells, droughts and warm intervals, which have had effects on human societies and ancient civilizations. Many such climate extremes have been well documented because they were disastrous for the human population. The climate of the Northern Hemisphere has been characterized by periods such as a warm "Roman" period (ca. 200 BC to AD 300), a generally colder "Dark Ages" (ca. AD 500 to AD 900), the Medieval Climatic Anomaly (MCA) (ca. AD 900–1200), and the Little Ice Age (LIA) (ca. AD 1300–1850). Of these, the MCA and the LIA are perhaps the best known and studied in the Northern Hemisphere. According to the review by Snowball et al. (2004), the temperatures in Fennoscandia were about 0.5–0.8°C warmer during the MCA and 1°C cooler during the LIA than at present.

There is convincing evidence that the climate in Fennoscandia and in continental northern Europe was generally warmer during the Middle Ages (Lamb 1965; Bradley et al. 2003). Vikings settled in Greenland at the end of the tenth century and their settlements existed until the beginning of the cold fourteenth century (Dansgaard et al. 1975). This is a good indication of warmer than present Greenland temperatures and probably easier ice conditions in northern parts of the Atlantic Ocean during the MCA. Briffa (2000) interpreted indications of warmer climatic conditions around AD 1000–1200 in the "northern average" tree-ring data, and Crowley and Lowery (2000) found evidence of three short warming periods during the MCA in their Northern Hemispheric temperature reconstruction. In central southern Finland, varved lake sediment records indicate diminished or even ceased spring flooding and catchment erosion probably caused by milder and wetter winters during ca. AD 1000–1200 than at present (Fig. 14.5) (Tiljander et al. 2003; Ojala and Alenius 2005). According to their varve records, the MCA was a two-stage event that was interrupted by a multi-decadal cold spell at the beginning of the twelfth century.

The trend from the MCA towards the cooler climate conditions of the LIA in the Northern Hemisphere began in the thirteenth century. The LIA was probably a global phenomenon, and not by any means a sustained multi-century long period of cooling, but rather a regionally asynchronous chain of climatic events (e.g. Bradley and Jones 1993; Meese et al. 1994). Glaciers expanded in Fennoscandia and in the

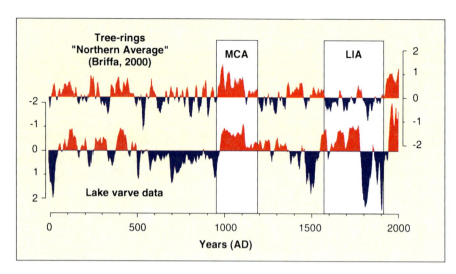

Fig. 14.5 The varve-based climate reconstruction for the period AD 1–1990 is compared with the tree-ring "Northern Average" by Briffa (2000). The Medieval Climate Anomaly (MCA) is clearly visible in both records. Changes in the studied lake varve records reflect winter climate fluctuations, where warming by a few degrees during the winter (positive NAO) causes considerable shortening or even the disappearance of the persistent snow cover in central and southern Finland. As a consequence, the spring discharge diminishes or disappears, which leads to a significant decrease in catchment erosion and sediment transportation into lakes (Tiljander et al. 2003; Ojala and Alenius 2005)

Alps, springs and summers were considerably colder causing damage to harvests around Europe, and winter snowfall was more common in continental Europe than ever since (e.g. Grove 1988; Nesje et al. 2001).

The complexity of the LIA anomaly is that at the hemispheric scale different palaeoproxy records seem to indicate contemporaneous cooling and warming patterns that are often highly seasonal in nature (Bradley and Jones 1993; Fisher et al. 1998). It seems certain that the LIA peaked with colder episodes several times between AD 1300 and 1850, during which the severe winters affected human societies, for example in the Fennoscandian region. Bradley and Jones (1993) argued that the coldest intervals were roughly between AD 1570 and AD 1730 in the Northern Hemisphere. The LIA was probably caused by decreased solar activity combined with increased volcanic activity on several occasions (e.g. Bradley and Jones 1993; Fisher et al. 1998; Wanner et al. 2008). A recent study of the past ca. 350 years by Lockwood et al. (2010) revealed that cold winter excursions have occurred more commonly in Europe at times when solar activity has been low. They emphasized, however, that this is a regional and seasonal effect rather than a global phenomenon.

Several cold spells during the LIA provide excellent examples of volcanic forcing of the climate at regional to global scales. For example, the coldest single year on record occurred in many places in Europe and North America in AD 1816 as a result of the eruption of Mount Tambora in Indonesia in AD 1815 (Stothers 1984). The year AD 1816 is known as the "year without a summer" because a substantial amount of atmospheric dust and sulfur dioxide from the eruption effectively prevented sunlight from passing through the atmosphere. More recently, the eruption of Krakatoa (in Indonesia) in 1883 caused a global average temperature fall of as much as 1.2°C, and temperatures did not return to normal until 5 years later (Self and Rampino 1981). Altogether, it has been calculated that more than one hundred significant volcanic eruptions occurred during the coldest phase of the LIA, which influenced summer temperatures in the Northern Hemisphere (Briffa et al. 1998).

Based on results from the northern Greenland ice core records, Fisher et al. (1998) argued that the LIA cooling was also accompanied by changes in the North Atlantic Ocean circulation pattern. Recently, Trouet et al. (2009) studied the influence of the NAO on the millennial-scale climate in the North Atlantic-European sector. Based on climate model results and palaeoproxy data, they suggested that a persistent positive NAO appeared during the MCA, while there was a clear shift to more negative NAO conditions in the Little Ice Age.

One approach to investigate past millennial-scale climate fluctuations is to use model simulations and quantified palaeoproxy records from various archives (e.g. Mann et al. 1999; Moberg et al. 2005, Mann et al. 2009). They allow larger-scale (hemispheric or global) reconstructions of temperature anomalies back into the recent past, but also provide a means to analyze the impact of human activities in relation to natural climate forcing factors during the past few centuries. A common feature in many of these reconstructions for the Northern Hemisphere is that they show a fairly rapid increase in temperatures after the Little Ice Age, ending at higher levels by the end of twentieth century than during the preceding 1,000 years. However, these multi-proxy temperature reconstructions to some extent give contradictory information on how the climate has fluctuated during the past millennia. The famous "Hockey-stick" curve, i.e. the average Northern Hemisphere temperature reconstruction of the past millennium by Mann et al. (1998, 1999) has attracted much discussion recently because it does not really exhibit any of the historically documented climate fluctuations. It only shows fairly steady temperatures until the last part of the twentieth century, and then a rapid increase. In another Northern Hemispheric mean temperature reconstruction for the past 2,000 years, Moberg et al. (2005) showed that recent decades and from about AD 1000–1100 have been warmer than the average temperature of AD 1961–1991. However, they found no evidence that the average Northern Hemispheric temperature was higher earlier during the last two millennia than is has been during the post-AD 1990 period. According to their estimation, the minimum temperature occurred around AD 1600. Moberg et al. (2005) concluded that there is large-scale natural climate variability, which may be larger than commonly thought and could be due to natural changes in radiative forcing, and which is likely to continue. However, the global warming of recent decades cannot be solely explained by these natural forcing factors (Moberg et al. 2005).

Crowley (2000) has estimated that variations in solar activity and volcanism have accounted for perhaps half of the mean temperature variability in the Northern Hemisphere during the past 1,000 years and preanthropogenic times (pre-1850), but only about 25% of the twentieth century temperature rise is due to natural variability. According to Bradley (2000), the causes of the twentieth century climate change are difficult to resolve, because instrumental observations do not extend before the period that has been characterized by considerable changes in the atmospheric composition. It seems that the uncertainties in past climate reconstructions can only be reduced by using a more thorough network of well-dated palaeoproxy records of the climate. However, not many high-quality palaeoproxies are available that would represent past temperatures with annual resolution, and those that exist are geographically sparse and mostly representative of summer or growing season temperatures.

References

Alley RB, Meese DA, Shuman CA, Gow AJ, Taylor KC, Grootes PM, White JWC, Ram M, Waddington ED, Mayewski PA, Zeilinski GA (1993) Abrupt increase in Greenland snow accumulation at the end of the younger Dryas event. Nature 362:527–529

Alley RB, Mayewski PA, Sowers T, Stuiver M, Taylor KC, Clark PU (1997) Holocene climatic instability: a prominent, widespread event 8200 yr ago. Geology 25:483–486

Alley RB (2000) The Younger Dryas cold interval as viewed from central Greenland. Quat Sci Rev 19: 213–226

Arppe L, Karhu J (2006) Implications for the Late Pleistocene climate in Finland and adjacent areas from the isotopic composition of mammoth skeletal remains. Palaeogeogr Palaeoclim Palaeoecol 231:322–330

Berglund BE (2003) Human impact and climate changes – synchronous events and a causal link? Quat Int 105:7–12

Björck S, Rundgren M, Ingólfsson O, Funder S (1997) The Preboreal oscillation around the Nordic Seas: terrestrial and lacustrine responses. J Quat Sci 12:455–465

Björck S, Muscheler R, Kromer B, Andresen CS, Heinemeier J, Johnsen SJ, Conley D, Koc N, Spurk M, Veski S (2001) High-resolution analysis of an early-Holocene climate event may imply decreased solar forcing as an important climate trigger. Geology 29:1107–1110

Bond G, Showers W, Cheseby M, Lotti R, Almasi P, deMenocal P, Priore P, Cullen H, Hajdas I, Bonani G (1997) A pervasive millennial-scale cycle in North Atlantic Holocene and glacial climates. Science 278:1257–1266

Bond G, Kromer B, Beer J, Muscheler R, Evans MN, Showers W, Hoffmann S, Lotti-Bond R, Hajdas I, Bonani G (2001) Persistent solar influence on North Atlantic climate during the Holocene. Science 294:2130–2136

Bradley RS (2000) 1000 years of climate change. Science 288:1353–1354

Bradley RS (2008) Holocene perspectives on future climate change. In: Battarbee RW, Binney HA (eds) Natural climate variability and global warming: a Holocene perspective. Wiley, Chichester, pp 254–268

Bradley RS, England JH (2008) The younger Dryas and the sea of ancient ice. Quat Res 70:1–10

Bradley RS, Jones PD (1993) 'Little Ice Age' summer temperature variations: their nature and relevance to recent global warming trends. Holocene 3:367–376

Bradley RS, Briffa KR, Cole J, Hughes MK, Osborn TJ (2003) The climate of the last millennium. In: Alverson KD, Bradley RS, Pedersen T (eds) Palaeoclimate, global change and the future. Springer-Verlag, Berlin, pp 105–141

Briffa KR (2000) Annual climate variability in the Holocene: interpreting the message of ancient trees. Quat Sci Rev 19:87–105

Briffa KR, Jones PD, Schweingruber FH, Osborn TJ (1998) Influence of volcanic eruptions on Northern Hemisphere summer temperature over the past 600 years. Nature 393:450–455

Chen DL, Hellstrom C (1999) The influence of the North Atlantic Oscillation on the regional temperature variability in Sweden: spatial and temporal variations. Tellus A51:505–516

Crowley TJ (2000) Causes of climate change over the past 1000 years. Science 289:270–277

Crowley TJ, Lowery TS (2000) How warm was the Medieval warm period. Ambio 29: 51–54

Dansgaard W, Johnsen SJ, Reeh N, Gundestrup N, Clausen HB, Hammer CU (1975) Climatic changes, Norsemen and modern man. Nature 255:24–28

Davis BAS, Brewer S, Stevenson AC, Guiot J, Contributors D (2003) The temperature of Europe during the Holocene reconstructed from pollen data. Quat Sci Rev 22:1701–1716

Ehlers J, Gibbard PL (eds) (2004) Quaternary glaciations: extent and chronology 2: part II North America. Elsevier, Amsterdam

Eronen M, Hyvärinen H, Zetterberg P (1999) Holocene humidity changes in northern Finnish Lapland inferred from lake sediments and submerged scots pines dated by tree-rings. Holocene 9:569–580

Field M, Huntley B, Müller H (1994) Eemian climate fluctuations observed in a European pollen record. Nature 371:779–783

Fisher H, Werner M, Wagenbach D (1998) Little ice age clearly recorded in northern Greenland ice cores. Geophys Res Lett 25:1749–1752

Funder S (1989) Quaternary geology of the ice-free areas and adjacent shelves of Greenland. In: Fulton RJ (ed) Quaternary geology of Canada and Greenland. Geological Survey of Canada, Ottawa

Giesecke T, Miller PA, Sykes MT, Ojala AEK, Seppä H, Bradshaw RHW (2010) The effect of past changes in inter-annual temperature variability on tree distribution limits. J Biogeogr 37:1394–1405

Grove JM (1988) The little ice age. Methuen, London

Hannah JL, Bekker A, Stein HJ, Markey RJ, Holland HD (2004) Primitive Os and 2316 Ma age for marine shale: implications for paleoproterozoic glacial events and the rise of atmospheric oxygen. Earth Planet Sci Lett 225:43–52

Helmens KF, Räsänen M, Johansson PW, Jungner H, Korjonen K (2000) The last Interglacial-Glacial cycle in NE Fennoscandia: a nearly continuous record from Sokli (Finnish Lapland). Quat Sci Rev 19:1605–1623

Huybrechts P (2002) Sea-level changes at the LGM from ice-dynamic reconstructions of the Greenland and Antarctic ice sheets during the glacial cycles. Quat Sci Rev 21:203–231

Hyvärinen H, Alhonen P (1994) Holocene lake-level changes in the Fennoscandian tree-line region, western Finnish Lapland: diatom and cladoceran evidence. Holocene 4:251–258

Imbrie J, Hays JD, Martinson GD, McIntyre A, Mix AC, Morley JJ, Pisias NG, Prell WL, Shackleton N (1984) The orbital theory of Pleistocene climate: support from a revised chronology of the marine ^{18}O record. In: Berger A (ed) Milankovitch and climate. Reidel, Dordrecht, pp 269–305

Isarin RFB (1998) Permafrost distribution and temperatures in Europe during the younger Dryas. Permafr Periglac Process 8:313–333

Isarin RFB, Bohncke SJP (1999) Mean July temperatures during the younger Dryas in northwestern and central Europe as inferred from climate indicator plant species. Quat Res 51:158–173

Johnsen SJ, Dahl-Jensen D, Gundestrup N, Steffensen JP, Clausen HB, Miller H, Masson-Delmotte V, Sveinbjörnsdotter AE, White J (2001) Oxygen isotope and palaeotemperature records from six Greenland ice-core stations: camp century, Dye-3, GRIP, GISP, Renland and NorthGRIP. J Quat Sci 16:299–307

Jouzel J, Masson-Delmotte V, Cattani O, Dreyfus G, Falourd S, Hoffmann G, Minster B, Nouet J, Barnola JM, Chappellaz J, Fischer H, Gallet JC, Johnsen S, Leuenberger M, Loulergue L, Luethi D, Oerter H, Parrenin F, Raisbeck G, Raynaud D, Schilt A, Schwander J, Selmo E, Souchez R, Spahni R, Stauffer B, Steffensenm JP, Stenni B, Stocker TF, Tison JL, Werner M, Wolff EW (2007) Orbital and millennial antarctic climate variability over the past 800,000 years. Science 317:793–796

Karlén W (1988) Scandinavian glacial and climate fluctuations during the Holocene. Quat Sci Rev 7:199–209

Kaufman DS, Ager TA, Anderson NJ, Anderson PM, Andrews JT, Bartlein PJ, Brubaker LB, Coats LL, Cwynar LC, Duvall ML, Dyke AS, Edwards ME, Eisner WR, Gajewski K, Geirsdóttir A, Hu FS, Jennings AE, Kaplan MR, Kerwin MW, Lozhkin AV, MacDonald GM, Miller GH, Mock CJ, Oswald WW, Otto-Bliesner BL, Porinchu DF, Rühland K, Smol JP, Steig EJ, Wolfe BB (2004) Holocene thermal maximum in the western Arctic (0–180°W). Quat Sci Rev 23:529–560

Kirschvink JL (1992) Late Proterozoic low-latitude global glaciation: the snowball Earth. In: Schopf JW, Klein C (eds) The proterozoic biosphere: a multidisciplinary study. Cambridge University Press, Cambridge, pp 51–52

Lamb HH (1965) The early medieval warm epoch and its sequel. Palaeogeog Palaeoclim Palaeoecol 1:13–37

Lisiecki LE, Raymo ME (2005) A Pliocene-Pleistocene stack of 57 globally distributed benthic $\delta^{18}O$ records. Paleoceanography 20:PA1003

Lockwood M, Harrison RG, Woollings T, Solanki SK (2010) Are cold winters in Europe associated with low solar activity? Environ Res Lett 5, doi:10.1088/1748-9326/5/2/024001

Mäkilä M, Saarnisto M (2008) Carbon accumulation in Boreal peatlands during the Holocene. In: Strack M (ed) Peatlands and climate change. International Peat Society, Saarijärvi, pp 24–43

Mann ME, Bradley RS, Hughes MK (1998) Global-scale temperature patterns and climate forcing over the past six centuries. Nature 392:779–787

Mann ME, Bradley RS, Hughes MK (1999) Northern hemisphere temperatures during the past millennium: inferences, uncertainties, and limitations. Geophys Res Lett 26:759–762

Mann ME, Zhang Z, Rutherford S, Bradley RS, Hughes MK, Shindell D, Ammann C, Faluvegi G, Ni F (2009) Global signatures and dynamical origins of the Little Ice Age and Medieval Climate Anomaly. Science 326: 1256–1260

Marmo JS, Ojakangas RW (1984) Lower proterozoic glaciogenic deposits, eastern Finland. Geol Soc Am Bull 98:1055–1062

Mayewski PA, Rohling EE, Stager JC, Karlén W, Maasch KA, Meeker LD, Meyerson EA, Gasse F, van Kreveld S, Holmgren K, Lee-Thorp J, Rosqvist G, Rack F, Staubwasser M, Schneider RR, Steig EJ (2004) Holocene climate variability. Quat Res 62:243–255

Meese DA, Gow AJ, Grootes P, Mayewski PA, Ram M, Stuiver M, Taylor KC, Waddington ED, Zielinski GA (1994) The accumulation record from the GISP2 core as an indicator of climate change throughout the Holocene. Science 266:1680–1682

NGRIP Members (2004) High resolution climate record of the Northern hemisphere back into the last interglacial period. Nature 431:147–151

Milankovitch M (1941) Kanon der Erdbestrahlungen und seine Anwendung auf das Eiszeitenproblem. Royal Serb Sci Spec Publ 132, Section of Mathematical and Natural Sciences, 33, Belgrade

Moberg A, Sonechkin DM, Holmgren K, Datsenko NM, Karlén W (2005) Highly variable Northern hemisphere temperatures reconstructed from low- and high-resolution proxy data. Nature 433:613–617

Nesje A, Matthews JA, Dahl SO, Berrisford MS, Andersson C (2001) Holocene glacier fluctuations of Flatebreen and winter-precipitation changes in the Jostedalsbreen region, western Norway, based on glaciolacustrine sediment records. Holocene 11:267–280

Ojala AEK, Alenius T (2005) 10 000 years of interannual sedimentation recorded in the Lake Nautajärvi (Finland) clastic-organic varves. Palaeogeogr Palaeoclim Palaeoecol 219:285–302

Ojala AEK, Heinsalu A, Kauppila T, Alenius T, Saarnisto M (2008a) Characterizing changes in the sedimentary environment of a varved lake sediment record in southern central Finland around 8000 cal. yr BP. J Quat Sci 23:765–775

Ojala AEK, Alenius T, Seppä H, Giesecke T (2008b) Integrated varve and pollen-based temperature reconstruction from Finland: evidence for Holocene seasonal temperature patterns at high latitudes. Holocene 18:529–538

Rapp D (2009) Ice ages and interglacials, measurements, interpretations and models. Praxis Publishing LDT, Chichester

Renssen H, Seppä H, Heiri O, Roche DM, Goosse H, Fichefet T (2009) The spatial and temporal complexity of the Holocene thermal maximum. Nat Geosci. doi:10.1038/NGEO513

Ruddiman WF (2003) The anthropogenic greenhouse era began thousands of years ago. Climatic Change 61:261–293

Saarnisto M, Lunkka JP (2004) Climate variability during the last interglacial-glacial cycle in NW Eurasia. In: Battarbee RW, Gasse F, Stickley CE (eds) Past climate variability through Europe and Africa. Developments in Paleoenvironmental research 6. Springer, Dordrecht, pp 443–464

Saarnisto M, Eriksson B, Hirvas H (1999) Tepsankumpu revisited – pollen evidence of stable Eemian climates in Finnish Lapland. Boreas 28:12–22

Self S, Rampino MR (1981) The 1883 eruption of Krakatau. Nature 294:699–704

Seppä H, Birks HJB, Giesecke T, Hammarlund D, Alenius T, Antonsson K, Bjune AE, Heikkilä M, MacDonald GM, Ojala AEK, Telford RJ, Veski S (2007) Spatial structure of the 8200 cal yr BP event in northern Europe. Clim Past 3:225–236

Seppä H, Bjune AE, Telford RJ, Birks HJB, Veski S (2009) Last nine-thousand years of temperature variability in Northern Europe. Clim Past 5:523–535

Shacleton NJ, Backman J, Zimmerman H, Kent DV, Hall MA, Roberts DG, Schnitker D, Baldauf JG, Desprairies A, Homrighausen R, Huddlestun P, Keene JB, Kaltenback AJ, Krumsiek KAO, Morton AC, Murray JW, Westberg-Smith J (1984) Oxygen isotope calibration of the onset of ice-rafting and history of glaciation in North Atlantic region. Nature 307:620–623

Snowball I, Zillén L, Gaillard M-J (2002) Rapid early-Holocene environmental changes in northern Sweden based on studies of two varved lake-sediment sequences. Holocene 12:7–16

Snowball IF, Korhola A, Briffa K, Koç N (2004) Holocene climate dynamics in Fennoscandia and North Atlantic. In: Battarbee RW, Gasse F, Stickley CE (eds) Developments in Paleoenvironmental research: climate variability in Europe and Africa. Springer, Dordrecht, pp 465–494

Stothers RB (1984) The great Tambora eruption in 1815 and its aftermath. Science 224:1191–1198

Stuiver M, Braziunas TF (1993) Sun, ocean, climate and atmospheric $^{14}CO_2$: an evaluation of causal and spectral relationship. Holocene 3:289–305

Svendsen JI, Alexanderson H, Astakhov VI, Demidov I, Dowdeswell JA, Funder S, Gataullin V, Henriksen M, Hjort C, Houmark-Nielsen M, Hubberten HW, Ingólfsson Ó, Jakobsson M, Kjær KH, Larsen E, Lokrantz H, Lunkka JP, Lyså A, Mangerud M, Matiouchkov A, Murray A, Möller P, Niessen F, Nikolskaya O, Polyak L, Saarnisto M, Siegert C, Siegert MJ, Spielhagen R, Stein R (2004) Late Quaternary ice sheet history of northern Eurasia. Quat Sci Rev 23:1229–1271

Tarasov PE, Peyron O, Guiot J, Brewer S, Volkova VS, Bezusko LG, Dorofeyuk NI, Kvavadze EV, Osipova IM, Panova NK (1999) Last glacial maximum climate of the former Soviet Union and Mongolia reconstructed from pollen and plant macrofossil data. Clim Dyn 15:227–240

Tiljander M, Saarnisto M, Ojala AEK, Saarinen T (2003) A 3000-year palaeoenvironmental record from annually laminated sediment of Lake Korttajärvi, central Finland. Boreas 26:566–577

Trouet V, Esper J, Graham NE, Baker A, Scourse JD, Frank DC (2009) Persistent positive North Atlantic Oscillation mode dominated the medieval climate anomaly. Science 324:78–80

Väliranta M, Kultti S, Seppä H (2006) Vegetation dynamics during the younger Dryas-Holocene transition in the extreme northern taiga zone, northeastern European Russia. Boreas 35:202–212

Väliranta M, Birks HH, Helmens K, Engels S, Piirainen M (2009) Early Weichselian interstadial (MIS 5c) summer temperatures were higher than today in northern Fennoscandia. Quat Sci Rev 28:777–782

Veski S, Seppä H, Ojala AEK (2004) Cold event at 8200 yr B.P. recorded in annually laminated lake sediments in eastern Europe. Geology 32:681–684

Walker MJC, Bohncke SJP, Coope GR, O'Connell M, Usinger H, Verbruggenet C (1994) The Devensian/Weichselian late-glacial in Northwest Europe (Ireland, Britain, North Belgium, The Netherlands, Northwest Germany). J Quat Sci 9:109–118

Wanner H, Beer J, Bütikofer J, Crowley TJ, Cubasch U, Flückiger J, Goosse H, Grosjean M, Joos F, Kaplan JO, Küttel M, Müller SA, Prentice C, Solomina O, Stocker TF, Tarasov P, Wagner M, Widmann M (2008) Mid- to late Holocene climate change: an overview. Quat Sci Rev 27:1791–1828

Zachos J, Pagani M, Sloan L, Thomas E, Billups K (2001) Trends, rhythms, and aberrations in global climate 65 Ma to present. Science 292:686–693

Chapter 15
Aerosols and Climate Change

Markku Kulmala, Ilona Riipinen, and Veli-Matti Kerminen

15.1 Short Introduction to Aerosol Particles

Each cubic centimeter of atmospheric air contains typically hundreds or thousands of small liquid or solid particles. The sizes of these aerosol particles range from a few nanometers to hundreds of micrometers. They originate from both natural and anthropogenic sources, such as volcanoes, combustion processes, deserts and biota (Pöschl 2005). The atmospheric life times and the effects of aerosol particles on atmospheric chemistry depend on their concentration, composition and size.

Atmospheric aerosol populations can be characterized by their size distribution – e.g. the mass or number concentrations of particles as a function of particle size. The number size distribution is usually dominated by sub-1 μm particles, while the large end of the size distribution often contributes significantly to the mass size distribution. The aerosol size distribution is typically divided to four modes: nucleation (3–20 nm in diameter), Aitken (20–100 nm), accumulation (100 nm–1 μm) and coarse (>1 μm) mode. Recently we have managed to observe also smaller particles or atmospheric clusters, in the size range 1–3 nm (Kulmala et al. 2007). The concentration and composition of atmospheric aerosol particles are highly variable depending on time and environment. The aerosol size distributions at clean marine environments are dominated by particles consisting of sea salt and sulfate that originate from emissions from the ocean. At urban environments, on the other hand, a large fraction of aerosols originates from anthropogenic activities

---

M. Kulmala (✉) • I. Riipinen
Department of Physics, Univesity of Helsinki, P.O. Box 64 (Gustaf Hällströminkatu 2), FI-00014 Helsinki, Finland
e-mail: Markku.Kulmala@helsinki.fi; ilona.riipinen@helsinki.fi

V.-M. Kerminen
Finnish Meteorological Institute, P.O. Box 503, FI-00101 Helsinki, Finland
e-mail: veli-matti.kerminen@fmi.fi

such as traffic and combustion processes. These include soot, sulfate, organics and road dust particles in the coarse mode. The air at remote continental sites is often dominated by particles emitted by natural sources such as the vegetation or ground.

Atmospheric aerosol particles are in constant interaction with the gas phase. During the recent years it has become evident that a large fraction of atmospheric particulate matter has been formed in the atmosphere by condensation of vapours, and dynamically alternates between the gaseous and condensed phases, depending on atmospheric conditions. Presenting these dynamic interactions between aerosol particles and gas phase compounds in an accurate way is a major challenge for current computational climate models.

15.2 Aerosol Particles and Radiative Balance

The atmospheric aerosol particles affect the quality of our life in many ways. First, they influence the Earth's radiation balance directly by scattering and absorbing solar radiation, and indirectly by acting as cloud condensation nuclei (CCN). The interaction between atmospheric aerosols and climate system is a dominant uncertainty in estimating the present-day radiative forcing, hindering seriously our ability to predict the future climate change (Forster et al. 2007; Myhre 2009; Quaas et al. 2009; Schwartz et al. 2010). The effects of aerosols on temperature and precipitation patterns can be seen mainly over regional and continental scales. Second, aerosol particles deteriorate both human health and visibility in urban areas (Pope and Dockery 2006; Hand and Malm 2007; Anderson 2009). Third, aerosol particles modify the intensity and distribution of radiation that reaches the earth surface, having direct influences on the vegetation and its interaction with the carbon cycle (Gu et al. 2002; Wang et al. 2008). Human actions, including emission regulations, forest management and land use change, along with various natural feedback mechanisms have substantial impacts on the complicated couplings between atmospheric aerosols, trace gases, air quality and climate (Brasseur and Roeckner 2005; Arneth et al. 2009; Jacob and Winner 2009; Raes et al. 2010). A better understanding of the various effects in the atmosphere requires detailed information on how different sources and atmospheric transformation processes modify the properties of aerosol particles and the concentrations of trace gases.

One example of how aerosol particles affect the regional climate and hydrological cycle is the deposition of soot particles on mountain tops and glaciers, where the temperature is well below zero. The carbon-containing soot particles on the snow surface absorb solar radiation, increase temperature locally and enhance the melting of the snow, subsequently affecting the planetary albedo (Flanner et al. 2007).On the other hand, increase in the aerosol particle number concentration can cause the cloud droplet concentration to increase as well. Consequently, the clouds will survive longer in the atmosphere and be optically thicker (Quaas et al. 2009). Both phenomena will cool the climate.

It has been estimated that the global average impact of aerosol particles has been cooling during the industrial period (Shindel and Faluvegi 2009; Schwartz et al. 2010). However, the exact radiative effect of aerosols is very difficult to estimate. This is due to the facts that (1) the climate effects of aerosols depend on the aerosol concentration but also on their size and chemical composition, (2) the behavior of aerosol size distribution is very non-linear, and it is practically impossible to make a difference between anthropogenic and natural contribution, and (3) the scientific understanding of the interactions between aerosols and clouds is incomplete.

The uncertainties related to radiative effects of aerosol particles make it difficult to test climate models. However, today, thanks to the increasing understanding on aerosol processes, aerosols can be included in global climate models much more reliably than 5 years ago. On the other hand, the atmospheric life times of aerosol particles typically range from days to weeks and are therefore much shorter than the life times of green house gases, particularly CO_2. This means that in future, the aerosol concentrations probably cannot keep up with the rising levels of CO_2 and other greenhouse gases, and their cooling effect on climate will smoothly lose the game – particularly taking into account the societal pressure to reduce aerosol emissions due to their air quality impacts. Therefore the general future trend of the climate will very likely remain warming, although several open questions still remain.

The effect of natural aerosol emissions is one of the open questions in current climate science. For instance, Charlson et al. (1987) proposed that the sulfur emissions from oceans will increase in warming climate, resulting in increased aerosol concentrations over the oceans – thus compensating for the warming effect of CO_2. This possible thermostat has been studied during the last 25 years, and it still remains open. Recently Kulmala et al. (2004a) proposed that biosphere, particularly forest, is able to produce increased number of fresh particles when climate is warming. Although these feedbacks involving oceans and forests may give us more time to adopt to a new climate, they will not able to prevent climate from changing.

As a summary one can say that there are good and bad aerosol particles from the climate perspective. Examples of the good particles are fine particles produced by the biosphere and sea salt particles emitted by the oceans, since these particles tend to cool the climate system via their direct scattering and their indirect cloud effects. Many types of soot particles produced by combustion processes could be viewed as bad because they tend to warm the atmosphere via absorption of solar radiation (Kopp and Mauzerall 2010).

15.3 Formation and Growth of Aerosol Particles

One of the key phenomena associated with the atmospheric aerosol system is the formation of new atmospheric aerosol particles from condensable vapours. Atmospheric aerosol formation consists of a complicated set of processes including the

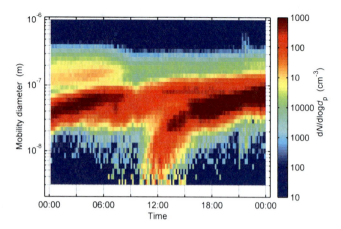

Fig. 15.1 A typical regional scale nucleation event, as recorded at the SMEAR II field station (see Hari and Kulmala (2005) for details) in Hyytiälä, southern Finland on July 5th 2006. The *color scale* indicates the aerosol concentration

production of nanometer-size clusters from vapours, the growth of these clusters to detectable sizes and the removal of growing clusters by coagulation with the pre-existing aerosol particle population (Kulmala 2003). Gas-to-aerosol conversion is one of the most important, yet still poorly understood, sources of atmospheric particles, and will thus be the focus of this subsection.

Direct evidence of aerosol formation has been observed in various environments in the continental boundary layer (Kulmala et al. 2004b). Figure 15.1 depicts a particle formation event measured at the SMEAR II station in Hyytiälä, Finland. During such events, particle number concentrations increase over areas of hundreds to thousands of square kilometers. As seen from Fig. 15.1, the formation of new particles and their growth continuous relatively smoothly for several hours. In that day the formation of new particles starts in the morning and the growth continues during the whole day. Before midnight particles have grown to 20–30 nm in diameter. Similar regional nucleation events have been observed all around the world (Fig. 15.2).

At the SMEAR II station in Hyytiälä, there are several new particle formation days per year (Fig. 15.3). The number of event days varies between 60 and 120 (Dal Maso et al. 2005). Annual variation depends mainly on meteorological conditions. Although the record shows a hint of a decadal peridiocity similar in length to the solar cycle, this may be a coincidence. The yearly variation of the nucleation frequency cannot be explained by cosmic ray induced ionization (Kulmala et al. 2010).

Seasonal variation is shown in Fig. 15.4. The frequency of nucleation events is highest in spring, with another maximum in September. Very probably these maxima are related to condensable vapour concentrations or changes in ecosystem functioning. The winter time minimum is related to lack of solar radiation.

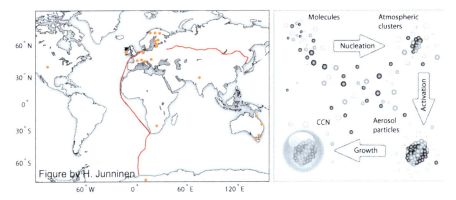

Fig. 15.2 Schematic picture of processes related to new particle formation and their growth to CCN (*right panel*, see also Kulmala 2003), and a map displaying the locations at which our atmospheric cluster measurements have been conducted so far – including a cruise from European coast to Antarctica and a train journey along the trans-Siberian railroad (see Kulmala and Tammet 2007)

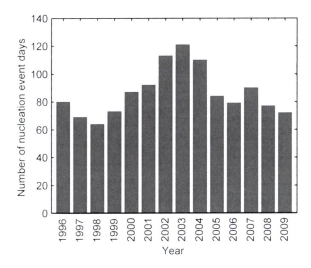

Fig. 15.3 Annual number of nucleation event days at SMEAR II station Hyytiälä, Finland

15.4 Forests as Source of Small Aerosol Particles

Forests, particularly Boreal forests, have been shown to be important aerosol sources and carbon sinks, and therefore they play an important role in the ecosystem-climate feedbacks. Figure 15.5 shows a schematic of one potential coupling between processes in forest ecosystems (photosynthesis and biogenic volatile organic compound, BVOC, emissions), aerosols and climate (Kulmala et al. 2004a). Photosynthesis drives plant gross primary production (GPP), which together with the total ecosystem respiration (TER) determines the net exchange of

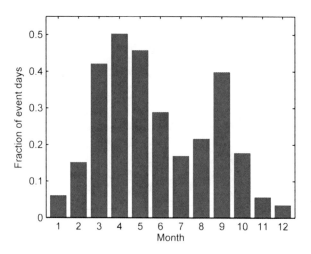

Fig. 15.4 Median seasonal variation of nucleation events in Hyytälä, Finland, during years 1996–2009

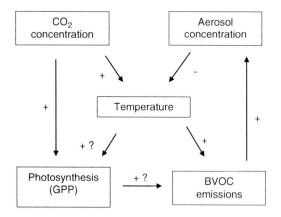

Fig. 15.5 Schematic figure of coupling of atmospheric CO_2 concentration, assimilation of carbon by photosynthesis (ecosystem gross primary production GPP), emission of biogenic volatile organic compounds (BVOCs), and aerosol particle concentration with atmospheric temperature (Kulmala et al. 2004a). Increased CO_2 concentration will increase temperature (+) and photosynthesis (+). Increased temperature will enhance BVOC emissions (+) and probably also photosynthesis (+?). Increased photosynthesis may enhance BVOC emissions (+?). Increased BVOC emissions will enhance aerosol formation and growth and therefore also enhance CCN concentrations (+). Enhanced aerosol and CCN concentrations will decrease temperature (−) due to increased reflection of sunlight from low clouds back to space. In addition, they increase diffuse radiation, which has a positive influence on photosynthesis (Gu et al. 2002)

CO_2 between the atmosphere and the ecosystem. In the boreal zone, photosynthesis occurs predominantly in sunlight during the growing season and is inhibited in winter. Forest ecosystems are usually sinks of CO_2, and a direct negative feedback (the higher the CO_2 concentration, the higher the rate of photosynthesis) exists between increasing atmospheric CO_2 concentrations and photosynthesis. On the other hand, a positive feedback exists between ecosystem respiration and temperature. With higher temperatures water becomes more important factor influencing both GPP and TER. Terrestrial vegetation contributes substantially to emissions of a variety of BVOCs, possibly as side products of photosynthesis (the purpose of the VOC emissions from the perspective of plant life is still unclear and may vary between plant species). Newly-formed aerosol particles in forested areas have been found to contain large amounts of organic material (e.g. Tunved et al. 2006). The ratio of BVOC emission to carbon assimilation is generally a few percent (Grace and Rayment 2000), and if increased CO_2 concentrations enhance photosynthesis, formation and emissions of several BVOCs may increase and possibly modify the aerosol particle formation routes.

The mechanism described above linking photosynthesis and aerosol forcing in a novel way is only one among other connections and feedbacks. The climatic and other effects of aerosols and trace gases are coupled together via both human actions, such as emission regulations and land use change, and via various natural feedback mechanisms involving the biosphere and atmosphere. For example, reduced aerosol loads due to actions aiming to improve air quality already seem to have affected climate in certain regions of the world (Brasseur and Roeckner 2005). Forestation enhances carbon sinks but at the same it decreases the surface albedo of the Earth, which tends to compensate for the effect of carbon sink feedback (Gibbard et al. 2005). It is also very probable that some new feedbacks or new pathways for components of biogeochemical cycling may appear in the future (e.g., Keppler et al. 2006).

References

Anderson HR (2009) Air pollution and mortality: a history. Atmos Environ 43:142–152
Arneth A, Unger N, Kulmala M, Andreae MO (2009) Clean the air, heat the planet. Science 326:672–673
Brasseur GP, Roeckner E (2005) Impact of improved air quality on the future evolution of climate. Geophys Res Lett 32. doi:10.1029/2005GL023902
Charlson RJ, Lovelock JE, Andreas MD, Warren SG (1987) Oceanic phytoplankton, atmospheric sulphur, cloud albedo and climate. Nature 326:655–661
Dal Maso M, Kulmala M, Riipinen I, Wagner R, Hussein T, Aalto PP, Lehtinen KEJ (2005) Formation and growth rates of fresh atmospheric aerosols: eight years of aerosol size distribution data from SMEARII, Hyytiälä, Finland. Boreal Environ Res 10:323–336
Flanner MG, Zender CS, Randerson JT, Rasch PJ (2007) Present-day climate forcing and response from black carbon in snow. J Geophys Res 112:D11202. doi:10.1029/2006JD008003
Forster P, Ramaswamy V, Artaxo P, Berntsen T, Betts R, Fahey DW, Haywood J, Lean J, Lowe DC, Myhre G, Nganga J, Prinn R, Raga G, Schulz M, Van Dorland R (2007) Cheanges in atmospheric constituents and in radiative forcing. In: Solomon S, Qin D, Manning M, Chen Z, Marquis M, Averyt KB, Tignor M, Miller HL (eds) Climate change 2007: the physical science basis. Contribution of working group I to the fourth assessment report of the intergovernmental panel on climate change. Cambridge University Press, Cambridge and New York
Gibbard S, Caldeira K, Bala G, Phillips TJ, Wickett M (2005) Geophys Res Lett 32:L23705. doi:10.1029/2005GL024550
Grace J, Rayment M (2000) Respiration in the balance. Nature 404:819–820

Gu L, Baldocchi SB, Verma TA, Black T, Vesala T, Falge EM, Dowty PR (2002) Advantages of diffuse radiation for terrestrial ecosystem productivity. J Geophys Res 107:4050. doi:10.1029/2001JD001242

Hand JL, Malm WC (2007) Review of aerosol mass scattering efficiencies from ground-based measurements since 1990. J Geophys Res 112:D16203. doi:10.1029/2007JD008484

Hari P, Kulmala M (2005) Station for measuring ecosystem-atmosphere relations (SMEAR II). Boreal Environ Res 10:315–322

Jacob D, Winner DA (2009) Effect of climate change on air quality. Atmos Environ 43:51–63

Keppler F, Hamilton JGT, Brass M, Röckmann T (2006) Methane emissions from terrestrial plants under aerobic conditions. Nature 439:187–191

Kopp RE, Mauzerall DL (2010) Assessing the climate benefits of black carbon mitigation. Proc Natl Acad Sci 107:11703–11708. doi:10.1073/pnas.0909605107

Kulmala M (2003) How particles nucleate and grow? Science 302:1000–1001

Kulmala M, Tammet H (2007) Finnish-Estonian air ion and aerosol workshops. Boreal Environ Res 12:237–245

Kulmala M, Suni T, Lehtinen KEJ, Dal Maso M, Boy M, Reissell A, Rannik Ü, Aalto PP, Keronen P, Hakola H, Bäck J, Hoffmann T, Vesala T, Hari P (2004a) A new feedback mechanism linking forests, aerosols, and climate. Atmos Chem Phys 4:557–562

Kulmala M, Vehkamäki H, Petäjä T, Dal Maso M, Lauri A, Kerminen V-M, Birmili W, McMurry P (2004b) Formation and growth rates of ultrafine atmospheric particles: a review of observations. J Aerosol Sci 35:143–176

Kulmala M, Riipinen I, Sipilä M, Manninen H, Petäjä T, Junninen H, Dal Maso M, Mordas G, Mirme A, Vana M, Hirsikko A, Laakso L, Harrison RM, Hanson I, Leung C, Palmer R, Lehtinen KEJ, Kerminen V-M (2007) Towards direct measurement of atmospheric nucleation. Science 318:89–92

Kulmala M, Riipinen I, Nieminen T, Hulkkonen M, Sogacheva L, Manninen HE, Paasonen P, Petäjä T, Dal Maso M, Aalto PP, Viljanen A, Usoskin I, Vainio R, Mirme S, Mirme A, Minikin A, Petzold A, Hõrrak U, Plaß-Dülmer C, Birmili W, Kerminen V-M (2010) Atmospheric data over a solar cycle: no connection between galactic cosmic rays and new particle formation. Atmos Chem Phys 10: 1885–1898.

Myhre G (2009) Consistency between satellite derived and modeled estimates of the direct aerosol effect. Science 325:187–190

Pope CA, Dockery DW (2006) Health effects of fine particulate air pollution: lines that connect. J Air Waste Manage Assoc 56:709–742

Pöschl U (2005) Atmospheric aerosols: composition, transformation, climate and health effects. Angew Chem Int Ed 44:7520–7540

Quaas J, Ming Y, Menon S, Takemura T, Wang M, Penner JE, Gettelman A, Lohmann U, Bellouin N, Boucher O, Sayer AM, Thomas GE, McComiskey A, Feingold G, Hoose C, Kristjánsson JE, Liu X, Balkanski Y, Donner LJ, Ginoux PA, Stier P, Grandey B, Feichter J, Sednev I, Bauer SE, Koch D, Grainger RG, Kirkevåg A, Iversen T, Seland Ø, Easter R, Ghan SJ, Rasch PJ, Morrison H, Lamarque J-F, Iacono MJ, Kinne S, Schulz M (2009) Aerosol indirect effects – general circulation model intercomparison and evaluation with satellite data. Atmos Chem Phys 9:8697–8717

Raes F, Liao H, Chen W-T, Seinfeld JH (2010) Atmospheric chemistry-climate feedbacks. J Geophys Res 115:D12121. doi:10.1029/2009JD013300

Schwartz SE, Charlson RJ, Kahn RA, Ogren JA, Rodhe H (2010) Why hasn't Earth warmed as much as expected? J Climate 23:2453–2464

Shindel D, Faluvegi G (2009) Climate response to regional radiative forcing during the twentieth century. Nat Geosci 2:294–300

Tunved P, Hansson H-C, Kerminen V-M, Ström J, Dal Maso M, Lihavainen H, Viisanen Y, Aalto PP, Komppula M, Kulmala M (2006) High natural aerosol loading over boreal forests. Science 312:261–263

Wang K, Dickinson RE, Liang S (2008) Observational evidence on the effects of clouds and aerosols on net ecosystem exchange and evapotranspiration. Geophys Res Lett 35:L10401. doi:10.10292008GL034167

Chapter 16
Enhanced Greenhouse Effect and Climate Change in Northern Europe

Jouni Räisänen

16.1 The Natural and the Enhanced Greenhouse Effect

The phenomenon known as the atmospheric greenhouse effect has prevailed on the Earth for thousands of millions of years. Solar radiation energy, about a half of which is visible light, penetrates the atmosphere quite well particularly in cloud-free conditions. By contrast, a large fraction of the longer-wavelength thermal radiation emitted by the Earth's surface is absorbed in the atmosphere. Without this phenomenon, which is known as the atmospheric greenhouse effect, the global mean surface air temperature would be only about $-18°C$, in stark contrast with the observed value of $+14°C$. The greenhouse effect is due to so-called greenhouse gases, which occur in relatively small concentrations but absorb thermal radiation very efficiently. The most important of these is water vapor (H_2O) (that is gaseous water, not clouds), the second carbon dioxide (CO_2). Others include, for example, ozone (O_3), methane (CH_4) and nitrous oxide (NO_2) (Kiehl and Trenberth 1997).

Because of human activities, the atmospheric greenhouse effect is now being rapidly enhanced. The single largest cause of this is due to emissions of carbon dioxide, about 85% of which are currently caused by the use of fossil fuels such as coal, oil and natural gas (Le Quéré et al. 2009). After the industrial revolution in the eighteenth century, the average CO_2 concentration has increased from 280 to nearly 390 parts per million (ppm) in volume, with two thirds of this increase in the past 50 years and a current growth rate of about 2 ppm per year. The concentrations of other greenhouse gases such as methane, nitrous oxide and (in the lower atmosphere) ozone have also increased.

Observations also show an increase in atmospheric water vapor (Trenberth et al. 2007). This is not due to anthropogenic emissions of this gas, which are tiny in

---

J. Räisänen (✉)
Department of Physics, Division of Atmospheric Sciences, University of Helsinki, P.O. Box 64, FI-00014 Helsinki, Finland
e-mail: jouni.raisanen@helsinki.fi

comparison with the natural water cycle, but is rather a consequence of the ongoing warming. The maximum amount of water vapor that air can hold in gaseous form, before condensation to cloud droplets begins, increases strongly with temperature. This is an important example of a positive feedback that tends to amplify temperature changes, regardless of if they are initially forced by increases in greenhouse gases other than H_2O or, for example, by a volcanic eruption (Soden et al. 2002).

The globally averaged surface air temperature has increased by about 0.7°C since the beginning of the twentieth century (Trenberth et al. 2007). A large part of this increase is attributable to anthropogenic forcing, particularly the increase in greenhouse gases (Hegerl et al. 2007). Were it not for the cooling influence of increasing aerosol concentrations (Kulmala et al. 2012, this volume) that has partly counteracted the enhanced greenhouse effect, the rate of warming would probably have been even higher.

As discussed by Ojala (2012, this volume), climate also varies for natural causes. Part of the natural variability on inter-annual to centennial time scales is forced by variations in solar and volcanic activity, part is generated by the internal dynamics and interaction of the atmosphere and the ocean. However, when considering global climate change during the rest of this century or so, these natural factors are likely to be dwarfed by continuing anthropogenic changes in the atmospheric composition.

It is obviously a difficult task to predict greenhouse gas and aerosol-producing emissions far into the future. Therefore, scientists have constructed a number of alternative emissions scenarios, building on different assumptions on factors such as population growth and economical and technical development in the world (Nakićenović et al. 2000). As an example, six different scenarios for the development of CO_2 emissions during this century are shown in Fig. 16.1a.

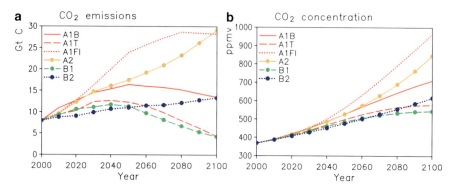

Fig. 16.1 Development of carbon dioxide (**a**) emissions and (**b**) concentration in the atmosphere under the twenty-first century according to six so-called SRES emission scenarios (Nakićenović et al. 2000; Houghton et al. 2001). The scenarios with the largest emissions (A2 and A1FI) represent a world with strongly increasing energy demand and continuing high reliance on fossil fuels, whereas the decrease in emissions in the B1 and A1T scenarios reflects both more efficient usage of energy and a shift toward non-fossil energy sources. See text for more details

For the next few decades, all scenarios indicate an increase in CO_2 emissions. The continued growth of the human population together with an expected increase in the average standard of living will nearly certainly be accompanied by increasing energy consumption. As far as fossil fuels serve for a lion's share (currently, about 80%) of the global energy production, this will also lead to an increase in CO_2 emissions. Towards the end of the century, the scenarios diverge. Some of them indicate a continued growth of greenhouse gas emissions driven by the increasing energy demand of the world, but others point towards a gradual decline in emissions allowed by reduced population growth, alternative energy sources and a shift towards a less energy-intensive economy. Nevertheless, the concentrations of the most long-lived greenhouse gases, particularly CO_2, keep rising even in those scenarios in which the emissions are reduced. Even under the so-called B1 scenario, which is the most optimistic in the group, the CO_2 concentration would increase from the present 390 ppm to about 540 ppm by the year 2100. For scenarios with larger emissions, the increase in atmospheric CO_2 concentration is naturally higher, with values up to over 900 ppm being possible in the end of this century.

In contrast to CO_2 and some other greenhouse gases, the concentrations of atmospheric aerosols are unlikely to increase much in the future. First, the negative environmental consequences of aerosols (acid rain, deteriorated air quality with adverse effects on human health) give a strong incentive for reducing the emissions. In Europe and North America, the emissions have already decreased during the past few decades, and the same is expected to happen in the developing world later in this century. Second, the lifetime of aerosol particles is short, of the order of 1 week. A decrease in the emissions would therefore lead to a virtually instantaneous decrease in aerosol concentrations. Consequently, greenhouse-gas induced warming is projected to become more and more dominant over aerosol-induced cooling.

16.2 Twenty-First Century Changes in the Global Climate

To find out how the projected changes in atmospheric composition are likely to affect climate, global climate models (GCMs) are used. These computer models attempt to describe, as accurately as possible, the physical processes that affect weather and climate in the atmosphere and in the ocean. With a time step of tens of minutes, they simulate the irregular fluctuation of atmospheric conditions, which can be aggregated to long-term mean values and other statistics that tell of the evolution of climate. Unfortunately, all the relevant phenomena cannot be simulated realistically in such models. For example, clouds in the real atmosphere generally have a scale of a few kilometers only, whereas the computing grid in most current GCMs has a resolution of 100–400 km. Thus, the models do not resolve individual clouds. What they do try to achieve is to estimate the total amount of cloudiness within a model grid box, but this can only be done approximately. Different models use different approximations when treating clouds and other small-scale phenomena. As a result, they also react differently to the forcing imposed by increasing greenhouse gas concentrations.

Anthropogenic climate change is often referred to simply as *global warming* (Harvey 2000; Houghton 2004). This refers to the expectation that increased greenhouse gas concentrations will warm up the climate practically everywhere in the world. Even for the B1 scenario with the smallest assumed greenhouse gas emissions, climate model simulations suggest a globally averaged warming of 1.1–2.9°C during this century (Fig. 16.2). The corresponding range for the A1FI scenario with the largest emissions is 2.4–6.4°C. These numbers reveal a considerable uncertainty in the magnitude of future global warming, both because of the unknown evolution of greenhouse gas and aerosol emissions and because of the unknown sensitivity of climate to these emissions. However, even a global mean warming of 1.1°C would be 50% larger than the warming experienced during the past century, whereas a warming of 6.4°C would mean a drastic change in the global climate. To put 6.4°C in context, the global mean temperature during the coldest phase of the last ice age about 20,000 years ago is estimated to have been 4–7°C lower than at present (Jansen et al. 2007).

Climate changes are not expected to be geographically uniform. One kind of "best estimates" of the annual mean temperature and precipitation changes that might take place during this century are given in Fig. 16.3. The climate changes in these maps are averaged over the results of 20 different GCMs (see Meehl et al. 2007b for documentation of the data set), with all the models forced with the "midrange" SRES A1B emissions scenario (see Figs. 16.1 and 16.2).

This averaging gives a global mean warming of about 2.6°C, which is in the midway of the wide range of possibilities indicated above. The simulated warming over nearly all land areas exceeds the global average, indicating that the global mean numbers actually understate the warming where humans live. The change over the oceans is more modest, with the important exception of the Arctic Ocean,

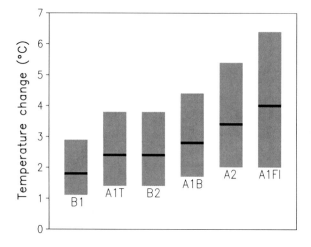

Fig. 16.2 Projected increase in the global mean temperature during this century (difference in mean temperature between the periods 2090–2099 and 1980–1999). The bars indicate the uncertainty range and the horizontal lines the best estimate of the warming for each of the six SRES emissions scenarios (based on numbers given by Meehl et al. 2007a)

a Temperature change (°C)

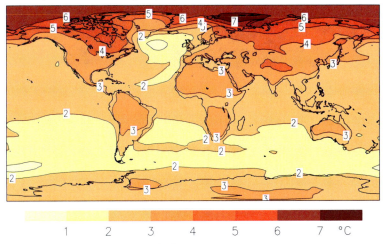

b Precipitation change (%)

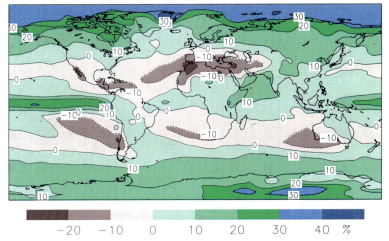

Fig. 16.3 "Best-estimate" changes in annual mean temperature (in °C) and precipitation (in percent) from the late twentieth century (mean of the years 1971–2000) to the late twenty-first century (mean of the years 2070–2099). The maps are based on the results of 20 climate models forced with the SRES A1B emissions scenario (see Meehl et al. 2007b for model documentation)

where the decrease and thinning of ice cover is projected to make the warming larger than anywhere else.

The amount of precipitation is projected to increase in high latitudes and in most of the equatorial tropics, but to decrease in the subtropical zones and to some extent in the lower mid-latitudes such as the Mediterranean area (Fig. 16.3b). Although not without exceptions, there is a tendency of dry regions to become even drier, and wet regions wetter, in a warmer climate.

16.3 Changes in Temperature and Precipitation in Northern Europe

A more detailed look at the projected climate change in the European area is given in Fig. 16.4. The annual mean warming in the Nordic countries varies from slightly less than 3°C to almost 5°C, increasing from southwest to northeast. As concretised

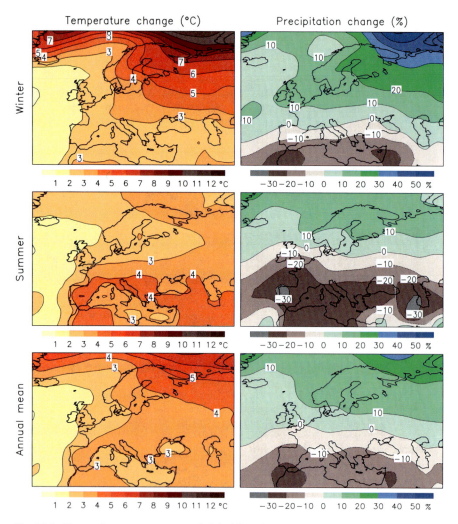

Fig. 16.4 Changes in mean temperature (*left*, in °C) and precipitation (*right*, in percent) from the late twentieth century (mean of the years 1971–2000) to the late twenty-first century (mean of the years 2070–2099) in Europe. From top to bottom winter (December-February), summer (June–August) and annual mean. The same model simulations were used as in Fig. 16.3

Table 16.1 Annual mean temperature at selected locations

Location	Coordinates	Observed 1971–2000 (°C)	Change (°C)	Scenario 2070–2099 (°C)
Utsjoki (Finland)	70°N, 27°E	−1.7	4.9	+3.2
Luleå (Sweden)	66°N, 22°E	+1.8	4.2	+6.0
Helsinki (Finland)	60°N, 25°E	+5.6	3.7	+9.3
Copenhagen (Denmark)	56°N, 13°E	+9.0	2.8	+11.8
Berlin (Germany)	52°N, 13°E	+9.7	2.9	+12.6
Paris (France)	49°N, 2°E	+12.1	2.7	+14.8

The observed values in 1971–2000 are based on data available from http://eca.knmi.nl/, and the change refers to the 20-model average warming from 1971–2000 to 2070–2099 under the SRES A1B scenario, as in Figs. 16.3 and 16.4. The resulting scenario for the mean temperature in 2070–2099 is shown in the last column

in Table 16.1, a warming of this magnitude would imply a dramatic northward shift in climatic zones. For example, the climate in Luleå in northern Sweden would become as mild as that in Helsinki at the south coast of Finland today, whereas Helsinki would become comparable with the present-day Copenhagen or Berlin, and the annual mean temperature in Copenhagen would approach that in Paris at present. However, geographic analogies like this should not be over-interpreted, as not all natural conditions will change together with temperature. For example, a warmer climate does not mean that more daylight would be available in high northern latitudes in winter.

The warming in northern Europe is likely to be largest in winter, when reduced snow and ice cover amplify climate changes in high latitudes (left column of Fig. 16.4). The warming in summer is expected to be smaller but still substantial, particularly when put in the context of the much smaller natural interannual variability of summer than winter temperatures in northern Europe. Further south in Europe, the seasonal cycle of the change is reversed, with the largest warming projected for the summer season in association with reduced precipitation.

Precipitation is projected to increase in northern Europe, with the largest increase in winter (right column of Fig. 16.4). Despite the increase in winter precipitation, the amount of snow is likely to decrease, because a larger fraction of precipitation comes as rain and melt episodes become more frequent in a warmer climate (Räisänen 2008a). For the summer, the models indicate a strong north-south contrast in precipitation change across the European region, with more precipitation in the north and less precipitation over central and southern Europe. However, the borderline between increasing and decreasing summer precipitation varies in location between the models, with some of them suggesting a slight decrease in summer precipitation as far north as in southern Finland.

As already noted, different models give different estimates of climate change even when forced with the same emissions scenario. As an example, annual mean temperature and precipitation changes in southern Finland in the 20 climate models used above are shown in Fig. 16.5. At this location, the simulated warming varies from 2°C to 6°C, and the increase in precipitation from just a few percent

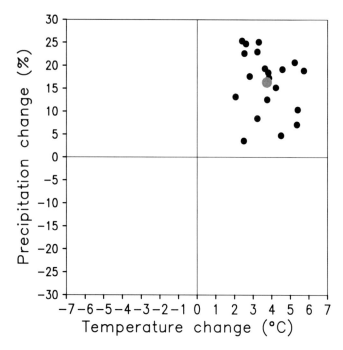

Fig. 16.5 Changes in annual mean temperature (in °C) and precipitation (in percent) in southern Finland (60°N, 25°E) from the period 1971–2000 to the period 2070–2099 in the 20 models used in Figs. 16.3 and 16.4. The larger *grey* dot shows the mean of the model results

to over 25%. Nevertheless, all 20 models agree on the sign of the changes. A claim frequently made in the public debate – that a greenhouse-gas-induced weakening of the North Atlantic ocean circulation would lead to a cooling of climate in northern Europe – is therefore not supported by these model simulations. In a few models, changes in ocean currents in fact lead to slight cooling to the south of Greenland, but this cooling is localized and does not extend to Europe.

16.4 The Interplay Between Greenhouse-Gas-Induced Climate Change and Natural Variability

Our focus this far has been on climate change to the end of the twenty-first century. A simple illustration of what might happen in the nearer future is given Fig. 16.6. In these diagrams, the first half of the time series up to the year 2000 shows the variation of temperature and precipitation as observed in Helsinki, Finland. For the twenty-first century, the interannual variability was assumed to be repeated as observed in the previous hundred years, but with linear trends of 4°C in temperature

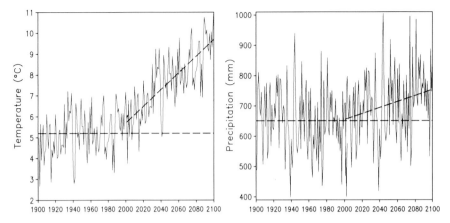

Fig. 16.6 Annual mean (**a**) temperature and (**b**) precipitation as observed in Helsinki, Finland, in the years 1901–2000, and a possible extension of the time series to the year 2100. The "forecast" assumes that, in the twenty-first century, the interannual variability would be repeated exactly as in the twentieth century, but the long-term average temperature and precipitation would increase by 0.4°C and by 1.5% per decade (*slanted dashed* lines, beginning in the year 2000 from the observed means of 1981–2000). The twentieth century mean temperature and precipitation are shown by horizontal dashed lines (Räisänen 2008b)

and 15% in precipitation superimposed. These assumed trends correspond approximately to the best-estimate long-term climate changes inferred from the 20 model simulations used above.

In reality, it is impossible to predict decades in advance, which individual years will be colder or warmer, or dryer or wetter, than their near neighbors. There is also substantial uncertainty in the magnitude of the long-term trends. Still, the figure gives a salient message that is not specific to this particular location but rather holds nearly everywhere in the world. In comparison with the natural interannual variability, the changes in temperature are much stronger than those in precipitation. If the forecast in Fig. 16.6 were realized, then the last year colder than the twentieth century mean would occur in Helsinki around the year 2040. By contrast, some individual years with below-average precipitation will still most likely occur in the end of this century, despite the projected gradual increase in the long-term mean. At lower latitudes, where inter-annual temperature variability is much smaller than it is in Finland but the variability of precipitation is at least equally large, the contrast between the two variables is even more pronounced.

16.5 Climate Change and the Frequency of Extremes

The winters 2007–2008 and 2009–2010 were dramatically different in northern Europe. The former was very mild, the latter cold and snowy. In Helsinki, Finland, the December–January–February mean temperature in winter 2007–2008 was

+1.3°C, the highest ever observed. By contrast, winter 2009–2010 with a mean temperature of −7.4°C was the coldest since 1987. How do such variations fit within the expected effects of the ongoing climate change?

It is impossible to directly link the weather conditions in an individual winter to global, long-term climate change. However, the issue can be approached from a probabilistic perspective. In a warming climate, mild winters are expected to become gradually more common and cold winters less common. In Fig. 16.7, the magnitude of this change is estimated using methods described in Räisänen and Ruokolainen (2008a, b).

The first curve in the figure presents the distribution of winter mean temperatures as observed in the years 1901–2005. The peak of the curve at −4°C indicates the most common level of winter temperatures during the 105-year period. The second curve, representing the estimated present-day climate, is shifted toward slightly milder temperatures, to take into account the climate change that has already occurred. A third, more tentative curve gives an estimate of the winter climate in Helsinki in the middle of this century, assuming that the warming of climate proceeds at the rate indicated by the average of model results.

The distribution for the observed climate suggests that a winter as mild as or milder than 2007–2008 should only have a probability of about 0.5% (this number is equal to the area below the curve for 1901–2005 and to the right of +1.3°C). Thus, a winter like this should only occur once in two centuries. The coldness of winter

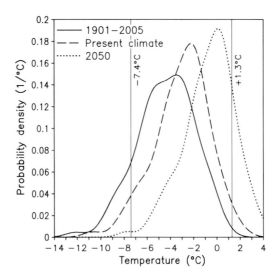

Fig. 16.7 Probability distributions of winter (December–January–February) mean temperature in Helsinki, Finland. The solid line shows the distribution observed in the years 1901–2005, the dashed line the distribution representing the actual present-day climate (around the year 2010), and the dotted line a projection for the distribution around the year 2050. The mean temperatures observed in the winters 2007–2008 (+1.3°C) and 2009–2010 (−7.4°C) are shown with vertical lines. The vertical axis gives the probability density, that is, the probability that the observed temperature falls within a given 1°C temperature increment

2009–2010 appears much less unusual against the long-term observed record, which gives a probability of about 13% (once in 8 years) for winters at least as cold ($-7.4°C$) as this one.

The situation changes when climate change is taken into account. For the actual present-day climate, the probability of winters at least as mild as 2007–2008 is estimated to be about 3%, which is six times more than the value derived directly from the observations. On the other hand, a winter mean temperature of $-7.4°C$ or colder is now estimated to have a probability of only 5–6%, less than a half of what was inferred from the observations for 1901–2005. Thus, the probability of occurrence of both very mild and cold winters is quite sensitive to even relatively small shifts in climate.

If climate change follows model predictions, this development will continue in the future. In the middle of the century, winters as mild as 2007–2008 are projected to have a probability of about 20%, thus repeating on the average twice in a decade. By contrast, winters similar to or colder than 2009–2010 seem to be facing a gradual extinction, with a probability of less than 1% in the middle of this century.

16.6 Impacts and Mitigation of Climate Change

The impacts on greenhouse-gas-induced climate change in northern Europe are unlikely to be catastrophic. In broad terms, climatic conditions in the area seem to be becoming more like the conditions currently observed in central Europe. Considering the world as a whole, however, the problems caused by anthropogenic global warming clearly seem to outweigh the benefits (Parry et al. 2007). The largest damages are anticipated in developing countries, but parts of the industrialized world such as the Mediterranean region may also be facing substantial adverse impacts. Even those areas where the direct consequences of climate change are modest will likely be affected indirectly, for example via disruptions in the global economy and by an increasing number of environmental refugees.

To reduce the rate and magnitude of future climate change, the international community has set the goal to reduce greenhouse gas emissions. The Kyoto Protocol obliges the industrialized world to reduce its emissions by 5% from the level of 1990 to the years 2008–2012. However, regardless of if this target will be reached, the Kyoto Protocol is insufficient to effectively mitigate climate change. The modest emission reductions required from the industrialized countries have been more than offset by the rapid growth of emissions in the developing world, particularly China. In fact, global carbon dioxide emissions from fossil fuel combustion and cement manufacture increased by about 41% from 1990 to 2008 (Le Quéré et al. 2009).

If the goal is, as it should be, to significantly mitigate anthropogenic climate change, much larger emission reductions and stronger political agreements will be necessary in the coming decades. This will be a tremendous challenge, not only technically but also because of the level of international co-operation required.

Perhaps the most difficult political question is how to make emission reduction targets acceptable for developing countries, which already produce a majority of the global emissions but have much smaller *per capita* emissions than the industrialized world. Progress on this issue seems unlikely unless industrial countries demonstrate a strong willingness to substantially reduce their own emissions.

References

Harvey LDD (2000) Global warming: the hard science. Prentice Hall, Harlow
Hegerl GC, Zwiers FW, Braconnot P, Gillett NP, Luo Y, Marengo Orsini JA, Nicholls N, Penner JE, Stott PA (2007) Understanding and attributing climate change. In: Solomon S, Qin D, Manning M, Marquis M, Averyt K, Tignor MMB, LeRoy Miller H Jr, Chen Z (eds) Climate change 2007: the physical science basis. Cambridge University Press, Cambridge/New York, pp 663–745
Houghton JT (2004) Global warming: the complete briefing, 3rd edn. Cambridge University Press, Cambridge/New York
Houghton JT, Ding Y, Griggs DJ, Noguer M, van der Linden PJ, Dai X, Maskell K, Johnson CA (eds) (2001) Climate change 2001: the scientific basis. Cambridge University Press, Cambridge/New York
Jansen E, Overpeck J, Briffa KR, Duplessy J-C, Joos F, Masson-Delmotte V, Olago D, Otto-Bliesner B, Peltier WR, Rahmstorf S, Ramesh R, Raynaud D, Rind D, Solomina O, Villalba R, Zhang D (2007) Palaeoclimate. In: Solomon S, Qin D, Manning M, Marquis M, Averyt K, Tignor MMB, LeRoy Miller H Jr, Chen Z (eds) Climate change 2007: the physical science basis. Cambridge University Press, Cambridge/New York, pp 433–497
Kiehl JT, Trenberth KE (1997) Earth's annual global mean energy budget. Bull Am Meteorol Soc 78:197–208
Kulmala MT, Riipinen I, Kerminen VM (2012) Aerols and climate change. In: Haapala I (ed) From the Earth's core to outer space. Lecture notes in Earth system sciences 137. Springer, Berlin/Heidelberg, pp 219–226
Le Quéré C, Raupach MR, Canadell J, Marland G et al (2009) Trends in the sources and sinks of carbon dioxide. Nat Geosci 2:831–836. doi:10.1038/ngeo689
Meehl GA, Stocker TF, Collins WD, Friedlingstein P, Gaye AT, Gregory JM, Kitoh A, Knutti R, Murphy JM, Noda A, Raper SCB, Watterson IG, Weaver AJ, Zhao Z-C (2007a) Global climate projections. In: Solomon S, Qin D, Manning M, Marquis M, Averyt K, Tignor MMB, LeRoy Miller H Jr, Chen Z (eds) Climate change 2007: the physical science basis. Cambridge University Press, Cambridge/New York, pp 747–845
Meehl GA, Covey C, Delworth T, Latif M, McAvaney B, Mitchell JFB, Stouffer RJ, Taylor KE (2007b) The WCRP CMIP3 multimodel dataset: a new era in climate change research. Bull Am Meteorol Soc 88:1383–1394
Nakićenović N et al (2000) Emissions scenarios. A special report of working group III of the intergovernmental panel on climate change. Cambridge University Press, Cambridge/New York
Ojala AEK (2012) The late Quaternary climate history Northern Europe. In: Haapala I (ed) From the Earth's core to outer space. Lecture notes in Earth system sciences 137. Springer, Berlin/Heidelberg, pp 199–218
Parry ML, Canziani OF, Palutikof JP, van der Linden PJ, Hanson CE (eds) (2007) Climate change 2007: climate change impacts, adaptation and vulnerability. Cambridge University Press, Cambridge/New York

Räisänen J (2008a) Warmer climate: less or more snow? Climate Dyn 30:307–319
Räisänen J (2008b) Kasvihuoneilmiö, ilmastonmuutos ja vaikutukset (Greenhouse effect, climate change and impacts: in Finnish). Lecture compendium, Department of Physics, University of Helsinki
Räisänen J, Ruokolainen L (2008a) Estimating present climate in a warming world: a model-based approach. Climate Dyn 31:573–585
Räisänen J, Ruokolainen L (2008b) Ongoing global warming and local warm extremes: a case study of winter 2006–2007 in Helsinki, Finland. Geophysica 44:45–65
Soden BJ, Wetherald RT, Stenchikov GL, Robock A (2002) Global cooling after the eruption of Mount Pinatubo: a test of climate feedback by water vapor. Science 296:727–730
Trenberth KE, Jones PD, Ambenje P, Bojariu R, Easterling D, Klein Tank A, Parker D, Rahimzadeh F, Renwick JA, Rusticucci M, Soden B, Zhai P (2007) Observations: surface and atmospheric climate change. In: Solomon S, Qin D, Manning M, Marquis M, Averyt K, Tignor MMB, LeRoy Miller H Jr, Chen Z (eds) Climate change 2007: the physical science basis. Cambridge University Press, Cambridge/New York, pp 235–336

Chapter 17
Will There Be Enough Water?

Esko Kuusisto

17.1 Introduction

Water is the most important natural resource on the Earth. It has two important properties, which all other resources are lacking:

- In a certain location, the amount of water varies considerably as a function of time. There is often a strong seasonal rhythm, but multiannual periodicities and trends are also possible. In addition, there is also a stochastic component always in the water-related time series, sometimes leading to wrong conclusions.
- Water moves unaided to new locations, crossing easily e.g. country borders. Ores and soils are not capable to do the same, neither do agricultural or silvicultural products have this ability. Game and fish naturally move by themselves but their migration does not normally have major consequences.

Time-domain change in water resources is often measured by coefficient of variation, i.e. the ratio of standard deviation to the average. In Finland the coefficient of variation of annual flows is typically 0.20–0.25, in Arctic rivers only 0.10–0.15, but in dry areas it may be larger than one.

In Finland, the ratio of the largest and smallest mean monthly flow is below two in watersheds with abundant lakes, but in excess of ten in many coastal rivers. Even the latter value is small when compared to the rivers of those climatic zones, where there is a distinct rainy season and dry season. In world's drylands, rivers may lose all the water for several months. Because of man's high water demand, major rivers like the Nile, Yellow River and Amy Darya may run dry in their lower reaches.

The number of river basins shared by at least two countries is over 250. They cover over one half of the global land area, inhabiting almost one half of the mankind. Fourteen river basins stretch to at least five countries, the most

E. Kuusisto (✉)
Finnish Environmental Institute, P.O. Box 140, FI-00251 Helsinki, Finland
e-mail: esko.kuusisto@ymparisto.fi

Fig 17.1 Egypt is fully dependent on external water resources, imported by the Nile. In addition to food production, water is used in Egypt to afforestation. However, it is not easy to convert a desert into a forest

international are the Danube (17), followed by the Nile and Niger (10 for both). Five countries receive over 90% of their water resources from abroad; Egypt, Turkmenistan, Mauritania, Hungary and Bulgaria (Fig. 17.1).

17.2 The Largest Material Cycle on the Earth

The hydrological cycle is the largest material cycle on the Earth. One measure for this cycle is the amount of renewable water resources on land areas, 41,000 km^3/year, of which 2,000 km^3 in frozen form. The annual worldwide water consumption is slightly over 4,000 km^3. Of this amount, the share of domestic water use is 8%, industrial 23% and agricultural 69%.

Is it possible to talk on water scarcity, when only 10% of the resource is in use? Yes, it is, mainly for two reasons:

- One half of the water resources is tied to floods, whose water volume cannot be stored. This water is mainly harmful to man, while it is valuable to many ecosystems.
- About one fourth is flowing in regions, where there are only a few people living.

When these "wasted waters" and ice flows are subtracted from the total, the renewable water resources are down to 9,000 km^3/a. Almost half of this amount is in use. The distribution of this exploitable resource in time and space also deviates from the expectations of mankind. For example, rain may fall in excess in wintertime, while the demand of water peaks in summer.

Rapid urbanization has increased the challenges for sufficiency of water. The urban half of the mankind has jammed to an area of the size of Finland, having local water resources of 50 L per capita per day. The rural half occupies an area 400 times the size of Finland and has over 30,000 L per capita per day available.

The water resources of different countries also range wildly. An Icelander has 1.6 million litres water per day, a man in Kuwait only 30 L. A Finn is born with a pretty big barrel in the armpit – 60,000 L. It is the largest in the EU, followed by Sweden and Slovenia (Kuusisto 1998).

The water use per capita also varies immensely between countries. According to the statistics by FAO, in Turkmenistan water is used at the average rate of 15,700 L per capita per day. This is due to inefficient, large-scale irrigation systems; domestic water consumption is well below 1% of the total. At the other end of the range lies Haiti, with the total use of 20 L per capita per day.

17.3 Climate Change: Water Will Be Redistributed

Climate change is going to have an impact on the total amount of world's water resources, but the impact on their areal distribution will be even more pronounced. The message of the Intergovernmental Panel on Climate Change (IPCC) is harsh; most areas with plentiful water resources will get more, those with scarcity will be losers. There are some exceptions, aerially largest being Sahara Desert, where some models suggest a runoff increase of 30–50% by the mid-twenty-first century. This sounds good but doesn't help much – the percentage change should be in excess of one thousand in Sahara to have a practical value.

If climate change is going to decrease the water resources of a country, this can be equated with the traditional forms of water use. Let a country have 110 km^3 of renewable water resources per year today, and let the climate change cause a reduction of 19% by the year 2050. The new hidden sector – climate change – will consume 21 km^3 of water in 2050, leaving only 89 km^3 to the other sectors.

Table 17.1 includes nine countries or areas with a total population of over half a billion. All of them have fairly high water consumption, particularly to irrigation. All of them will also be significant water losers in the future due to climate change. The example given above is the first country in the table, Spain.

In all countries described in Table 17.1, climate change will consume more water than the domestic water supply consumes today. Climate change will surpass the agricultural water consumption only in Turkey, where also the absolute reduction of water resources will be highest. On the other hand, the present water stress is lowest in Turkey, implying that the country is not as vulnerable to water losses as the others.

Which of the countries in Table 17.1 will be the most vulnerable? One candidate will be Spain, where the dry south tried to get help from the north, but the large-scale water transfer did not go ahead. It is probable that agricultural water use needs to be reduced in Spain in the future. New desalination plants will be constructed along the southern coasts of the country, but they will mainly relieve the thirst of a tourist than the thirst of a tomato seedling.

Table 17.1 Internal renewable water resources and sectoral water use in some countries

Country	Internal water resources (km³/a)	Degree of use today (%)	Use in different sectors (km³/a)			
			Domestic	Industrial	Agricultural	Climate change in 2050
Spain	110	33	4.3	9.4	22.3	20.9
Italy	159	37	8.3	16.2	34.8	19.1
Ukraina	53	49	4.7	13.5	7.8	5.8
Morocco	30	50	0.6	3.6	11.0	4.8
South Africa	45	38	2.9	1.9	12.2	4.5
Turkey	196	17	2.8	2.0	28.8	33.0
Iran	128	46	3.5	1.5	54.0	21.0
California	101	51	9.3	0.7	41.8	24.0
Mexico	357	25	4.8	7.2	77.5	28.6

Data are from different sources from recent years, the reductions due to climate change correspond to a mid-range scenario

Fig 17.2 A scanty rain-fed harvest can be produced in some parts of the Atlas mountains in Morocco. When climate change proceeds, this might not be possible any more

The population of California is still increasing rapidly. The Los Angeles region would never have developed into a megapolis without huge water transfer projects. An intense debate about future use of water is ongoing in California, fuelled by the challenges of climate change (Vicuna et al. 2007) (Fig. 17.2).

The "Climate Glutton" will test the countries of Table 17.1 with a fierce hand. On the other hand, many countries will get more water. A large majority of climate models suggest an increase of runoff almost everywhere north of the latitude 60°N. The same is true in many tropical regions, but not in the Amazon Basin. Likewise, eastern parts of North America, northeast China and Argentina will be gaining regions.

In Finland, the scenarios give one of the highest increases of runoff in the world. According to the ECHAM4 scenario (EEA 2007) the increase in 2070 is over 30%, in Eastern Finland locally even more than 50%. No other country in Europe has an increase exceeding 20% in this scenario.

However, an increased runoff is not necessarily a good piece of news. It may imply higher floods, in some cases both higher floods and more intense droughts. Both of these extremes may become worse e.g. in some parts of Central Europe and Eastern Africa.

Controversial modeling results have been obtained e.g. for China and India. These giants have today 11% of world's water resources, but over one third of the population. Most climate models seem to have still major problems with the monsoon precipitation. More water to southern China, less to water-scarce central China? India's dry West may get even drier, wet Northeast may get even wetter? But some models disagree.

17.4 The Economy Does Not Grow Without Water

The changes of water use in the traditional sectors were not dealt with in Table 17.1. This topic was discussed in detail by e.g. Alcamo et al. (2007). They estimated that domestic water use will grow globally up to four- to five-fold by the year 2055. The increase will be fastest in developing countries, paralleled with the economic growth. The role of their population growth will be considerably smaller in terms of water consumption.

The industrial water use will also increase, but only about twofold worldwide by the year 2055. In Europe and North America it will decrease. The relative growth will be highest in Africa, but the share of that continent will still be only about 5% of the industrial water use in the world. The climate change itself will naturally have a direct impact on sectorial water consumptions. As to the domestic and industrial sectors, the impact will be minor, but in agriculture it will be significant.

At present, the irrigated area in the world is around 270 million hectares. This is 18% of total agricultural land, but 40% of the harvest volume, in developing countries up to 60%. The irrigation efficiency is only 50% – thus there is plenty of room for rationalization.

Fischer et al. (2007) estimated that the irrigated area will be around 390 million hectares by the year 2080. The efficiency will be 70%. Without the direct impact of climate change, the water need would grow by roughly one fourth. Climate change will, however, increase the water need by roughly another fourth, 670–725 km^3 in different scenarios. If the mitigation efforts can lead to SRES B1-scenario, water savings in irrigation will be 220–275 km^3.

17.5 Glaciers Shrink, Snow Melts Away

The area of world's glaciers without the ice caps of the Antarctic and Greenland is around 530,000 km^2, and their volume 51,000–133,000 km^3 (Solomon et al. 2007). The area is known rather accurately, but not the volume. If all these glaciers would melt completely, the level of the world ocean would rise by 15–37 cm. In the period of 1960–2005, the volume of glaciers decreased by about 7,000 km^3 or 5–14%. The share of Alaska was almost one third, the Arctic islands contributed over one fourth. Half of the losses occurred after 1990, indicating a doubling of the melt rate (Solomon et al. 2007).

It should also be noted that the "secondary" ice caps of the Antarctic and Greenland have an area of at least 200,000 km^2. They are not included in the estimates above. In the Antarctic, they might have lost up to 3,000 km^3 of their volume after 1960 (de Woul 2008), and perhaps half of that in Greenland.

As to the water supply, the mountain glaciers in Asia, Europe and the Andes are important. Almost 1 billion people use their meltwaters. In the mountains of Asia, the shrinking rates have varied considerably, in Karakorum region some glaciers have even gained more mass. Glaciers in Central Europe melted in 2003 at an unprecedented rate, four times the average of the period 1980–2001 (Solomon et al. 2007). In the Andes some small glaciers have completely disappeared; e.g. the water supply of the Peruvian capital, Lima has endangered (Leavell 2007).

Salomon et al. (2007) estimated that glaciers in some mountain regions in the Northern Hemisphere may lose half of their volume by the year 2050 (Fig. 17.3). This implies increasing summertime flows in the next few decades. Thereafter, a severe flow reduction can be anticipated e.g. in large rivers having their sources in the Himalayas (Mall et al. 2006).

The NH snow cover of land areas contains some 2,200 km^3 of water in February. The share of Asia is 46%, North America 40% and Europe 15% (Kim et al. 2008). In Finland, the average maximum storage is 45 km^3, 2% of the global total.

According to two different models (ECHAM5, HadCM3) and scenarios (A1B, B1), the snow storage in February will be 1,790–2,000 km^3 in the period 2071–2100. The reduction will thus be only 10–20%, because the increase of wintertime precipitation partly compensates the effect of warming. In Europe the estimated reduction will be as high as 33–47% (Kim et al. 2008).

During the last four decades, the maximum areal extent of snow cover in the Northern Hemisphere has decreased by over 10%, about 15 times the area of Finland (Salomon et al. 2007). This trend is going to continue, but the rate varies considerably between different scenarios. In general, snow will disappear from many areas that are presently having a thin snow cover, while more ample covers may get thicker.

In Southern Finland, the losses of snow have already been significant (Table 17.2). After the year 1990, mean maximum water equivalent of snow has decreased by 25–35% in the south (the first three river basins), while in the north an increase of 5–15% has been observed.

Fig. 17.3 The tongues of the Columbia glacier in the Canadian Rockies have receded by up to 2 km during the last century. A similar trend has been observed in most mountain glaciers in North America

17.6 Water Problems in Megacities

When a city grows, local water resources become insufficient. An extreme example of that is the Valley of Mexico, now hosting a megapolis of 23 million people in an area of 8,000 km^2. In the valley there is a huge aquifer, which has been exploited for many centuries. There are over 10,000 wells and groundwater is pumped at an average rate of 45 m^3/s. Water level has lowered in the aquifer by up to 100 m, leading to soil subsidence of 15 m in some areas.

Table 17.2 The mean maximum water equivalent of snow cover in selected river basins in Finland during the periods 1961–1990 and 1991–2010

River basin	L_{max} (mm) 1961–1990	L_{max} (mm) 1991–2010	Δ (%)
Vantaanjoki	109	73	−33
Aurajoki	86	61	−30
Kyrönjoki	92	68	−26
Vuoksi	146	145	−1
Oulujoki	162	177	+9
Kemijoki	175	186	+6
Paatsjoki	149	173	+13

The water consumption of Mexico City is now as high as 63 m³/s, some 240 L per capita per day. The need for additional water became evident already in the 1930s and the first water transfer system, 60 km long, was built. When the population passed the 6 million line around 1970, new water sources were again needed. A system where water is pumped 1,000 m uphill was constructed, having 102 pumping stations and 17 tunnels. The transfer distance is 150 km. Now there are plans for an even longer distance and a pumping head of almost 2 km.

Varis (2008) compares the water supply of a mega-city to blood circulation, which keeps the human hive alive. Much too often this circulation gets additional substances, which are harmful or even fatal. It is a mega-challenge to solve the water supply problems of a mega-city – a challenge which will not be easier when climate change proceeds.

17.7 Bioenergy or Food: Competition on Water

The production of bioenergy is an essential item in the palette of climate change mitigation. This leads to a growing competition on both water and land. The price of food has already increased and poor people in developing countries are the first to suffer.

The cultivation of biomass suitable for energy production typically requires 400 L of water per 1 kg of dry-weight. Ambitious bioenergy targets for the year 2050 will imply an additional water use of around 2,000 km³, almost half of the present use of blue water by mankind. It is clear that this would be a critical stress factor to world's water resources.

Most suitable areas for bioenergy production plants would be abandoned agricultural lands, rangelands and wastelands. In many rangelands, the food production per water unit is quite low. On the other hand, the demand of rangeland food, which is mainly meat, is growing steadily. One of the most promising bioenergy plants, *Jatropha curcas*, grows relatively well on wastelands. These have been taken into use with this plant e.g. in India. Wastelands may be, however, an important source of firewood – and even food – for the poorest people.

17.8 The Virtual Rivers of Global Trade

In addition to Nature's water cycle, there is now a virtual water cycle created by man. Water embedded in various products is flowing from country to country in the global trade. When a Finn buys an Indian T-shirt, he also buys 1,000 L of water from India.

Some 1,600 km^3 of virtual water is crossing country borders annually (Roth and Warner 2008). The share of agricultural products is around four fifths, industrial products making the last fifth. The largest amounts of virtual water are flowing in the trade of beef, followed by soybean, wheat, cacao, coffee and rice (Hummel et al. 2006). United States (150 km^3) is the largest exporter of virtual water, the next ones being Australia, Canada and Thailand. The largest importer is Japan, close to 100 km^3. Israel has the highest dependency from virtual water import, 74% of all consumption (Hummel et al. 2006).

The globalization has extended the virtual water trade and this trend is going to continue. Climate change has also entered the stage; e.g. the drought in Australia has decreased agricultural production and the export of virtual water. In most of the countries of Table 17.1, the virtual water balance will probably change towards negative direction. For example Spain and Italy are already virtual water importers, although they still are large exporters of agricultural products.

All in all, the water problems could be relieved if water intensive products would not be cultivated or manufactured in countries where water resources are scarce. There are still obstacles to this development, although the opening of global trade and weakening of political tensions have works to the other direction (Fig. 17.4).

17.9 Is the Time of Large Dams and Water Transfers Already Over?

At present, there are over 120 large dams in the world having a reservoir capacity of at least 1 km^3 of water. Two of them are located in Finland, Lokka and Porttipahta. The number of world dams with a crest height exceeding 15 m is 38,000. The global water storage of all reservoirs is over 7,000 km^3. This is some 30 times the volume of Finnish lakes.

The controversies related to large dams have been an essential part of environmental disputes since the 1960s. A trigger was the construction of the Assuan Dam in Egypt, shifting the multi-millennial irrigation practice to a new stage with higher production – and higher problems.

Particularly controversial have been the reservoirs constructed in tropical rain forests in South America and Africa. In Brazil, the hydro schemes of Tucurui and Balbina may, among other things, have released huge amounts of greenhouse gases from the decomposing organic material. In addition to power production, large dams have been built particularly for flood protection and irrigation water supply.

Fig. 17.4 A paddy field in China. Over 100 km^3 of virtual water is flowing in the world trade, although the share of domestic consumption is larger for rice than to most other staple foods

One of the major disputes has been Sardar Sadovar project in Narmada River in India. The plan was to create a reservoir with a length of 210 km and produce 1,450 MW of power. The network of irrigation canals would have covered an area larger than all the agricultural lands in Finland. The original plan was, however, cancelled.

The Three Gorges in China is the largest hydropower plant in the world, large enough to cover the electricity demand of the whole Finland. It is supposed to lower considerably the threat of a catastrophic flood in the lower reaches of the Yangtze River, and provide water for extensive irrigation. On the other hand, 1.3 million people had to be relocated and some of the world's most magnificent sceneries were lost.

The largest inter-basin water transfers are located in California and Central Asia. The California State Aqueduct (CSA) is equivalent of transferring 300 m^3/s of water from the northernmost tip of Finland to the southernmost, over a distance of 1,000 km. In the Soviet era, the Kara-Kum Canal in Central Asia beat the CSA both in distance and in water amount, at the same time heavily kicking down the surface of the Aral Sea.

World's largest groundwater transfer system is located in Libya. During the last Ice Age, heavy rains stored up to 35,000 km^3 of water in the sands of the present Central Libya. The surface water resources of the country are now less than 1 km^3/year, the annual water use below 3 km^3. Water was pumped earlier from small coastal aquifers, with a consequent salinization. The transfer capacity of the double pipeline with a length over 1,000 km is almost 6 million cubic meters per day.

Is there still place for large dams and gigantic water transfer systems? Yes, but social and environmental aspects will play a much more essential role than earlier. Some old dams will also be torn down, but restoring the former biotopes of reservoir bottoms might not be possible. As to the water transfers, construction of China's South–North water transfer plan will be by far the largest project in the world within the next two decades. India also has some transfer plans, from Himalayan rivers towards the south.

17.10 New Gimmicks to Water Scarcity

Non-traditional water sources include e.g. desalination, international water trade, icebergs and rainmaking. Although rainwater harvesting is an age-old method, new technological innovations might be an excuse to include it to this category, too.

In the beginning of 2010, there were almost 15,000 desalination plants with a capacity of over 100 m^3/day. Their total capacity was 60 million m^3/day. In three decades, the capacity has grown almost 50 times.

The largest capacities by country are those of Saudi-Arabia (27%), United States (13%), UAE 11% and Kuwait (9%). Although these four countries have 60% of the capacity, there are desalination plants in over 120 countries or island groups.

The cost range of distilled seawater is around 0.2–2 €/m^3. Methodological development may decrease the costs, while energy prices may increase. The latter may not be a concern in some oil-producing countries, which can utilize energy that would otherwise be wasted.

Rainwater has been collected for thousands of years. Some relatively large systems were built during last century, e.g. the water supply of Gibraltar was based on the collection of rainwater from rocky slopes in 1903–1984. Representatives of a new generation include collection systems at airports, stadiums and other public buildings. Much more important, however, are projects in developing countries; e.g.

in the Chinese countryside where at least 7 million pools are suitable for rainwater harvesting (Shah et al. 2000).

There have been some examples of small-scale international water trade in recent decades. Having abundant water resources, Finland has also been mentioned when discussing the future development of the water trade. However, suitable water sources are available in the vicinity of countries, where traded water will be needed – Finland simply is locater too far north.

Towing of icebergs would be one option that is technically possible. If the target has an area of 1 km^2 and thickness of 200 m, the trip from Antarctic waters to Arabian Peninsula would take about half a year. Heat insulation for the icy storage would be needed. A chopping plant should be built off-shore, with pipelines to land.

Rainmaking has been scientifically studied for more than half a century. Silver iodide is a good chemical to create large numbers of condensation nuclei. There has been some success in enhancing winter precipitation e.g. in Colorado. Political issues easily prevent these activities close to country boundaries.

17.11 Conclusions

Human population on the Earth grew fourfold, from 1.5 billion to 6 billion during the last century. At the same time, domestic water consumption grew 21-fold, industrial 28-fold, agricultural sixfold. All these four growth rates were unprecedented – and will never be repeated.

The growth of water consumption is, however, continuing. In the 2050s, it may already be 6,500–9,000 km^3. The upper limit of this range could be reached, if irrigation systems for bioenergy production will be extensive. The highest growth rates of water consumption will occur in countries, where population growth is the most rapid. Most of these countries will also be water losers due to climate change.

How much are global water resources going to change by the year 2050? Climate models do not give reliable answers; even the direction of change is unclear. A change higher than ±10% on the global level is, however, not probable. Countrywide changes may be much larger, perhaps as much as −40% to +30%.

The growth of water consumption might necessitate the construction of large-scale water transfers and major reservoirs in the first half of this century. Rainwater harvesting may increase substantially, the same will apply to desalination and waste-water reuse. It is possible that some icebergs will be towed to Australia or Middle East by the year 2050.

International water conflicts have occurred, even some water wars. The international climate convention, UNFCCC, has got a watery sister, UNFCWS (United Nations Framework Convention on Water Solidarity). It will guide water allocation between nations and solve disputes between them and different user groups.

References

Alcamo J, Flörke M, Märker M (2007) Future long-term changes in global water resources driven by socio-economic and climatic changes. Hydrol Sci J 52(2):247–275

de Woul M (2008) Response of glaciers to climate change: mass balance sensitivity, sea level rise and runoff. Ph.D. dissertation, Department of Physical Geography and Quaternary Geology, Stockholm University, Sweden

EEA (European Environmental Agency) (2007) Europe's environment. The fourth assessment. Chapter 3: Climate change. EEA, Copenhagen, pp 145–174

Fischer G et al (2007) Climate change impacts on irrigation water requirements: effects of mitigation, 1990–2080. Technol Forecast Soc Change 74(8):1083–1107

Hummel D et al (2006) Virtual water trade. Documentation of an international expert workshop. Institute for Social-Ecological Research, Frankfurt am Main

Kim H et al (2008) Impacts of climate change on long-term global water balance. http://www.cahmda3.info/ab_files/CAHMDA3_Kim.pdf

Kuusisto E (1998) International river basins and the use of water resources. A report to the Ministry for Foreign Affairs. Finnish Environment Institute, Helsinki

Leavell D (2007) The impacts of climate change on the mountain glaciers of the central Andes, and the future of water supply in Lima, Perú. In: Heinonen M (ed) Proceedings of the 3rd international conference on climate and water, Helsinki, Finland, 3–6 September 2007, pp 290–295. Finnish Environmental Institute (SYKE), Helsinki. http://www.ymparisto.fi/download.asp?contentid=732908.lan=en

Mall RK et al (2006) Water resources and climate change: an Indian perspective. Curr Sci 90:1610–1626

Roth D, Warner J (2008) Virtual water: virtuous impact? The unsteady state of virtual water. Agric Hum Values 25:257–270

Shah T, Molden D, Sakthivadivel R, Seckler D (2000) The global groundwater situation: overview of opportunities and challenges. International Water Management Institute, Colombo. http://www.publications.iwmi.org/pdf/H025885.pdf

Solomon S, Qin D, Manning M, Chen Z, Marquis M, Averyt KB, Tignor M, Miller HL (eds) (2007) Climate change 2007: the physical science basis. Contribution of Working Group I to the fourth assessment report of the Intergovernmental Panel on Climate Change. Cambridge University Press, Cambridge/New York

Varis O (2008) Maailman megakaupungit ja niiden vesiongelmat. Tieteessä tapahtuu 5:16–20

Vicuna S et al (2007) The sensitivity of California water resources to climate change scenarios. J Amer Water Resour Assoc 43:482–498

Part IV
Planet Earth, Third Stone from the Sun

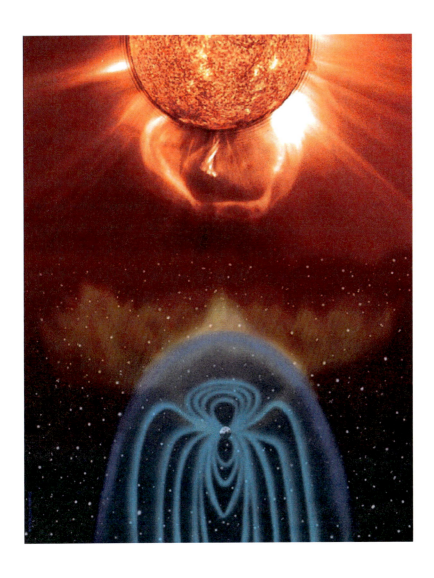

Cover page: Solar wind and particle streams of the space. Above the Sun, below the Earth's magnetosphere Credit: ESA

Chapter 18
Trends in Space Weather Since the Nineteenth Century

Heikki Nevanlinna

18.1 Introduction

Space weather is the concept of changing conditions in the physical state in the near-space of the Earth. Changes occur mainly in the Earth's ionosphere and magnetosphere. Varying space weather phenomena are connected with electromagnetic processes as well as particle precipitations in the space. Most familiar manifestations of space weather are the northern lights. Space weather is ultimately governed by solar activity, varying solar wind and transient solar eruptions like flares and coronal mass ejections.

Changes of the intensity of space weather phenomena are also connected with long-term variations of the solar activity in the course of the 11-year sunspot cycle. During sunspot maximum years space weather conditions are usually stormier than during sunspot minima. There are also slower variations in space weather in centennial time frame of several successive sunspot cycles. These are space climatic variations.

Information about space weather conditions is obtained by e.g. satellite monitoring of solar particles and processes. However, ground based measurements, like ionospheric radars, auroral imagery, and magnetic field recordings, give important space weather data. When studying space climate variations back in time to the eighteenth and nineteenth centuries, systematic magnetic observatory recordings and observations of auroral occurrence frequencies play important role in long-term data acquisition.

In this study we represent space weather and climate data series since the nineteenth century and analyze decadal trends and power spectra of the data. A simple forecast is given for the amplitude of the sunspot number of the next sunspot maximum 2013–2014.

H. Nevanlinna (✉)
Finnish Meteorological Institute, P.O. Box 503, FI-00101 Helsinki, Finland
e-mail: heikki.nevanlinna@gmail.com

18.2 Magnetic Recordings

In space weather studies utilizing geomagnetic observations, standardized magnetic indices are widely used for characterizing magnetic changes in different time scales from hours to days and longer. An index is a function of the magnetic field amplitude in the time interval chosen. Indices are local, calculated from observations at a single site, or global, based on data from observatory networks. Indices contain information from both ionospheric and magnetospheric disturbances, and are thus proxy data for the solar activity. In space climate studies, an important requirement is the long time coverage and the homogeneity of the index series.

The longest available global index series is the 3-h *aa*-index belonging to the so-called *K*-index family (Menvielle and Berthelier 1991). It is based on two almost antipodal observatories in Australia and the UK. The *aa*-series covers the years since 1868. However, an extension of the original *aa*-index series for about two solar cycles to 1844 was introduced by Nevanlinna and Kataja (1993). In this analysis old magnetic recordings from the Helsinki (Finland) magnetic observatory (Lat. 60.2°N; Lon. 25.0°E) were utilized. Later the data analysis was completed using data from other magnetic observatories in the nineteenth century (Nevanlinna and Häkkinen 2010).

Another long space weather activity index series is from the Sodankylä observatory (Lat. 67.4°N; Lon. 26.6°E) in Finland. It covers almost 100 years since 1914 (Fig. 18.1). Sodankylä series in the longest available space weather data source inside the Arctic Circle.

Recently, new type of magnetic indices have been introduced that can be easily calculated from present day and historical magnetic data if available in numerical form. One of these is the daily IHV-index (Inter-Hourly Variability) calculated

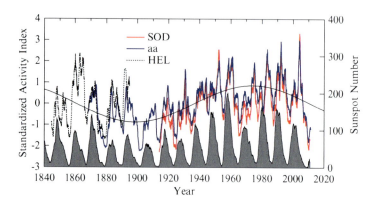

Fig. 18.1 Smoothed (365 days) daily means of *Ak*-indices from the Sodankylä (SOD) 1914–2010 and Helsinki (HEL) 1844–1897 and global *aa*-indices 1868–2010. Sunspot numbers are depicted by *grey area*. The smooth curve is a sinusoidal fit of the *aa*-values showing an apparent period of about 150 years

from hourly variations of the magnetic field during night. Because the daytime ionospheric variations and other regular field changes are absent in the night, the IHV-index describes solar wind related activity. It is a useful tool in studies of the long-term behavior of global geomagnetic variations in connection with solar activity reaching back to the 1830s (e.g., Svalgaard and Cliver 2007). One of the advantages of the IHV-index compared with the traditional hand-scaled activity numbers (K-indices) is its objective derivation.

Figure 18.1 shows a summary of the magnetic activity series of smoothed (365 days) daily index from Sodankylä (1914–2010) and Helsinki (1844–1897), and global aa (1868–2010). Daily indices in Fig. 18.1 have been standardized to z-scores by subtracting from each series the mean value and dividing the difference by the standard deviation. One can see that the time variations of all overlapping index series are rather similar and follow the 11-year sunspot variation. The long-term behavior of the activity was slightly decreasing in the late nineteenth century. There was a minimum around 1900 and a maximum in the mid of the twentieth century, after which the general trend of the magnetic activity has been slightly decreasing since about 1985. The peak values of the sunspot numbers in (Fig. 18.1) follow closely the slow trend in the geomagnetic activity. Similar trends have been reported in auroral magnetic storm occurrence frequencies during the late nineteenth century and early decades of the twentieth century (Nevanlinna 2004b).

As is well known the highest peaks in the activity time series are usually occurring in years just before or at the sunspot maximum and a couple of years after it during the descending phase of the cycle (Usoskin 2008). The index series have thus a dual peak structure during the 11-year sunspot cycle. The latter activity peak is usually stronger than the previous one as can be seen in Fig. 18.1 showing 15 solar cycles in 1844–2010.

The activity between two such peaks can be plunged deeper than in the preceding sunspot minimum as was the case in 1980.

The time period from about 1920 to 2010 represent an epoch called "Grand Maximum" (e.g., Usoskin 2008; Lockwood 2010) because solar activity has been probably greater at that time than centuries to millennia before. The overall level of solar activity will probably decrease significantly during the next sunspot cycles. A signal of this solar weakening is seen in the progress of the present sunspot cycle which has activated much slower than on average.

According to statistics of sunspot numbers at NOAA, the present cycle 24 started in December 2008. As can be seen in (Fig. 18.1), the global magnetic activity (aa-index) had its minimum about 1 year later as is the usual lag between sunspot and magnetic activity. After the minimum, magnetic activity has been slowly increasing. The minimum in late 2009 in the time series of the aa-index was deepest since the beginning of the twentieth century in 1901 (solar cycle 14). The next solar maximum is predicted to occur in 2013–2014. During the last sunspot cycle (1996–2008) solar activity, as measured by geomagnetic disturbances, was at highest level in autumn 2003 since the start of activity recordings 1844. On the other hand in the beginning of the new solar cycle in 2009, magnetic activity was at

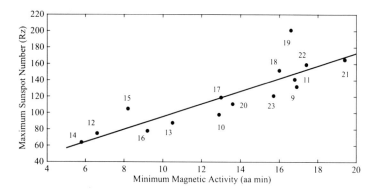

Fig. 18.2 The precursor method for predicting smoothed sunspot maximum (Rz) from the preceding magnetic activity minimum ($aa\ min$) (e.g., Kane 2010). The linear fit between $aa\ min$ and Rz for the past 15 solar cycles (1844–2008) has the linear correlation coefficient 0.89. The prediction for the next solar maximum (24) in 2013–2014 is $Rz = 79$

lowest level for about 100 years predicting that the present sunspot cycle (24) will be much less intensive than its predecessors in the late twentieth century.

There is an empirical connection between the maximum value (Rz) of the next sunspot number and the minimum value ($aa\ min$) of the preceding geomagnetic activity. Both Rz and $aa\ min$ are expressed as 12 months running means (averages?). This is the so-called precursor method (e.g., Kane 2010). When applied to the 15 aa minima between 1844 and 2008 shown in Fig. 18.1, the linear expression between $aa\ min$ and Rz is as follows

$$Rz = 18.2(\pm 17.3) + 7.7(\pm 1.3) aa\ min,$$

where the linear correlation coefficient is high, 0.89 (Fig. 18.2), and the $\pm$ error is one standard deviation. The last minimum of the smoothed aa-index was 8.3, which means that the estimate of the next maximum of the sunspot number will be about 82 ($\pm$28) according to the linear equation above. This is about 30% lower than the previous maximum in 2000, and about the same level as the maxima of the sunspot cycles 13 and 14 in the turn of the nineteenth and twentieth centuries. The recent (Sep 2010) estimate for the maximum Rz given here is higher than that published by NOAA (Space weather Predictions Center) (64).

18.3 Power Spectra of Activity Time Series

Figure 18.3 shows an example of the power spectra of activity index time series after a Fourier-transformation. Given are the spectra based on daily activity indices aa (1868–2010) and Sodankylä Ak (1914–2010). For comparisons between geomagnetic activity and their solar forcings, a similar power spectrum from the solar

18 Trends in Space Weather Since the Nineteenth Century

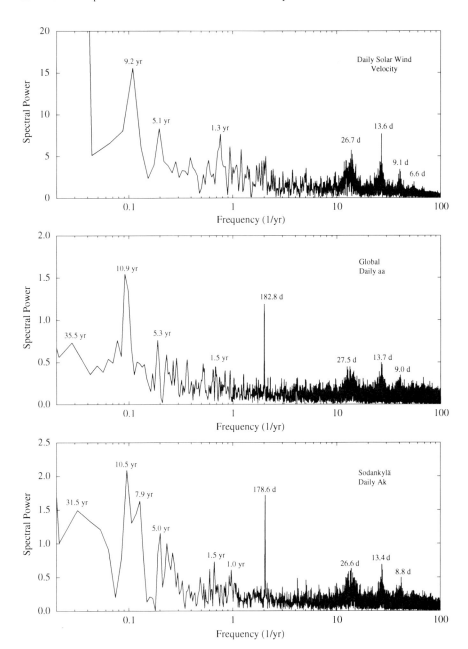

Fig. 18.3 Power spectrum of daily solar wind velocity (*upper panel*), global *aa*-index (*mid-panel*), and Sodankylä *Ak*-numbers (*lower panel*). Prominent spectral lines are attached by the period of the wave

wind at 1 AU (data source: OMNIweb Data Explorer, http://omniweb.gsfc.nasa. gov/form/dx1.html) has been given for 1965–2010. The power spectra have been calculated for the whole time periods although a division according to 11-year sunspot cycles with ascending and descending sunspots would be a meaningful analysis too, as has been done by e.g. Kane (1997). However, the spectral spikes found here and shown in (Fig. 18.1) represent signatures in the activity data that are stable throughout several sunspot cycles.

All three time series show major spectral spikes at the Schwabe sunspot cycle (9.2–10.9 years), 5.0–5.3 years, 1.3–1.5 years, 26.6–27.5 days, 13.4–13.7 days, and 8.8–9.1 days reflecting basic characteristics in the solar activity behavior although the periods obtained here differ slightly from each other in the spectra. This is partly due to the different lengths and different spatial coverage of the original time series of activity indices. However, the periods found here correspond well to the results obtained from similar spectral analysis of activity index series (e.g., Fraser-Smith 1972; Delouis and Mayaud 1975; Clua De Gonzales et al. 1993; Nevanlinna 2004a).

In the aa- and Ak-spectra there are weak signals for periods longer than the double-sunspot cycle 22 year similarly to that found by e.g. Kane (1997). The spectral line at about 5 years is connected with the dual-peak structure of the magnetic activity during the course of the sunspot cycles as shown in (Fig. 18.1). In the dual activity behavior the magnetic storms occur in at least two episodes with time separation of about 4–6 years around the sunspot maximum year.

The spectral region between about 3 years to 1 year is very rich of lines in the geomagnetic activity spectral series as well as in the solar wind spectrum. The period 1.3–1.5 year is found here and in other studies and also in many heliospheric parameters (e.g., solar wind speed, interplanetary magnetic field; e.g., Richardson et al. 1994; Clua De Gonzales et al. 1993; Mursula et al. 2003) activity indices and auroral occurrence frequency (Silverman and Shapiro 1983). These lines are quasi-periodic in such manner that they exist not through the entire solar cycle and are weaker than average during low-activity sunspot cycles (Kane 1997; Mursula et al. 2003). Clua De Gonzales et al. (1993) claimed that this periodicity could be traced back to the sector structure of the interplanetary magnetic field.

There is a rather weak annual line visible in the Sodankylä data, which is absent in the aa-series. The annual variation is cancelled in the aa-index series because it is based on two antipodal observatories where the annual magnetic variation is at antiphase.

The strongest spectral line after the sunspot signal is the semiannual wave seen only in geomagnetic activity data (Fig. 18.3). The existence of the semiannual periodicity has been known since the early days of space weather studies in the beginning of the nineteenth century. It is caused mainly by the varying tilt of the Earth's magnetic dipole axis relative to the Earth-Sun line in the Russell-McPherron model (Russell and McPherron 1973). The semiannual wave has its minima at solstices and maxima at equinoxes.

In addition to the solar cycle and semiannual periodicities, the solar rotation recurrence peak around 27 days is one of the most important features in the geomagnetic activity variation as well as in the solar wind data. Actually, in the short period part of the spectrum there are three clusters of spectral lines at about 27, 13, and 9 days, respectively. The amplitude of the 13 day line is slightly greater than the two others.

Historically, the existence of the 27 day solar rotation periodicity of magnetic disturbances was recognized as early as in the 1850s and the first writings about the 14 days periodicity appeared in the early twentieth century (e.g., Chapman and Bartels 1940). Statistics of the 27 day and 14 day variations as revealed in the magnetic recordings at Sodankylä were analyzed by Sucksdorff in his thesis about magnetic activity (Sucksdorff 1942).

In the 1930s Bartels introduced M-regions as hypothetical (M for mysterious) solar sources of the recurrent magnetic activity (Bartels 1934). Satellite missions in the 1970s revealed that M-regions are coronal holes from which enhanced solar wind streams escape into the interplanetary space causing disturbances in the magnetosphere of the Earth (e.g., Crooker and Cliver 1994).

A recurrent two-sector structure in the sun is typically associated with an emerging spectral peak close to 27 day while the 14 day modulation becomes more important during intervals corresponding to four sectors per solar rotation. The recurrence tendency of two high-velocity streams in a solar rotation seems to reinforce the relative importance of 14 day variation.

18.4 Conclusions

Information about geomagnetic activity on hourly basis exists since the mid-nineteenth century. Magnetic indices derived from these data sources give important proxy material for determining variations in solar activity in time periods from hours to years and decades. In the long-term run the geomagnetic and solar activity was in deep minimum in early decades of the twentieth century when the global activity was on average about half of that during the last two sunspot cycles in 1986–2009 (23 and 24). However, the prolonged duration of the sunspot cycle 23 (1996–2008) and the slow progress of the solar activity during the present cycle 24, indicate that the Sun will be in less active state during the coming decade. The average magnetic activity in 2008–2009 was lowest for about 100 years.

References

Bartels J (1934) Twenty-seven day recurrences in terrestrial-magnetic and solar activity, 1923–33. J Geophys Res 39:201–202

Chapman S, Bartels J (1940) Geomagnetism, Vol. I & II. Oxford Clarendon Press, Oxford

Clua De Gonzales A, Gonzales G, Dutra SL (1993) Periodic variation in the geomagnetic activity: a study based on the Ap index. J Geophys Res 98:9215–9231

Crooker NU, Cliver EW (1994) Postmodern view of M-regions. J Geophys Res 99:23383–23390

Delouis H, Mayaud PN (1975) Spectral analysis of the geomagnetic activity index *aa* over a 103-year interval. J Geophys Res 80:4681–4688

Fraser-Smith AC (1972) Spectrum of the geomagnetic activity index *Ap*. J Geophys Res 77:4209–4220

Kane RP (1997) Quasi-biennial and quasi-triennial oscillations in geomagnetic activity indices. Ann Geophys 15:1581–1594

Kane RP (2010) Size of the coming solar cycle 24 based on Ohl's precursor method. Ann Geophys 28:1463–1466

Lockwood M (2010) Solar change and climate: an update in the light of the current exceptional solar minimum. Proc R Soc A 466:303–329

Menvielle M, Berthelier A (1991) The *K*-derived planetary indices: description and availability. Rev Geophys 29:413–432

Mursula K, Zieger B, Vilppola JH (2003) Mid-term quasi-periodicities in geomagnetic activity during the last 15 solar cycles: connections to solar dynamo strength. Sol Phys 212:201–217

Nevanlinna H (2004a) Historical space climate data from Finland: compilation and analysis. Sol Phys 224:395–405

Nevanlinna H (2004b) Results of the Helsinki magnetic observatory 1844–1912. Ann Geophys 22:1691–1704

Nevanlinna H, Häkkinen L (2010) Results of Russian geomagnetic observatories in the 19th century: magnetic activity 1841–1862. Ann Geophys 28:917–926

Nevanlinna H, Kataja E (1993) An extension of the geomagnetic activity series *aa* for two solar cycles (1844–1868). Geophys Res Lett 20:2703–2706

Richardson JD, Paularena KI, Belcher JW, Lazarus AJ (1994) Solar wind oscillations with a 1.3 year period. Geophys Res Lett 21:1559–1962

Russell CT, McPherron RL (1973) Semi-annual variation of geomagnetic activity. J Geophys Res 78:92–108

Silverman SM, Shapiro R (1983) Power spectral analysis of auroral occurrence frequency. J Geophys Res 88:6310–6316

Sucksdorff E (1942) Die erdmagnetische Aktivität in Sodankylä in den Jahren 1914–1934. Sanan Valta, Kuopio

Svalgaard L, Cliver EW (2007) Interhourly variability index of geomagnetic activity and its use in deriving the long-term variation of solar wind speed. J Geophys Res 112:A10111–A10143

Usoskin IG (2008) A history of solar activity over Millennia. Living Rev Solar Phys 5:1–88

Chapter 19
Space Weather: From Solar Storms to the Technical Challenges of the Space Age

Hannu Koskinen

19.1 Northern Lights: A Nordic View to Space Weather

The term space weather became widely used in the 1990s when several organizations in the United States interested in space environmental conditions formulated a common National Space Weather Strategy. In this context space weather was defined through its harmful effects. As a natural phenomenon space weather had, of course, been studied long before its negative consequences were even known.

The most glorious appearance of space weather, the Northern Lights, has occupied the imagination of scientists for centuries. Its Latin name, *Aurora Borealis*, was most likely coined by Galileo Galilei around the year 1619. He had seen a magnificent reddish light display on the northern sky of Tuscany and he named it according to the Greek goddess of the dawn, Aurora. Obviously, nobody in the northern countries would have made an association with the northern lights and the dawn.

An important step toward the understanding of the underlying physics was taken much later by Anders Celsius and Olof Hiorter in Uppsala, Sweden, in the 1790s. They noticed that during an auroral display the compass needle can deviate several degrees from the north–south direction. Consequently, the auroras were somehow able to disturb the magnetic field on the ground. Celsius did not yet have tools to understand the reason for this behavior. Today we know that the visible auroras at altitudes of about 100 km are associated with strong and rapidly changing electric currents. Celsius and Hiorter had observed the magnetic perturbations caused by these currents.

In Finland measurements of magnetic disturbances started in 1840s when observations at the Magnetic-Meteorological Observatory founded in 1838 started

---

H. Koskinen (✉)
Department of Physics, University of Helsinki, P.O. Box 64, FI-00014 Helsinki, Finland
e-mail: hannu.e.koskinen@helsinki.fi

in Kaisaniemi in the present downtown Helsinki. The observations continued there until the early twentieth century when the electrification of the city made magnetic observations impossible with the accuracy required for scientific purposes. Later the observatory evolved to become the Finnish Meteorological Institute, which continues to make magnetic observations at the Nurmijärvi Observatory in Southern Finland and also manages the extended magnetometer chain from Estonia to Spitzbergen known as IMAGE. One of the strongest sectors of the Institute's space activities concentrates on the effects of solar perturbations near the Earth and other terrestrial planets.

The first prominent Finnish scientist trying to understand the physics of northern lights was Selim Lemström (1838–1904), Professor of Physics at the University of Helsinki. He developed a theory of the auroras based on atmospheric electricity. In his theory the auroras and lightning were considered as two faces of a common phenomenon. To test his theory Lemström started in 1870 a series of experiments at the mountains of Finnish Lapland. He set up constructions resembling a circle of lightning rods with the goal of creating artificial auroras. However, Lemström's theory proved to be wrong. Among the reasons for the failure were Lemström's own erroneous estimates of the altitude of the auroras. He did not have sufficiently good cameras to make reliable altitude determination by triangulation and did not realize that the auroral light comes from the altitudes of 100 km and above (Fig. 19.1).

Despite of the fate of his auroral theory, Lemström's expeditions to Lapland brought important scientific activities to the Northern Finland. During the First Polar Year 1882–1883 numerous geophysical observations were conducted in Sodankylä and in the recently established gold mining village Kultala on the bank of the Ivalo River. In 1913 the Finnish Academy of Sciences and Letters

Fig. 19.1 Sketch of Selim Lemström's experiment in which he believed to have produced artificial auroras (Biese 1885; Lemström and Biese 1886)

established the Sodankylä Geophysical Observatory to continue the observations interrupted in Helsinki. The observatory was destroyed during the Second World War, but was rebuilt and is today part of the University of Oulu.

19.2 Storms on the Sun

Galileo actually opened the correct path leading to the modern explanation for the origins of northern lights and space storms. He was one of the developers of the first optical telescopes who pointed their devices toward the Sun and noticed dark stains on it, the sunspots. After all, the Sun was not a divine perfect body. Instead there were signs of activity on its surface.

The Sun radiates the energy released through nuclear fusion in its central core with the power of 3.84×10^{26} W. This corresponds to about 60 million one thousand MegaWatt power plants for each human inhabitant of the Earth. Within about the last quarter of the path from the center of the Sun to the surface the energy transfer takes the form of turbulent convection where hot gas rises up, cools down and the cool gas sinks downward. In addition, the Sun rotates around its axis such that at the equator one rotation takes about 26 days whereas close to the poles the rotation is slower and takes more than 30 days. This differential rotation and up-and-down convective motion form a generator that creates a magnetic field. This field was first observed by George Hale at sunspots in 1908.

While the energy production in the solar core is steady, the division of energy on the surface is not. The sunspots look as dark specks because they are cooler than their surroundings. This is due to the magnetic field that has higher concentration in the spots and locally inhibits the upward flow of the hot ionized gas, the plasma. When there are a lot of sunspots, the variations of the magnetic field are strongest and we say that the Sun is in its active phase. This happens in a cyclic manner and one sunspot cycle from one activity minimum to the next takes about 11 years. However, both the exact length as well as the intensity of the maximum activity varies from one cycle to another. The solar cycles are enumerated from the cycle that peaked in 1760. The cycle reaching its minimum in 2008 was number 23, after which a very long and deep minimum followed and cycle 24 did not really start until 2010. The next maximum is expected to occur in 2013, more than 12 years after the previous maximum.

The sunspots are a signature of the overall solar activity. The most important stormy phenomena of the Sun are the solar flares and coronal mass ejections. A flare was observed for the first time independently by Carrington and Hodgson in 1859. Only 17 h later the ground based magnetometers registered one of the strongest magnetic storms in history (e.g., Cliver and Svalgaard 2004, and a series of articles in *Advanced Space Research* 38/2, 2006, pp. 115–388). This event led to the idea that solar flares would be the primary cause of terrestrial magnetic storms. No physically acceptable mechanism that would transport the effects of the flare to the Earth was, however, found for a long time.

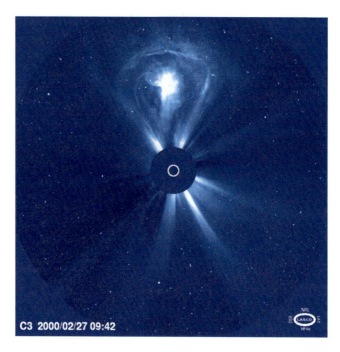

Fig. 19.2 A coronal mass ejection observed on February 27, 2000, by the LASCO coronagraph aboard the SOHO spacecraft of the European Space Agency (ESA) and the National Aeronautics and Space Administration (NASA). The size of the Sun is shown by the small *white circle* in the middle of the occulting disc. Credit: ESA

Starting from the 1970s the explanation of the flares as drivers of space storms turned out to be questionable (Gosling 1993). In 1973 a massive ejection of solar matter was observed for the first time with a coronagraph onboard the satellite OGO 7. Coronagraph is an instrument in which an artificial solar eclipse is created using an occulting disk. These eruptions became known as coronal mass ejections. Today we know that the strongest space storms in the near-Earth space are caused by coronal mass ejections directed toward the Earth (Fig. 19.2).

19.3 Solar Activity and Geomagnetism

Connection between sunspots and geomagnetic disturbances on the ground was found by Edward Sabine in 1852 (Fig. 19.3). Similarly as in case of flares, any explanation of what would carry the information from the Sun to the Earth was not found until a century later. It was suspected that some kind of corpuscular ejections would hit the Earth, but it was highly unclear how to extract enough particles from the very strong gravitational field of the Sun. The escape velocity from the solar

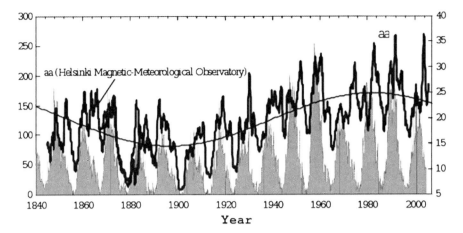

Fig. 19.3 Geomagnetic activity (thick line, so-called aa index, scale on the right) follows the sunspot number (*grey* background; scale on the left). The official global aa index is available since 1868; it has been extended to 1844 by using the local observations in Helsinki during 1844–1890 (Figure by courtesy of Heikki Nevanlinna, Finnish Meteorological Institute)

surface is namely 618 km/s, which corresponds to thermal motion of protons at a temperature of about 20 million degrees.

The explanation for the magnetic connection between the Sun and the Earth came from a direction that may sound a little surprising. A German astronomer Ludwig Biermann studied cometary tails in the 1950s. The most visible dust tail is caused by the radiation pressure, which blows the gas around the comet slowly outward. When the comet moves with a high velocity slightly sideways with respect to the Sun, the dust tail looks more or less bent depending on the direction to the observer. Biermann noticed that in addition to the dust tail there is another less clear stream from the comet directed almost radially outward from the Sun. Consequently, particles forming this stream are moving much faster than the dust tail. Biermann concluded quite correctly that this stream consists of plasma ionized from the dust by solar ultraviolet radiation. These particles are picked up by the magnetized solar wind plasma and rapidly accelerated to the ambient solar wind speed. The existence of this continuously flowing plasma wind originating from the Solar surface was confirmed by direct in situ observations with the first spacecraft sent toward Venus and Mars, i.e., outside the terrestrial magnetosphere.

The solar wind blows continuously and carries about one million tons of matter per second from the Sun. The flow is made possible by the very high temperature, more two million degrees of the solar corona. The heating of the corona is apparently associated with wave-like magnetic perturbations, but the question how to sustain such a hot gas above the surface of 6,000° only, has not yet been satisfactorily explained. The solar wind brings the disturbed solar magnetic field through the whole solar system. The speed, density, and temperature of the flow, as well as the strength and direction of the magnetic field are highly variable when the flow interacts with the magnetic shield of the Earth. These variations shake the magnetosphere and cause the magnetic storms. The previously mentioned coronal mass ejections belong to the strongest perturbations in the solar wind and also cause the strongest magnetic storms in the near-Earth space.

19.4 Effects of Space Weather in Space

Particularly dangerous for technological systems and humans in space is corpuscular radiation enhanced by space storms. A solar flare may directly accelerate charged particles to very large energies. Coronal mass ejections are even more efficient accelerators. Frequently the ejecta propagate in the solar wind faster than the local sound speed pushing shock fronts in front of them. These shocks accelerate large numbers of protons and other nuclei to such high energies that they easily cross through the magnetic shield of the Earth and penetrate deep into the near-Earth space being stopped only by the atmosphere.

Being accelerated to energies of tens of millions of electronvolts and above can penetrate deep into spacecraft structures. Hitting, for example, a spacecraft's control unit they can cause changes in the contents of a memory unit leading to a command with fatal consequences to the spacecraft operations. The radiation is also strongly ionizing. Astronaut suit does not provide much protection against these particles. For example, the astronauts that once walked on the Moon were lucky that there were no strong solar particle eruptions at the time they were on the lunar surface.

Only a small fraction of the solar wind penetrates to the Earth's magnetosphere, but both fast solar wind and in particular shock fronts and magnetic clouds of coronal mass ejections shake the magnetosphere strongly causing electromagnetic perturbations in the space environment. These perturbations accelerate a fraction of magnetospheric electrons to nearly the speed of light. These relativistic electrons are sometimes called "killer electrons" because the enhanced flux of high-energy electrons appears in many cases have been associated with damages on satellites in orbit. Relativistic electrons introduce charge accumulation in the electronic components that can cause internal discharges, which in the worst case may break the device irrecoverably.

Space is never a safe place, not even during good weather. The Earth is surrounded continuously by radiation belts consisting of energetic charged particles. These were found for the first time by the team of James Van Allen with their Geiger counter onboard of the U.S. satellite *Explorer 1* in 1958 (Fig. 19.4). Spacecraft must always be designed to survive the radiation to be encountered on their orbit. Space storms pose particular design challenges. One must be able to estimate the strength of strongest particle events and the total radiation in the orbit during the phase of the solar cycle of the mission. It is also useful if it is possible to switch-off the current to particularly vulnerable subsystems during bad space weather without simultaneously causing too much problems for the primary tasks of the spacecraft.

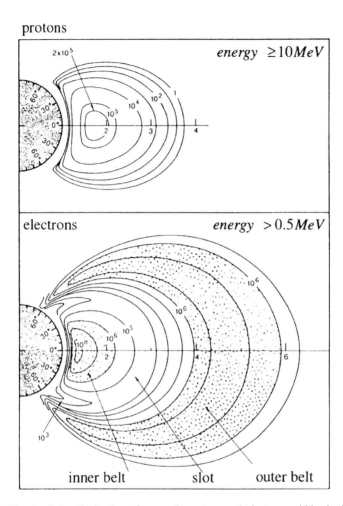

Fig. 19.4 Sketch of the distribution of energetic protons and electrons within the Van Allen radiation belts (Adapted from the textbook of Kivelson and Russell 1995). In the inner belt the energetic particles are both electrons and protons whereas in the outer belt mostly electrons. The outer belt reaches up to the geostationary orbit where most commercial satellites are located. The orbits of positioning satellites in turn move deep into the outer belt

Space storm perturbations together with stronger than average solar ultraviolet radiation also heat the uppermost layers of the atmosphere. This leads to increased drag on spacecraft at low-altitude orbits. The friction decelerates to vehicle and thus lowers its orbit. When the Russian MIR station was finally to be de-orbited, even "good" space weather was a problem. The procedure was first to let the orbit decay by the atmospheric friction before final braking with rocket engines. However, during that period the space environment was more quiet than expected and the whole operation took more time than planned.

19.5 Aviation, Telecommunication and Satellite Positioning

The solar ultraviolet radiation ionizes the upper atmosphere from about 100 km altitude upward and an electrically conductive ionosphere is formed around the Earth. There is considerable day-night variation in the properties of the ionosphere even during quiescence, to which space weather variations introduce their contributions, which are difficult to predict. The existence of the ionosphere was found during the first years of the twentieth century when Marconi succeeded to send radio signals across the Atlantic Ocean. This is possible because the radio wave propagates in a wave guide between the surface and the ionosphere. This is still utilized by the AM radio transmitters at long and medium wavelengths. Also the shortwave (HF) signal is reflected from the ionosphere if it is transmitted in a sufficiently large angle with respect to the vertical direction.

Economically important space weather related communication problems are encountered at commercial airline routes across the polar regions. Poleward of the auroral regions the aircraft cannot communicate reliably through geostationary satellites and become dependent on HF contacts with the aviation control. A strong space storm disturbs the auroral ionosphere and the reliability of the HF links is lost. The only way out of this problem is to fly around the polar region. This adds to fuel costs and increases flight times.

Another aviation issue is, of course, the enhanced energetic particle radiation, which also is strongest in polar regions owing to the quasi-dipolar structure of the terrestrial magnetic field. The problem can be mitigated somewhat by flying at a lower altitude to allow more air mass above the aircraft to damp the radiation. For an occasional traveler the radiation doses remain small, but crews crossing the polar cap frequently must be monitored not to get doses above the safety limits. The European Union has defined a very strict limit to the highest acceptable dose of 6 mSv (millisievert) per year. For pregnant women the limit is 1 mSv for the remaining time until delivery, which may in principle be reached already by being above the polar region during one strong space storm. Furthermore, the ever more miniaturized electronics is becoming increasingly more sensitive against particle radiation and this forces space engineers to include several spare units in the most sensitive and critical systems.

When the frequency of the radio waves increase, the signal can propagate through the ionosphere, which is essential for satellite communications. The communication usually takes place at such high frequencies that the space weather disturbances in the ionosphere do not impose actual direct problems for telecommunication satellites. However, the recent rapid expansion of satellite positioning applications is much more sensitive to space weather disturbances, particularly when maximum position accuracy is sought. Although the positioning satellite signal propagates through the ionosphere and atmosphere without apparent problems, it may during space storms be disturbed enough that the calculation of the receiver's position can lead to errors of tens of meters. The problem can be mitigated by using two different receiver frequencies, but this is not always a practical solution, for example, in cell phone positioning applications. Neither can the aviation rely completely on satellite positioning during space storms.

In radio communications forecasting disturbances due to space weather has long history. In fact, according to radio amateurs, the radio weather is good when space weather is most stormy, because then it is possible to obtain very long connections through ionospheric reflections. Sometimes it is possible even to receive TV signals from the opposite hemisphere.

19.6 Effects on the Ground

Storms in space introduce strong variations in the magnetic field measured on the ground. Consequently, all means of using the magnetic field for orientation are sensitive to space weather perturbations. Particularly important sources of disturbances are electric currents flowing in the auroral ionosphere at the altitude of about 100 km. This phenomenon is not limited to arctic regions, because during strong magnetic storms the auroral activity extends to mid-latitudes reaching central European latitudes.

Besides the deviation of the magnetic needle, which of course can be a serious problem for applications using magnetic orientation, an electric current of a few million amperes at the altitude of 100 km is a rather harmless phenomenon. Serious problems, however, arise from rapid variations of this current during a space storm. This leads to electromagnetic induction, which creates an electric field on the ground. The induced electric field in turn drives strong currents in extended good conductors such as electric power transmission networks, gas and oil pipelines, or submarine telecommunication cables. Thus the space weather effects reach all the way under the ground and bottom of the oceans (Fig. 19.5).

The first technological problems associated with a magnetic storm were identified around 1850 when problems in wired telegraph connections were observed during strong auroras. The first reports on serious problems in an electric power network are from a strong storm in March 1940 in the United States and Canada (Davidson, 1940). During that storm also 80% of long distance telephone network in Minneapolis were out of order.

Thus far the most serious system failures took place in March 1989 in North America. In that storm the electric power distribution network in Quebec collapsed in a few seconds and it took 9 h to get it up again. Smaller disturbances are observed frequently, for example, at the end of October 2003 there was an electric power black-out for about 1 h in the region of Gothenburg, Sweden, as a consequence of a space storm.

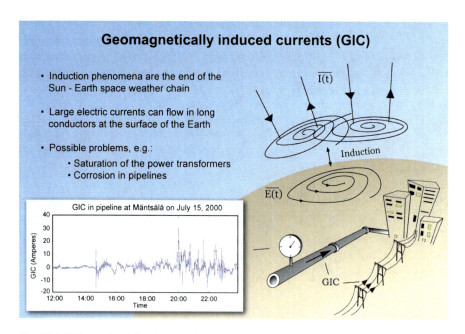

Fig. 19.5 Effects of rapidly changing ionospheric electric currents on the ground (Courtesy of Finnish Meteorological Institute)

19.7 Forecasting is Difficult

There evidently is a need for reliable space weather forecasts in order to mitigate problems and to inform the public of upcoming beautiful auroral displays. To forecast space weather is, however, far more difficult than to predict the atmospheric weather.

A space storm in January 1997 received significant publicity. Just over a year earlier the joint ESA–NASA *Solar and Heliospheric Observatory (SOHO)* had been launched to the Lagrange point between the Sun and the Earth, where the spacecraft remains near the Sun–Earth line as the Earth moves around the Sun (Fig. 19.6). In *SOHO*'s coronagraph pictures a faint mass ejection toward the Earth was observed on January 6, 1997. Scientists contacted the space weather service of NOAA in Boulder, Colorado (http://www.swpc.noaa.gov/). No storm warning was, however, issued because no other space weather indicator pointed to approaching severe weather. The ejection was weak and not particularly fast, and caused only a medium-size storm in the near-Earth space about 4 days later. As a consequence of the storm, the flux of killer electrons in the magnetosphere increased and on January 11, 1997 the telecommunication satellite Telstar 401 was permanently damaged. Next day space weather was front page news in various media. The storm sequence has been analyzed in a number of articles in *Geophysical Research Letters*, vol 25 (14) 1998.

19 Space Weather: From Solar Storms to the Technical Challenges of the Space Age 275

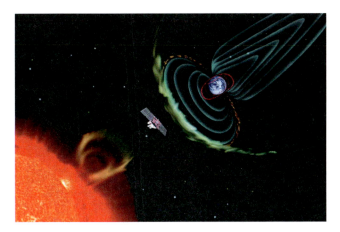

Fig. 19.6 SOHO moves around the Lagrange point between the Sun and the Earth. It observes the sources of space storms on the Sun and the Earthward flowing solar wind and energetic particles. Credit: ESA

Another similar but in that particular case harmless event took place in the spring 2000. On April 4 a relatively weak flare with an associated fast ejecta was observed on the western limb of the Sun, and the probability of the ejecta hitting the Earth was uncertain. The forecast issued by NOAA in the evening of April 5, 2000 gave a probability of 35–40% for a small storm and only 6–11% for a big or severe storm. The shock front driven by the mass ejection hit the Earth's magnetosphere very strongly only 18 h after the too conservative forecast was issued. This led to one of the strongest storms of the solar cycle 23. The storm has been analyzed thoroughly by Huttunen et al. (2002).

There are several time scales associated with predictions. For example, we know that statistically the geomagnetically most active periods occur just before the solar maximum and a couple of years thereafter. When an eruption is observed on the solar surface it is sometimes possible to predict a high probability for a storm to hit the near-Earth space 2–3 days later but predictions for the primary solar eruptions are still in their infancy.

Short term forecasts are much better. Once a spacecraft with appropriate instrumentation observes a shock driven by a mass ejection and measures the direction and magnitude of the associated magnetic field, it is possible to make a somewhat reliable forecast of a magnetic storm starting within a few hours. The case is similar if fast solar wind is observed for several hours where the direction of the magnetic field is conducive for a storm, i.e., antiparallel with the geomagnetic field. Even after the commencement of a storm it may not be too late to take measures against damages because a storm lasts from a few hours to more than a day and, for example, the production of killer electrons to the orbits of geostationary or global positioning satellites takes up to a day after the start of the storm main phase. Thus a satellite operator has some time, for example, to switch-off the electricity from components whose risk for damage is largest. To make such a decision is, however, somewhat awkward because some storms do produce relativistic electrons, whereas others do not. We do not yet really know why it is so.

International space weather activities are often motivated by the need for better forecasts. Unfortunately, such activities sometimes are rather short-sighted: The aim is to produce something very quickly neglecting the unpleasant fact that we do not yet sufficiently well understand the physical processes starting from the Sun and resulting in distortion of the satellite navigation signal, for example, at a busy airport.

19.8 Science of Space Weather Versus Meteorology

The scientific challenges of predicting space weather may be useful to compare with meteorology, which is at a much more advanced level in forecasting.

The medium of space weather is very tenuous ionized gas, plasma. Everyone who has studied plasma physics knows that plasma is a very difficult state of matter to handle both experimentally and theoretically. The hydrodynamics of the atmosphere is, of course, also difficult, but the difficulties are different. Collisions between the atmospheric constituents drive the gas to a local thermodynamic equilibrium at a given local temperature. On the contrary, space plasmas are practically never in thermal equilibrium. Instead in a given volume there can be several particle populations at different temperatures moving to different directions, which can cause numerous electromagnetic instabilities. These instabilities give rise to processes that lead, for example, to the acceleration of the previously mentioned killer electrons or to auroral breakups. The magnetic fields in space guide electrically charged particles and, on the other hand, the electric currents carried by these particles deform the magnetic field. Consequently, there is significant nonlinear feed-back in the space weather system.

The spatial scales of space weather processes strongly couple to each other. In the ionosphere the individual particle motion is characterized by length-scales of the order of 1 m. This motion is simultaneously coupled to the Sun–Earth connection in the scale of 150 million kilometers. This makes local space weather prediction very difficult, contrary to standard meteorology, where local weather for tomorrow or the day after tomorrow can be forecast with great accuracy.

In meteorology the temperature varies within a few tens of degrees. In space weather the temperatures reach from $1,000°$ in the ionosphere (i.e., 0.1 eV in terms energy) to one million degrees (100 eV) in the solar wind, and the energies of protons and nuclei accelerated in solar eruptions coronal mass ejection shocks reach up to energies of billions of electronvolts. Also the time scales vary from microseconds corresponding to radio-frequency perturbations to the sunspot cycle of 11 years.

Finally, space weather research is empirical science, where accurate and comprehensive observations are essential. This is actually the sector where space weather activities are most behind meteorology. The domain to be covered by observations is vast, from the Earth to the Sun, and there are only few well-equipped observing systems. In situ observations from space extend only about 50 years back and time series are short and often discontinuous. For operative space weather forecasting the situation is even worse than for basic research. For example, the Halloween storm in 2003 allows excellent material for research but cannot be used to forecast space weather for the Halloween 2013.

19.9 Challenges for Tomorrow

The modern society is becoming increasingly dependent on space infrastructure and this will unavoidably increase the need for space weather services. For example, the global positioning satellites are in orbits passing through the most intense radiation environment of the outer Van Allen radiation belt. The *GPS* satellites form the presently most used navigation constellation. They have been designed according to military standards and presumably have high radiation tolerance. It is more uncertain if sufficient attention has been paid to space weather hazard in the design of the European *Galileo* satellites. For example, while there is a lot of radiation expertise within ESA, the organization has only recently started to build up a coherent space weather activity within the newly established Space Situational Awareness Programme.

It is evident that space tourism will become a growing activity in the future, and it is likely that someday manned spaceflights to the Moon, and perhaps further, will return to the agenda. Astronauts residing for long periods on the Moon will certainly request for better and more comprehensive space weather forecasts than they can get today.

The biggest challenge of all is, however, our inadequate understanding of the physics underlying space weather. This is due to both insufficient observational data and complicated physical processes. It is not enough for services that we know the big pictures of solar-terrestrial physics. The Devil is in the details and the complicated interaction chains must be understood much better than today. Not nearly enough material or intellectual resources have been directed to this quest. Computer models and simulation tools keep on improving, but the road is long, and the models need input data from sufficiently versatile and comprehensive observation networks.

References

Biese E (1885) Expeditionens fortgång och vigtigaste resultat. Bidrag till kännedom af Finlands natur och folk 42
Cliver EW, Svalgaard L (2004) The 1859 solar-terrestrial disturbance and the current limits of extreme space weather activity. Sol Phys 224:407–422
Davidson WF (1940) The magnetic storm of March 24, 1940 – effects in the power system. Edison Elecric Institute Bulletin, July 1940, pp. 365–366 and 374
Gosling JT (1993) The solar flare myth. J Geophys Res 98:18937–18949

Huttunen KEJ, Koskinen HEJ, Pulkkinen TI, Pulkkinen A, Palmroth M, Reeves GD, Singer H (2002) April 2000 magnetic storm: solar wind driver and magnetospheric response. J Geophys Res 10(A12):1440. doi:10.1029/2002JA0099154

Kivelson MG, Russell CT (1995) Introduction to space physics. Cambridge University Press, Cambridge/New York

Lemström S, Biese E (1886) Exploration international des régions polaires 1882–83 et 1883–1884. Expédition polaire fnlandaise. Le Gouvernement Finlandais – la Societé des Sciences de Finlande, Helsingfors

Chapter 20
Space Geodesy: Observing Global Changes

Markku Poutanen

20.1 Before Sputnik

The idea to measure the precise size and shape of the Earth is very old but before the space age there were no means to do such measurement accurately. Especially oceans were insurmountable hindrances. Sun, Moon, planets and stars have been used since ancient times for mapping and navigation but the accuracy remained quite modest.

Although there were some earlier attempts to determine the size of the Earth (e.g. by Aristotle), the first actual space geodetic measurement is usually credited to Eratosthenes of Cyrene in c. 200 BC. Eratosthenes was also the first to calculate the tilt of the Earth's axis. He determined the circumference of the Earth with remarkable accuracy using the observations of the Sun at two sites on the summer solstice (see Fig. 20.1).

Eratosthenes had heard that on the summer solstice the Sun shone to the bottom of a deep well in the town of Syene (now Aswan in Egypt). At the same time he measured at Alexandria that the Sun at midday casted a shadow, which was equivalent to 1/50th of a circle. Assuming that Syene is directly south of Alexandria and estimating the distance between the two cities, he concluded that the circumference of the Earth is 50 times the distance of Syene and Alexandria.

In fact, Syene is not directly south of Alexandria, neither it is not on the Tropic of Cancer so the Sun was not exactly at zenith on the Summer Solstice. Distance between the cities was also quite uncertain. After all, it is surprising that the Eratosthenes' result deviates from the actual circumference of the Earth less than 1%. It is partly due to good luck and errors cancelling each other, but it also shows that the method itself was adequate. Unfortunately, Eratosthenes' results were partly forgotten during the next centuries, and later determinations about the size

---

M. Poutanen (✉)
Finnish Geodetic Institute, P.O. Box 15, FI-02431 Masala, Finland
e-mail: markku.poutanen@fgi.fi

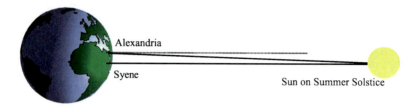

Fig. 20.1 Eratosthenes method to determine the circumference of the Earth

of the Earth were less accurate, some even badly erroneous. These may have had some consequences to the great journeys of exploration in the early centuries of modern times.

Latitude determination is relatively simple using the Sun or measuring altitude of stars. These quantities are absolute and change with the distance to the equator. Longitude, however, is much more difficult to determine because it requires ability to measure time and time differences accurately. There is no logical zero for the longitude, contrary to the latitude where the equator forms the natural zero. Therefore, one can speak only longitude differences between two sites. There was no single zero meridian but many sites have been used for this purpose during centuries: Rhodes, Alexandria, Kap Verde or Canary Islands at the western border of the known world in antiquity, or e.g., Paris, Pulkovo or Washington in modern times. Our habit to use Greenwich as the zero meridian is only a bit more than 100 years old. In 1884, representatives of 25 countries agreed at a conference in Washington, USA, that Greenwich meridian would be adopted as the zero line.

Longitude difference of two sites is actually the same as the difference of their local times. The Earth rotates 360° in 24 h, i.e., 15° in 1 h. If a site is 15° east of another place, the local time there is 1 h ahead of the latter one. Before modern clocks this was too demanding task to measure. In ancient maps the difficulty is clearly visible: maps are quite precise in north–south direction but in east–west direction distortions can be remarkable. The difficulty continued long in modern times; King Louis XIV in France complained at the end of the seventeenth century that he has lost more land to his astronomers than to his enemies. The reason was that due to increased accuracy of longitude determination, the west coast of France moved quite a distance to the east.

One possibility to measure longitude differences was to use some celestial event, such as the lunar eclipse, which is visible simultaneously on several places. This offers a possibility to measure the local time at the same moment at both sites. The method was known already in antiquity but results were very modest and errors in longitude were even more than 10°. Combined with the erroneous estimations of the circumference of the Earth gave rise to the totally false assumption of the distances between continents.

This false assumption may have had some consequences even in the great journeys of exploration. In Columbus time the size of the Earth was considered too small and the distance from Europe to China across the Atlantic Ocean was believed to be less than 1/4 of the circumference of the Earth. It was Columbus' luck that a new continent happened to be in between. Columbus' expeditions used the lunar eclipse method in 1494 and 1503 to determine the longitude of the new continent, but both determinations were badly erroneous, giving 20° and 40° too western longitudes. Only more than two centuries later precise portable chronometers finally solved for the problem of longitude determination.

Meanwhile terrestrial techniques were developed. The Dutch mathematicians Gemma Frisius and Willebrord Snellius, and the Danish astronomer Tycho Brahe developed triangulation in the sixteenth and seventeenth centuries for mapping and surveying purposes, although similar technique was used already in China in the second century BC. In triangulation a chain of adjacent triangles are measured. If one measures two angles in each triangle and the length of one side, all other angles and sides can be calculated. This allows each observing site (corners of triangles) to be positioned relative to the others. Measuring angles is much faster than a precise distance measurement; therefore precise mapping of large areas became possible. However, one needs visibility between two points of observations and oceans remained insurmountable.

Triangulation allowed a more precise determination of the shape of the Earth. The question if the Earth is flattened or prolonged was argued in the early eighteenth century when the French Academy of Sciences decided to send two expeditions to solve the question. The expedition of Pierre Louis Moreau de Maupertuis travelled far north to Lapland to the River Tornio and expedition of Charles Marie de La Condamine and Pierre Bouguer went to Peru (nowadays territory of Ecuador), near the equator. The question was solved already in 1736–1737 when measurements of Maupertuis expedition showed that the length of one-degree-long meridian arc is longer at the north than in France. Some years later the result was confirmed when the Peruvian expedition returned back to France.

Much more extensive works were conducted in the nineteenth century. The most famous one was initiated in 1816 by the German-born astronomer Friedrich Georg Wilhelm Struve. The measurements were completed in 1852. The chain of triangles stretches from the Black Sea to Hammerfest in Norway; a total of 2,820 km. Struve conducted the work first from Dorpat Observatory (Tartu in Estonia) where he lived at the time, and later on from Pulkovo Observatory near St. Petersburg, the first director of which he became in 1839. The value of the measurement is recognized even today, and the Struve Geodetic Arc was inscribed on the UNESCO World Heritage List in 2005.

The famous Swiss mathematician and expert in celestial mechanics, Leonhard Euler (1707–1783), mentioned an idea to use solar eclipses in distance measurements. When starting times of an eclipse are observed at two sites along the zone of totality, the distance between the sites can be computed if the speed of the shadow on the Earth surface is known. There were no technical possibilities to perform such a measurement at Euler's time but totally new innovations were needed: radio and a movie camera. This happened two centuries later. With a radio one can get precise time signals simultaneously at both sites and the eclipse is recorded with a movie camera together with the time signal markings on the film. The measurement was made successfully in 1947 when expeditions of T.J. Kukkamäki and R.A. Hirvonen of the

Finnish Geodetic Institute measured the distance of sites in South America and Africa with an accuracy of 141 m (Kukkamäki and Hirvonen 1954), see Fig. 20.2. This was a remarkable increase in accuracy and a proof of the potential of space geodetic techniques.

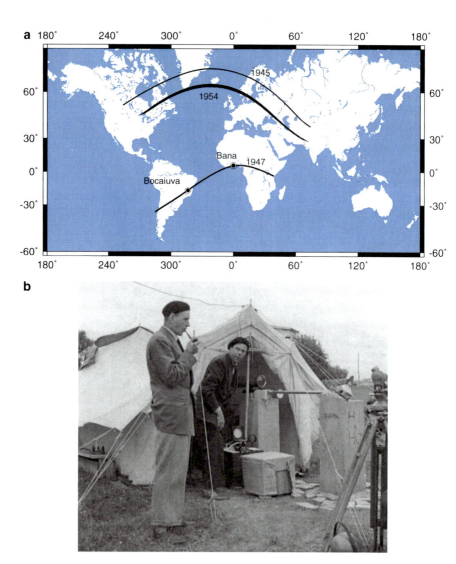

Fig. 20.2 (a) Total solar eclipses were used for distance measurements in mid-twentieth century. Only the measurement of 1947 was successful. (b) R.A. Hirvonen and T. Honkasalo of the Finnish Geodetic Institute during the total solar eclipse on Öland Islands in 1954. Clouds prevented observations (Finnish Geodetic Institute)

20.2 Under the Clouds

Solar eclipses are rare and even rarer to have a favourable path of totality to connect two continents. In 1954 a large international campaign was organized to observe the great solar eclipse ranging from Northern USA and Canada to Scandinavia, Soviet Union, Iran, Pakistan and India. Mother Nature showed her bad side to all expeditions: the eclipsed Sun was behind clouds at all sites! This was the last trial to use eclipses for distance measurement. First artificial satellite *Sputnik* was launched to orbit the Earth 3 years later, in 1957. Finally the eternal dream of geodesists became true. Satellites enabled measurements between continents.

First positioning experiments were made almost immediately based on optical observations of satellites using stellar triangulation. The principle of stellar triangulation was introduced in Finland much earlier by Academician Yrjö Väisälä. He described the method already in 1946 and later on tested an application where a flashlight is lifted with a balloon up to 25–30 km above the ground. When the flashes are photographed simultaneously at two sites against the starry background, one can compute the distance of the two sites. A five-point network was measured by Juhani Kakkuri in Southern Finland in late 1960s – early 1970s (Kakkuri 1973). Longest distances were more than 200 km.

With satellites, one is not limited to couple of hundreds of kilometres but can measure distances of even thousands of kilometres thus allowing intercontinental measurements. Especially the US launched several satellites for this purpose. The most famous one was *Pageos* (Passive Geodetic Earth Orbiting Satellite, launched in 1966), a 30-m balloon-like satellite which was easy to observe when sunlit. Already in mid-1970s as a result of an international co-operative project there existed the Worldwide Satellite Triangulation Network of 45 points where the mean accuracy of points was better than 5 m (Schmid 1974).

Weather was a severe problem also with optical satellites and cloudy skies delayed observations. Additionally, measuring the position of a satellite relative to the stars limited the accuracy and there were no means to observe the distance of a satellite. Optical observations were replaced by radio waves which can be detected through the clouds and they allowed also the precise distance measurements.

First Transit Doppler satellites were launched in the 1960s. The system was primarily used by the U.S. Navy to provide accurate location information to submarines and ships, but satellites were also used for geodetic purposes. Using the Doppler delay of several passes of a Transit satellite, one achieved a relative accuracy of a few decimetres in geodetic positioning. During the two decades many international Doppler campaigns were arranged, until at the end of the 1980s the GPS (Global Positioning System) superseded Doppler. The Transit Doppler system was officially ceased in 1996.

GPS allowed precise measurements in hours (nowadays in minutes), it is completely weather-independent, and receivers are inexpensive and simple to use. GPS applications have increased rapidly and small handheld navigators are now familiar to most of us. Today everybody can get 1-m accuracy in minutes, which was mere a dream in 1960s. With geodetic GPS receivers one can follow plate tectonics or even minor local deformations almost in real-time.

20.3 Gravity Wags the Satellite

Most satellites orbit the Earth at the altitude of a few hundred or a few thousand kilometres. Weather and communication satellites are on geostationary orbits, about 36,000 km above the equator. Orbital period of these satellites is the same as the rotation period of the Earth and therefore they seem to stay above the same spot on the Earth. Navigation satellites (GPS or the European Galileo) are in somewhat lower orbits, about 20,000 km above the ground.

In addition to positioning one can use satellites to determine the shape of the Earth's gravity field. Bumps and depressions of the gravity potential field dictate the orbit and orbit changes of a satellite. Only 1 year after the first *Sputnik* the flattening of the Earth was computed from the orbits of satellites, and a few months later the asymmetry between the North and South Pole was obtained. These were already known quite well, but with redetermination of the parameters the new technique showed its speed and usability.

The closer the satellite orbits, the smaller are the details they can detect. Geostationary satellites are useless in this respect because they remain above the same point on the Earth. GPS satellites are also too far to reveal much more details. Satellites orbiting a few hundred kilometres above the ground are most useful for gravity field observations but the air drag becomes a problem. In course of the 50-year history of satellites, perhaps the biggest leap was taken after year 2000 when three gravity satellites were launched to orbit the Earth.

CHAMP (CHAllenging Minisatellite Payload) was launched in 2000, followed by *GRACE* (Gravity Recovery And Climate Experiment) in 2002 and the European Space Agency (ESA) *GOCE* (Gravity field and steady-state Ocean Circulation Explorer) satellite in 2008. They have revolutionized the gravity measurements (Fig. 20.3).

There are two innovations that make this possible. First one is a precise accelerometer which can detect very small changes in acceleration. The second one is a GPS receiver that onboard the satellite allows a cm-range positioning.

The shape of the Earth, the geoid, can be determined within couple of centimetres in the spatial resolution of 100 km. This makes possible to create a global unified height system where sea level changes and glacier melting can be detected. Earlier such a precise global height system was not possible.

Especially *GRACE* time series show changes in geoid and sea surface. Monsoon or dry seasons change the ground water level in tropical and subtropical areas (e.g. Amazonas), which is clearly visible. Even much smaller changes can be detected, like annual groundwater change in Finland or water balance of the Baltic Sea. Melting of the Greenland or Antarctic glacier is a more complicated question, and

Fig. 20.3 GOCE satellite of the European Space Agency (ESA) was launched in year 2009. GOCE is mapping global variations in the gravity field with extreme detail and accuracy. Credit: ESA

for this we'll need the full selection of space geodetic techniques. But what are those modern techniques and what is the scope of geodesy?

20.4 Three Pillars of Geodesy

Historically, geodesy has been defined as a discipline of geosciences that is "concerned with measuring or determining the shape of Earth or a large part of its surface, or with locating exactly points on its surface". The definition originates from the famous German geodesist, Friedrich Robert Helmert (Helmert 1880). With modern techniques, the scope of geodesy has extended beyond Helmert's definition to include the dynamics and mass transport within the Earth system. At the same time, Heinrich Bruns, mathematician, astronomer and geodesist discussed a wider scope of geodesy, which is now known as the "three pillars of geodesy" (Bruns 1878): measuring the geometric shape of the Earth's surface and its kinematics; Earth rotation and position in space; and determining and monitoring the Earth's gravity field and its changes. In satellite era these tasks are still valid (Fig. 20.4).

Increased accuracy requires also more stable reference systems. Continental drift as well as local deformations degrade the accuracy by moving the ground-fixed benchmarks defining the reference frames, and these motions should be known better. One can compare measurements today with measurements 10 years from now only if the reference frame remains stable over decades. It is a central task of geodesy to create and maintain such systems, understand deformations, perform the most precise measurements and offer best possible coordinates and time series to users. Studying the sea level or glacier mass changes are examples of long term studies.

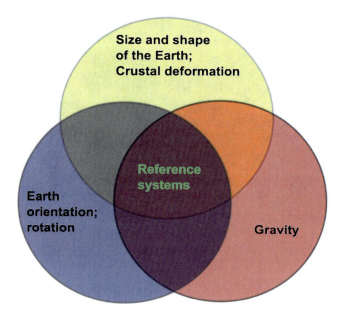

Fig. 20.4 Three pillars of geodesy

Navigation satellites orbit the Earth, and for positioning one needs to compute the distance between the observer and a satellite. For this we need precise orbits of the satellites but we have to know also the orientation of the whole Earth in space. Space geodetic techniques, especially VLBI (Very Long Baseline Interferometry) and SLR (Satellite Laser Ranging) are crucial in this. Without continuous observations VLBI observations of the Earth orientation in space, navigation satellites would not be able to be used in such high accuracy.

The third pillar, gravity, is needed when satellite orbits are computed. It is needed also in height determination. Traditional levelling gives heights above the sea level and it also tells the direction the water flows. Satellite measurements give only geometric quantities, distance from the centre of the Earth, and therefore one needs a gravity-based geoid model to convert heights from satellite measurements to heights obtained with levelling. Gravity satellites can also detect large scale mass changes like melting of glaciers but ground based gravity measurements are still needed for local details.

20.5 Space Geodetic Techniques and Global Networks

Accuracy of space geodetic measurements is amazing. Just imagine sitting behind a table with a sheet of paper on the table. Mark a cross on the paper and then measure the distance of the cross to the curtain on the window some meters away with an accuracy of a millimetre. Coming back next week and repeating the measurement most likely will give different result. The question is if the difference is caused by the measurement uncertainty, or if the sheet of paper has shifted on the table, perhaps the whole table has moved or the curtain has moved. Geodesists are regularly doing measurements where nothing can be considered fixed and where distances are not meters but thousands of kilometres. Our information, e.g., on global change is based on reliability, repeatability and accuracy of geodetic measurements.

Global Positioning System GPS is the best known space geodetic technique (Figs. 20.5 and 20.6). There are lot of applications from navigation to most precise geodetic measurements where even country-wide millimetre-sized motions can be detected over time span of years. GPS has superseded many traditional geodetic techniques and its speed and accuracy are superior to most of them. However, GPS alone is not sufficient to maintain global reference frames or follow the orientation

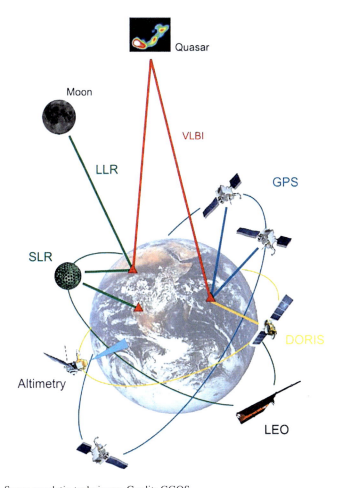

Fig. 20.5 Space geodetic techniques. Credit: GGOS

Fig. 20.6 GPS technique has fully changed geodetic measurements. Speed and accuracy allow precise measurements of large areas or networks in a short time (H. Koivula, Finnish Geodetic Institute)

of the Earth in space. The Russian navigation system GLONASS or the European Galileo will not change this situation albeit they will improve ordinary navigation or surveying measurements. We need also other space geodetic techniques.

VLBI is the most fundamental technique where large radio telescopes, often on different continents, observe distant celestial objects, quasars. Quasars form a fixed network of reference points where the orientation of the Earth can be measured. VLBI is the only technique that is independent of satellites, and it cannot be replaced with any other technique. GPS and other navigation satellites cannot function without the information of the Earth orientation from VLBI measurements.

SLR is another indispensable technique. Short laser pulses are shot towards a satellite equipped with a retroreflector and reflections are detected with a telescope. Distance of a satellite can be measured within millimetres using the flight time of a pulse. SLR is used for satellite orbit determination but also to compute the precise position of the mass centre of the Earth. Origin of our reference system is placed in the mass centre and therefore its position and especially temporal variation must be known. Annual variation is centimetres, and even 1 mm unknown change can cause an error in our reference frame where we try to retrieve millimetre-sized annual changes in sea level height.

One observing station is not enough but one needs a network of fundamental stations over the globe. There are more than 30 VLBI and SLR stations in the world and hundreds of GPS stations in the same network. These stations, however, are unevenly distributed. Oceans cover 2/3 of the surface of the Earth but also land areas are not equal. Most stations are in Europe and North America, some in Japan or Australia, but very few elsewhere (Fig. 20.7).

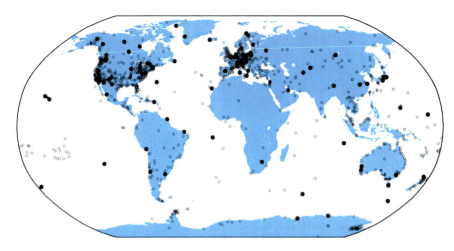

Fig. 20.7 Network of global geodetic stations which is used for maintenance of the global reference frame. Most of the points are permanent GPS stations. There are only 30–40 SLR and VLBI stations

Fig. 20.8 Metsähovi Fundamental Station is one of the global fundamental stations hosting all major space geodetic instruments (J. Näränen, Finnish Geodetic Institute)

Observations are coordinated and collected by services of the International Association of Geodesy (IAG). As a result of these observations global reference frames are maintained and data and results are freely available via Internet. Services and observing stations are maintained by national mapping authorities, research institutes and universities. All these rely on national funding. Political decisions, short-sighted savings or changes in the strategy of an institute threaten such structure. Stations and working groups are ceased arbitrarily and there is no guarantee on continuation.

It is surprising that such a fundamental service and infrastructure of the whole civilized world depends on a "voluntarily" network of geodetic stations (Fig. 20.8). It is estimated that about 500 person years per year are provided on a voluntary commitment to maintain the ground-based networks and to run data and analysis centres (Plag and Pearlman 2009). Investments on instruments and infrastructure are even bigger. It is one of the biggest challenges in the future to secure the continuation of global networks and to maintain geodetic knowhow. Geodesy's reply to this challenge is the Global Geodetic Observing System, GGOS (Plag and Pearlman 2009).

GGOS is a part of IAG, and it is being built on the existing IAG Services as a unifying umbrella. GGOS provides the links between IAG services and science groups or professionals needing such precise and reliable geo-data, United Nations or other international programs like the Group on Earth Observation (GEO) which is implementing the Global Earth Observation System of Systems (GEOSS), and ultimately to the great public who use space geodetic products and services. GGOS has also started an initiative at the international level to develop an intergovernmental agreement to secure existence of geodetic reference frames and high-quality network in the future.

20.6 The Challenge of the Global Change

We will now return back to the question of measuring sea level changes and crustal deformation. Continental drift is visible in satellite measurements, especially in time series obtained with continuously observing permanent GPS stations (Fig. 20.9). The speed is only a few centimetres per year, but during millions of years they can cause even high mountain ranges when continents collide each other. Violent earthquakes originate from the same motion. We do not fully understand

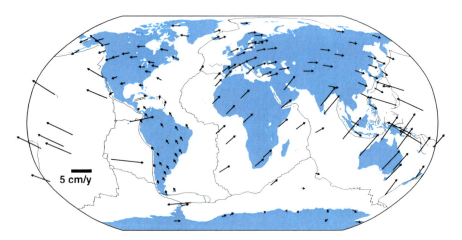

Fig. 20.9 Space geodetic observations can be used to observe motion of conti-nents. Tectonic motions are associated with earthquakes and volcanism

how all the motions happen, and we are far from predicting disastrous earthquakes accurately.

There are also other crustal deformations that happen more slowly and less violently. Postglacial rebound in Fennoscandia and Canada started after the last ice age, more than 10,000 years ago, when up to 2–3 km-high glaciers melted in these areas. Crust was depressed more than half a kilometre under the ice and the rebound is still continuing, about 1 cm/year near the maximum areas of the uplift. At the Gulf of Bothnia, a narrow and shallow sea area between Finland and Sweden, the uplift will continue almost 100 m more. Kvarken, the narrowest part of the gulf, and a site in the UNESCO list of World Heritage, will disappear in a few thousand years.

Crustal deformations are associated with gravity changes. The shape of the Earth is changed, mass distribution is changed, and there exist also slow mass flow in the mantle of the Earth. We know the relation between the uplift and gravity change locally in Fennoscandia but what happens in Greenland or Antarctica. In addition to crustal movements, there exist also glacier mass changes. How to separate these from each other?

A gravimeter on the ground feels gravity change due to two different effects: due to the mass change and due to the height change caused by the uplift or subsidence. If we lift the gravimeter higher, it feels smaller gravity. With repeated GPS observations during several years we can measure height changes; GPS is totally insensitive to changes in gravity.

A satellite orbiting the Earth feels only mass changes, like decreasing the mass due to melting of a glacier. It is insensitive to the height change of the crust. With gravity satellites it is possible to determine the geoid within centimetres but also detect its changes (Fig. 20.10). But what is the rate, how much and where the glacier melts and where the melt water is flowing? Are there any other causes for the mass change? Does the precipitation change and increase glacier in other areas? There are still a lot of open questions after such observations.

Let's consider the distribution of the melt water. When the mass of the melting glacier is decreasing, the gravity caused by the glacier is also decreasing. Melt water flows mostly to the opposite side of the globe increasing mass there, which attracts more water. As a result, the sea level is not rising in the vicinity of the melting glacier.

Radar altimetry satellites measure variations in sea surface height. If one ignores winds or sea currents, there are two other major effects, thermal expansion of the water and actual increase (or decrease) of the amount of the sea water. Thermal expansion can change sea surface height even a couple of decimetres over several years, as happens during the El Niño in the Pacific Ocean when the sea water temperature is changed by a few degrees. Another example is the Golf Stream where one sees a ridge of several decimetres because water is warmer there than in the surrounding ocean. What if the global 1 mm increase in the sea level height is just due to global warming? How to handle the uneven distribution of the melt water and where to measure the height change?

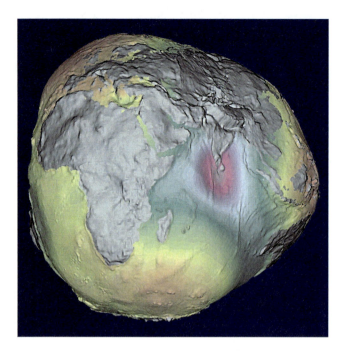

Fig. 20.10 Free sea surface determines the shape of the geoid. On continents the shape can be computed from gravity observations. With satellite observations one can compute precise geoid model everywhere on the Earth. Large depression in the geoid south of India is due to the mass deficiency in the upper mantle. It is a consequence of the Indian plate movement to the north. Slow motions of mantle material will gradually remove the deficiency. Vertical changes are greatly exaggerated; the actual depression is about 100 m. Credit: GFZ-Potsdam

Tide gauge observations at the coasts will not give the final answer either. When the amount of water is increased in the seas, the sea floor is pressed down. At the same time a slightly lighter continental crust is squeezed up, and the volume of the seas is also changed. The net effect of the sea surface rise at the coast is different from that computed directly from the amount of the melt water. We may get contradicting observations or even misinterpretations if only one method is used.

The only way to understand and measure the effect of global change is to use all our tools in space geodesy toolbox combined with terrestrial observations. With GPS one can measure changes in crustal height near the glacier. Radar altimetry satellites measure height changes of the glacier and sea surface and gravity satellites observe the total mass change. Satellite laser ranging is needed to calibrate the radar satellite orbits, and also to track any changes in the mass centre of the Earth. VLBI is needed for fixing the orientation of our reference frame and ensure the stability of the scale. All these must be done over a decade or more within an accuracy of millimetre.

After all, this is not so simple...

References

Bruns H (1878) Die Figur der Erde: Ein Beitrag zur europäischen Gradmessung. Publication des Königl Preussischen Geodätischen Institutes, Berlin

Helmert FR (1880) Die mathematischen und physikalischen Theorien der höheren Geodäsie. B.G. Teubner Druck und Verlag, Leipzig

Kakkuri J (1973) Stellar triangulation with balloon-borne beacons. Publ Finn Geod Instit 76:1–48

Kukkamäki TJ, Hirvonen RA (1954) The finnish solar eclipse expeditions to the Gold Coast and Brazil 1947. Publ Finn Geod Instit 44:1–71

Plag H-P, Pearlman M (2009) Global geodetic observing system. Meeting the requirements of a global society on a changing planet in 2020. Springer, Heidelberg

Schmid HH (1974) Worldwide geometric satellite triangulation. J Geophys Res 79:5349–5376

Chapter 21
Destination Mars

Risto Pellinen

21.1 Early History

The study of Mars is about to become the largest and the most systematic long-lasting research project of mankind. It was started by Galileo Galilei (1564–1642) almost exactly 400 years ago. Realising that telescopes were useful for astronomical observations, he was the first astronomer to make scientific observations of Mars. In 1610 he wrote: "I'm not sure I can see the phases of Mars, however, if I'm not mistaken, I believe I see it not perfectly round". Actually he had found the varying phases of Mars (Pellinen and Raudsepp 2006).

The next step was taken in 1659 by Christiaan Huygens (1629–1695), a Dutch astronomer. By observing gradual motion of a V-shaped dark area on the planet surface he discovered that the rotation period of Mars was about 24 h, very close to the presently known value (24 h 37 min 23 s). This observation was confirmed by the Italian-French astronomer Giovanni Domenico Cassini (1625–1712) who found that rotation period was 24 h and 40 min and by the German-British astronomer William Herschel (1738–1822) who was able to diminish the error to 2 min. Both Huygens and Cassini made the first observations of the southern white polar cap. By using stellar occultation techniques Cassini concluded that Mars had an atmosphere and Herschel in the late eighteenth century confirmed that the Mars atmosphere was tenuous and had clouds and seasons that deformed the polar ice caps.

Mapping of Mars started in 1830, when Mars was near Earth. Two German astronomers, Wilhelm Beer (1797–1850) and Johann von Mädler (1794–1874), managed with their own data to produce the first map of Mars in 1840.

In 1877–1890 the Italian astronomer Giovanni Virginio Schiaparelli (1835–1910) made continuous, detailed observations of Mars at the Milan astronomical observatory. In the most detailed map published in 1878 (refined in 1879) he used Latin and

R. Pellinen (✉)
Leipurintie 16 F, FI-00620 Helsinki, Finland
e-mail: risto.pellinen@presentor.fi

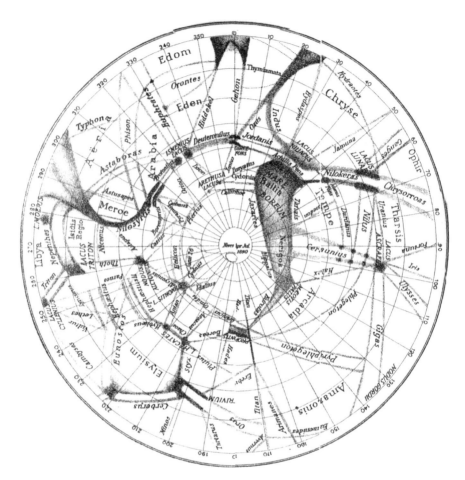

Fig. 21.1 One of Schiaparelli's original Martian maps (Copyright: Osservatorio Astronomico di Brera)

Mediterranean place names taken from ancient history, mythology, and the Bible (e.g. Elysium, Arcadia, Arabia). In 1877 he also started to use word "canali" (in Italian) of long straight features on the surface meaning "channels" that seemed to connect "seas" and "lakes". The term was famously mistranslated as "canals", which meant artificially produced landforms. They seemed to be produced by highly intelligent life forms to conduct water from polar regions to plains at the equator (Fig. 21.1).

The American astronomer Percival Lowell (1855–1916) continued Schiaparelli's work in 1903–1907 at his own observatory in Flagstaff, Arizona. He strongly believed that the canals were water pipes surrounded by vegetation zones. These areas had to be more than 80 km wide to be visible to Earth.

The Turkish-born astronomer of Greek descent, Eugène Michael Antoniadi (1870–1944), was given access to the largest telescope in Europe at the Meudon Observatory during the 1909 opposition of Mars. After careful observations he concluded: "The geometrical canal network is an optical illusion." This was the end of speculation; as a next step we had to go there (Hotakainen 2008).

21.2 First Missions to Mars

Since the launch of *Sputnik 1* in 1957 the launch vehicle fleet in the Soviet Union was growing quickly, mainly to meet planetary flight conditions. In 1959 Sergei Korolyov (1907–1966) had developed a 4-stage 'Molniya' planetary launch vehicle and a new, much improved multi-purpose modular spacecraft was quickly developed. The second launch in 1962, named *Mars 1*, succeeded but the mission was lost after 5 months.

The activities continued until 1973 with altogether 17 attempts were made of which many failed. Six launch windows (separated by 26 months) were used; only the one in 1967 was missed. Some data on upper atmosphere composition were collected; no magnetic field was observed; surface temperature variations were observed during different atmospheric conditions; CO_2 column densities were measured; and temperature and density profiles were obtained by occultation techniques, as were *in situ* pressure and temperature profiles by means of landing probes. Some images from *Mars 4* on flyby, and from *Mars 5* in orbit, were received. The key issue for future was testing and development of spacecraft and observing technologies. For this purpose a completely new heavy spacecraft and the new Proton launcher were designed and built in the years 1966–1969 (Pellinen and Raudsepp 2006).

After 1973 there was a period of 15 years before the next launch to Mars took place in the Soviet Union.

In November 1964 the first two U.S. spacecraft, *Mariner 3* and *4*, were launched towards Mars. *Mariner 3* failed at launch but *Mariner 4* was able to reach the planet and pass it at a distance close to 10,000 km. During the flyby on 15 July 1965 21 images of the planet surface were taken. These were the first close-range pictures of another planet taken by mankind. The images showed Mars to be heavily cratered and a very dry planet, with no channels or any other indication of life (Fig. 21.2).

The United States continued up to 1975 and eight attempts were made of which only two failed. Four launch windows were used; the windows in 1967 and 1973 were missed. The whole planet was mapped in rough scale. Two long-lived *Viking* landers were placed on Mars surface in 1976. No life was detected in the surface soil but the landers provided extensive long-duration sets of surface meteorology and climate data. The two landers were operated until 1980 and 1982, respectively. Atmospheric composition was measured with high accuracy. Elemental composition of the soil and rough mineral mapping was done on the surface and from orbit.

After 1975 there was a period of 17 years before the next launch to Mars took place in the United States.

Fig. 21.2 One of the first images (No. 11) of Mars taken by the Mariner 4 spacecraft (Credit: NASA)

21.3 Space Era History: "Bridging Phase"

In spring 1985 the Soviet government decided on a highly ambitious and massive new Mars programme. The Soviets also decided to invite international partners to the programme.

Two identical 2,600 kg (dry mass, total mass 6,220 kg) spacecraft of a completely new design were developed. The goal of these two missions was to study Mars from the orbit and to fly-by and to drop two landers on the Martian moon Phobos. Both Proton launches were successful but contact with *Phobos 1* was lost after 2 months due to an error in a steering command. *Phobos 2* reached Martian orbit in late January 1989 but was never able to reach the vicinity of the Phobos moon, its ultimate target. However, it was able to collect entirely new information of the atmospheric escape, a result of direct interaction between the upper atmosphere and the solar wind. This was made by means of the ASPERA – Finland's first space experiment provided together with Sweden (Pellinen and Raudsepp 2000).

Quite soon after the launches of the two Phobos missions, planning for a new and even more ambitious *Mars-92* dual mission was started. Originally two large orbiters should have carried 40 instruments, two landers, two penetrators, one balloon and one rover to be deployed on the surface and atmosphere of Mars. The program was renamed as *Mars-96* and modified to single launch on 16

November 1996, which failed due to a problem in the fourth stage booster of the Proton launcher.

Finland with an unusually large contribution was one of the international partners in this programme. Highly sophisticated new instruments were developed for the *Mars-96* mission, one of them in Finland. The miniaturized instruments built for the landers have been re-flown on the U.S. surface missions to Mars and the orbiter instruments are now producing new dramatic data on ESA's *Mars Express* mission. Without the challenge offered by the *Mars-96* mission these instruments might not have been developed at all.

U.S. scientists and institutes were also involved in the *Mars-96* mission. Simultaneously the United States also wanted to restart its own activity, which led to the *Mars Observer* mission (with seven highly sophisticated experiments) launched successfully in 1992. Unfortunately the spacecraft was lost on approach to Mars. As a consequence of this loss a new approach to distribute risk with smaller spacecraft over multiple launches was implemented in the U.S.

In the 1996 launch window Americans were able to test their new ideas and approach by sending a lander, *Mars Pathfinder*, with a small rover and a separate orbiter, *Mars Global Surveyor*, to Mars. The lander mission collected about 16,500 pictures, while the 10.5 kg rover *Sojourner* took 550 images of the rocks and soils and made 16 chemical studies of the rocks. During 3 months the rover traversed altogether about 100 m.

The *Mars Global Surveyor (MGS)*, was equipped with six primary experiments: high-resolution imager, laser altimeter, magnetometer, radio science experiment for gravity field measurements, thermal emission spectrometer to look at the surface composition and atmospheric properties and a data relay system to communicate with the surface experiments. The *MGS* high-resolution imager found clear soil layering and signs of water. The laser altimeter MOLA mapped the entire planet with 1–10 m altitude accuracy. The global images clearly indicate differences in the hemispherical terrain formations: southern highlands with mountains and a smooth almost flat northern hemisphere. The altitude variation across all of Mars is 31 km (Fig. 21.3).

In this phase, which we call "the Bridging Phase" three more missions were carried out before entering the current era of continuous, intensive, multi-spacecraft and multi-national exploration of Mars.

In the 1998–1999 launch window the U.S. launched two missions to Mars: *Mars Climate Orbiter* and *Mars Polar Lander*, both of which were lost at Mars. The reasons for the failures were discovered and careful measures were taken not to repeat the mistakes again. The U.S. programme was restructured by replacing the two planned 2001 landers with a single orbiter, and rescheduling future launches. Today the new approach seems to be highly successful.

The third mission to be mentioned in this phase is the Japanese *Nozomi*, launched in summer 1998 into a parking orbit around Earth, and after 6 months boosted towards Mars. Unfortunately, a malfunction in the propulsion system delayed its arrival at Mars until January 2004. High solar activity during the long interplanetary cruise time damaged some key electronics of the spacecraft, which,

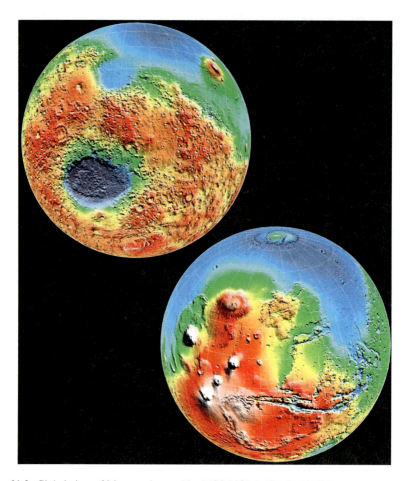

Fig. 21.3 Global view of Mars as observed by MGS MOLA (Credit: NASA)

ultimately, prevented *Nozomi* from being inserted into a Martian orbit and the mission was lost.

21.4 Space Era History: Mars Today

Today in late 2011 there are three operational satellites orbiting Mars, one rover still traversing the surface of Mars and another in a jammed position probably not able to operate any more. One spacecraft landed in May 2008 at a Martian arctic site and was successfully digging into the soil up to the end of October 2008.

The oldest still operational orbiter, *Mars Odyssey (MO)*, was launched by NASA in 2001. It has four instruments onboard to look for surface mineral composition and subsurface water ice and to monitor the radiation environment. *MO* has found that in certain areas subsurface water is frozen close to the surface. This observation is supported by the *Mars Express* PFS instrument, which has found both enhanced water vapour and methane fluxes in the same areas. Speculation about the existence of a simple form of underground biological activity in these areas is ongoing.

Mars Express (MEX) was launched by ESA in 2003 In addition to its main payload, it carried the *Beagle 2* lander to Mars, which failed during the landing phase. The instruments onboard the orbiter are taking color 3D-images, looking for surface composition and subsurface water, and measuring atmospheric composition. The *MEX* High Resolution Stereo (Colour) Camera has been imaging Mars since January 2004. At the moment the entire surface of the planet has been photographed in high resolution and in real colour (Pellinen and Raudsepp 2006).

The most recent mission, NASA's *Mars Reconnaissance Orbiter (MRO)* launched in August 2005, has been orbiting Mars since March 2006. *MRO* is equipped with cameras to allow for extreme close-up (30 cm resolution) photography of the Martian surface. Other science instruments are identifying surface minerals, looking daily global weather, doing atmospheric profiling and studying how dust and water are transported in the Martian atmosphere. *MRO* carries a radar sounder (quite similar to the one on *MEX*, but operated at a higher frequency) to look for subsurface water layers. Other main objectives of the mission are to search for future safe landings at scientifically interesting sites and to be the first dedicated link in a communications bridge back to Earth. The mission will also serve as a high-precision interplanetary lighthouse to guide incoming spacecraft as they approach Mars.

Two rovers were launched by NASA in 2003. The *Spirit* rover landed in the Gusev crater on 3 January 2004 and the *Opportunity* rover in Meridiani Planum on 24 January 2004. The rovers are identical carrying imaging and geochemical instruments to look for surface rock composition and thermal properties of the soil and atmosphere. Originally designed for 3 months operation they have now been traveling for eight years (four Martian years) and at least *Opportunity* is expected to continue for some time. *Spirit* is stuck into sand and had its last communication with Earth on 22 March 2010. The rovers have traversed 7.73 km *(Spirit)* and 34.46 km *(Opportunity)*. On their way they have stopped to photograph panoramas and to conduct geological investigations with seven different instruments. At both landing sites they have found physical and chemical evidence of water that earlier had existed at these sites (Squyres 2005).

Phoenix was launched in August 2007 towards Mars and landed on 25 May 2008 in the Martian northern polar region at 68.2 N and 125.7 W (Fig. 21.4). *Phoenix* operated successfully about 2 months longer than its planned 3-month mission. After arrival, *Phoenix* deployed its robotic arm to dig trenches up to half a meter into the layers of water ice. The selected soil samples were heated in a small oven to release volatiles that could be examined in a dedicated laboratory for their chemical composition and other characteristics. Signs of organic molecules necessary for life were of great interest. Like the earlier landed rovers, *Phoenix* carried a stereo camera located on the top of a 2-m mast. *Phoenix* was also scanning the Martian atmosphere vertically up to 20 km to study atmospheric processes such as the formation, duration and movement of clouds, fog, and dust plumes. It was also

Fig. 21.4 Phoenix landed successfully on Mars on 25 May 2008 (Credit: NASA)

carrying temperature and pressure (provided by Finland) sensors to observe local weather conditions. During its mission *Phoenix* identified calcium carbonate, which suggests occasional presence of thawed water. It found soil chemistry with significant implications for life and observed falling snow. The mission's biggest surprise was the discovery of perchlorate, an oxidizing chemical on Earth that is food for some microbes and potentially toxic for others (Hotakainen 2008).

It is important to know more about the internal structure of Mars. This can obviously be addressed only by seismological instruments. We would like to know what is the present status of volcanic activity, how it evolved earlier, and is the planet completely silent at the moment? We would like to know about fluvial and subsurface processes and how subsurface features are impacting atmospheric behaviour. We would also like to make an inventory of the subsurface water reservoirs. Maybe the low-frequency radars currently operated in Martian orbit will give an answer to these last questions. And, of course, we would like to know whether life ever existed on Mars or is there still some form of primitive life. *Beagle 2* was supposed to address these questions and it is important to repeat these experiments and ideas in the near future.

21.5 Near Future Plans

Three major space agencies (NASA, ESA and Roscosmos) are actively preparing or planning for near future missions to Mars. Also Japan, China and India have activities preparing for flights to Mars sometime in the future.

NASA's next Mars mission, the *Curiosity* rover *(Mars Science Laboratory)*, was launched on 26. November 2011, and arrival at Mars will be in August 2012. *Curiosity* will be substantially larger but its construction quite similar to the twin rovers now on the surface of Mars. It will carry a laser for vapourising a thin layer from the surface of a rock and remotely analysing the elemental composition of the underlying materials. It will be able to collect and crush rock and soil samples and distribute them to onboard test chambers in order to analyse them in greater detail than ever before, searching for organic compounds and environmental conditions that could have supported microbial life now or in the past. In addition, *Curiosity* will study the Martian atmosphere and determine the distribution and circulation of water and carbon dioxide in the atmosphere-ground system. The estimated life time of the mission is one Martian year. The payload for *Curiosity* will be provided by an international consortium including Finland.

The Russian Federal Space Agency (Roscosmos) has been working several years on the *Phobos Sample Return* mission named *Phobos-Grunt*. The spacecraft was launched into low Earth orbit on 8. November 2011. However, the main propulsion system did not ignite during the attempted firings to propel the spacecraft on a cource to Mars and the mission was lost. The remains of the spacecraft fell down to the Pacific Ocean on 15. January, 2012. The main goal of the mission was to collect 100 g soil samples from the Martian Phobos moon, and to deliver them to Earth for laboratory analysis. It is assumed that the samples would have provided insight into the origin and evolution of Mars, its moons and the whole solar system. The spacecraft was planned to land on Phobos and a special device had to drill into the moon, collect soil samples and place them into an airtight capsule. The capsule

Fig. 21.5 The *Phobos-Grunt* mission with two (probably only one) meteorological stations provided by Finland located in the cylinders on the *top* of the spacecraft (Credit: Lavochkin Association)

was then planned to be sent to Earth by a return vehicle with 300 days travel time. The rest of the landed spacecraft would have remained on Phobos and continued to study Phobos, the Martian environment and various surface processes on the planet. The spacecraft had a planned capacity to carry 100–150 kg of international equipment such as a sub-satellite provided by China and one to two Martian small meteorological stations provided by Finland. However, the meteorological station had to be removed from the spacecraft at a very late stage of the construction work (Pellinen and Raudsepp 2006) (Fig. 21.5).

In 2001 ESA began the development of a Mars program called "Aurora". The ultimate goal is to send European astronauts to Mars in the 2030 time frame. The first preparatory steps will be taken by science and technology demonstration missions that will be implemented in close cooperation with NASA. Roscosmos has also been invited to join this effort. A mission called *"ExoMars"* having two stages, to be launched in 2016 and 2018, will be the first major step. The main aims are to search for signs of past and present life on Mars, to study the water and geochemical environment as a function of depth in the shallow subsurface, and to investigate Martian atmospheric trace gases and their sources. The two mission elements described below will be modified if Russia is involved.

To achieve these objectives, the 2016 ESA-led mission includes a Mars Orbiter and an Entry, Descent and Landing Demonstrator Module (EDM). The Orbiter will carry scientific instruments to detect and study atmospheric trace gases, such as methane. The EDM will contain sensors to evaluate the lander's performance as it descends, and additional sensors to study the environment at the landing site.

The 2018 mission is a NASA-led mission and includes two rovers, one European and one American. Both rovers will be transported in the same aeroshell and will be delivered to the same site on Mars. The ESA Rover will carry a drill and a suite of instruments dedicated to exobiology and geochemistry research. The drill will be designed to penetrate the surface and obtain samples from well-consolidated (hard) formations, at various depths, down to 2 m.

After the *ExoMars* mission plans become more diffuse and vague. The first preparatory steps to sample return missions will be taken. Also new missions to test airborne mobility techniques on Mars might take place. It is clear that near-future missions will build on scientific discoveries from past and ongoing missions. The search for life is the dominating theme, and drilling deeper and having more miniaturized instruments are the main challenges.

Both NASA and ESA are planning sample return missions to be carried out soon after 2018. A *Mars Sample Return* mission would typically use robotic systems and a Mars ascent rocket to collect and send samples of Martian rocks, soils, and atmosphere to Earth for detailed chemical and physical analysis. According to the recommendation of the International Mars Exploration Working Group, a total weight of 500 g of the various samples would satisfy the needs of all dedicated laboratories on Earth. Scientists on Earth can measure chemical and physical characteristics much more precisely than in-situ instruments. On Earth, they have the flexibility to make changes as needed for intricate sample preparation, instrumentation, and analysis if they encountered unexpected results. In addition, for decades to come, the collected Mars rocks will yield new discoveries as future generations of scientists apply new technologies to study them.

21.6 The Remote Future: Human Exploration of Mars

Since the historical moment when Christopher Columbus approached the shores of America in 1492, mankind has continuously expanded its presence at earlier unknown territories on Earth. Major new steps in exploration seem to have taken place in approximately 60-year intervals. Around 1910 both poles were reached and in 1969 humans landed on the Moon. If we believe in this periodicity over the last 500 years, mankind should now be looking for the next step, which quite naturally could be landing a human on Mars in 2030. In fact, this process has already been started by various space agencies and there seems to be a strong political support to continue these efforts (Pellinen and Raudsepp 2006) (Fig. 21.6).

We are technically almost ready to take the next step today. Fifty years have passed since the first manned space flight in 1961. Since the launch of the first *Sputnik* in 1957, on the average two satellites per week have been launched to space. Today some 700 operational spacecraft serve the needs of the population on Earth. Humans have visited space (among them a few tourists) 800 times and spent altogether more than 20,000 days there. Twelve men have walked on the Moon. Humans have been living at low-Earth orbit (LEO) continuously about 25 years. The *International Space Station (ISS)* is almost completed and manned activities beyond the ISS era are under consideration. When President John F. Kennedy in 1961 declared that the USA would land a human on the Moon before the start of 1970, Americans were there half a year ahead of schedule on 21 July 1969. If mankind decides in 2015 to go jointly to Mars in 2030, there will be twice as much time for preparing for the launch as was ever allocated in the Apollo programme.

With the propulsion technologies presently available we require narrow "launch windows" for Mars, determined by celestial mechanics of planetary motions and temporally separated by approximately 26 months. After 2010 there are still nine launch windows available for preparatory missions before the first human mission to Mars in late 2030.

The preparatory missions must follow a certain logical order to support the final goal. Knowledge about Mars has to be enhanced and capabilities have to be developed through specially designed demonstration missions. The orbital mapping activities of Mars will soon be completed through the various instruments onboard *Mars Express* and the *Mars Reconnaissance Orbiter*. Safe (soft), accurate landing and long-duration mobility on the surface are the next steps to be demonstrated by missions planned for the 2011–2018 launch windows. Returning small samples to Earth will be demonstrated during the following launch windows. The sample return missions, besides bringing samples back to Earth for laboratory analysis,

Fig. 21.6 The Aurora programme of ESA (Credit: ESA)

can demonstrate rendezvous and docking techniques in Martian orbit and test ascent vehicle technologies (Fig. 21.6).

All these missions will be strongly science oriented. Typically, the search for life, water and geological structures beneath the Martian surface will be the main focus of the next phase of scientific studies. The orbital terrain-mapping accuracy will be on the order of a few 10s of cms; a meteorological network will be started; and drilling down to 2 m or even deeper will be performed. It has to be remembered, however, that both science and exploration goals characterise missions in the launch windows after 2010. In the following decade new observing techniques will also be tested with airplanes and balloons. Life support, telecommunication and positioning systems will be developed, and astronaut training on the *ISS* will be started.

Preparation for a manned mission to Mars should include a large sample return mission, where the remote collection and retrieval of samples can be tested. Forward and backward planetary protection are key issues to be addressed during this mission, as well as all the different stages for a round trip to Mars. In the same time frame human technology demonstrator experiments are to be carried out onboard *ISS*. These include assembly in orbit, life support, habitation and extravehicular activities (EVA), testing of docking systems and human operation in microgravity conditions. In addition, demonstration of re-entry vehicle technologies would be planned. It should demonstrate intermediate size technologies for aero-capture and aero-braking, solar electric propulsion, and controlled landing. Furthermore, setting up an *in situ* propellant production experiment is planned. This would also be the right moment to implement a telecommunication and remote sensing infrastructure on Mars.

Moon could serve as test bed for various activities to be performed later on Mars. A human rehearsal mission could be directed towards the Moon. First there would be a 180-day test flight in LEO, corresponding to the flight from Earth to Mars followed by transit to and landing on the Moon where living and EVA activities together with a new space suit would be tested. The mission would end with a return to Earth. Other Moon-related activities are also under consideration. The present goal of NASA and some other space agencies is to send humans to the Moon in the near future. This would be followed by the establishment of a permanent Lunar base in international partnership.

The manned mission to Mars will most probably carry six humans to be selected according to very careful criteria. The spacecraft will be assembled in orbit, which will require about 500–1,000 t of components to be carried to LEO. Assuming that an Energia-type of launcher is available, capable of carrying 80 t, about ten subsequent launches would be required over a few years' time. Assembly experience from *MIR* and *ISS* will be utilised. Construction work on the ground has to be started around 2020 to be ready in time. The size of the spacecraft will be comparable to the *MIR* station, with facilities for exercise, work, medical treatment, food storage, common gathering area, hygiene facilities and private areas for six individuals.

The first manned mission is planned to have two separate stages. A cargo mission to Mars should be sent in late 2030, with the humans following in early 2033. The manned round trip will last close to 3 years, with a stay of slightly more than 1 year (over half Martian year) on the surface. In the present scenario three humans would land on Mars, while the three others would remain in orbit, providing support to the ground-based activities.

On the surface the astronauts would have a carefully planned scientific and operational programme. The main task would be to assess the suitability of the planet for long-term human presence (e.g. habitability, resources availability and various engineering constraints). A return to Mars would be possible in the 2037 and 2044 launch windows, possibly with a start, if it proves feasible, of a permanent base and presence on the planet.

21.7 Epilogue

After several logical steps we have arrived at the moment when on a daily basis we are receiving new data both from Mars orbit and its surface. Every forthcoming launch window will be effectively utilized in order to rapidly gain more knowledge about some still missing blocks of information: the amount of water hidden by Mars and possibly existing life. Several space agencies are continuously developing their technical capabilities for more extensive flights beyond near-Earth orbits. A great space adventure is awaiting mankind; plans are made and the first firm and clear steps have been taken. The details in these plans and their timetables may still change, but we are definitely on a path with no turning back. The future of mankind is in the universe and Mars is the clear first milestone.

References

Hotakainen M (2008) Mars: from myth and mystery to recent discoveries. Springer, Berlin
Pellinen R, Raudsepp P (eds) (2000) Towards Mars! Raud Publishing, Helsinki
Pellinen R, Raudsepp P (eds) (2006) Towards Mars! Extra. Raud Publishing, Helsinki/Tallinn
Squyres S (2005) Roving Mars spirit, opportunity, and the exploration of the red planet. Hyperion, New York

Chapter 22
In Search of a Living Planet

Harry J. Lehto

22.1 Life: Its Characteristics, Requirements and Limits

Life is an orderly system. This order is manifested from molecules to cells and further on to the levels of individuals and ecosystems. This high level of ordering appears to be in conflict with the second law of thermodynamics, which says that any closed system tends to smooth out with time and to become a sort of homogeneous soup. Life does not function this way, and it tries to keep up the order by all means. The complex order is maintained by energy that flows though the system. The original source of this energy can be the central star of the planetary system, or really the fusion reactions inside the star. It can also be chemical energy or energy harvested from tidal friction or from the decay of radioactive isotopes.

Life needs a means for transporting internal information and energetic compounds. It needs a solvent and preferably also some sort of a limited volume in space. In Tellurian life, the only type of life we know of, the solvent is water and the space limiter is a membrane surrounding the cell.

One of the properties of life is the tendency to maintain various unstable states such as complex molecules and cellular structures. Another example of the unstable condition is the high level of oxygen in the atmosphere, which is a consequence of photosynthetic life. Would life dwindle from the Earth, oxygen would disappear in 10,000 years.

In addition to cellular structure and water solubility the known life has also other properties that can be considered universal. The genetic information transferred from one generation to the next is written in the structure of complex molecules. The DeoxyriboNucleic Acid (DNA) acts as the bank of information, which is then copied into RiboNucleic Acid (RNA) and translated into instructions for building

---

H.J. Lehto (✉)
Department of Physics and Astronomy, Tuorla Observatory, University of Turku, Väisäläntie 20, FI-21500 Piikkiö, Finland
e-mail: hlehto@utu.fi

Table 22.1 The limits of life have been shown for some relevant parameters

Temperature (active)	0°C	121°C
(survival)	−180°C	250°C
pH	pH 0	pH 13.2
Pressure	~mbar (vacuum)	1,100 bar
Salinity		32% concentration
Gamma radiation		20,000 Gray
UV radiation		<280 nm destroys DNA
Oxygen, CO_2	0%	100%

proteins. Proteins in turn maintain and regulate various functions of the cell, such as copying of the DNA into RNA. In addition, a large number of molecules take part in the complex molecular pathways of life. The folding of large molecules requires also a suitable solvent. All this takes place inside a cell membrane formed of lipids.

Life has adapted to almost every conceivable environment on Earth (Stan-Lotter 2007). The living beings that dwell in environments we consider extreme are called extremophiles. They are mainly Bacteria and Archaea, but extremophiles can be found also in the third domain of life, Eucharyota. If objectivity is of high importance, then beings that live in "normal" oxidized conditions are really extremophiles. Atmospheric oxygen appeared on Earth rather late. At that time it was indeed a very harmful substance for most of the life forms. Life adapted to the changing conditions and developed new protective and corrective mechanisms to deal with oxygen, which was important because oxygenic metabolism was vastly superior to other types of metabolisms. The limits of life turn out to be quite broad we may conclude that life has adapted to all possible environments on Earth except for molten lava (Table 22.1) (Stan-Lotter 2007; Roadcap et al. 2006; Kanervo et al. 2005).

Life on other planets or moons may have similar limits. It is quite possible that life will be quite different from what we are accustomed to. Because of adaptation to very different environments, extraterrestrial life may show a different set of limits than what we are accustomed to here on Earth.

22.2 Searching for Life in Our Solar System

The conditions in our Solar System outside the Earth are harsh, at least in the frame of known life.

In Mercury, the diurnal cycle is long, lasting 167 Earth days. A long sun-baked day alternates with an equally long very cold night. The temperature differences are not reduced the way they are at Earth, as Mercury has practically no atmosphere. The surface of Mercury is directly exposed to the vacuum of space, and to the energetic particles from both the Sun and the deep space. Furthermore, a large flux of high-energy electromagnetic radiation such as UV, x-ray and gamma radiation is present on the dayside. Mercury is possibly the most hostile place for life among planets and Moons within our solar system.

Venus, the sister planet of Earth, has an atmosphere, but the conditions can hardly be considered balmy. Venus is covered by a thick carbon dioxide atmosphere. The air pressure on the surface is 90 bars, which corresponds to the pressure in oceans at a depth of 900 m. The temperature is 464°C, which is enough to melt lead. It is too hot for all known life forms. The Venusian atmosphere has a potentially interesting zone. Around an altitude of 55 km the clouds have a temperature between 0°C and 100°C and a pressure of 1 bar (Seiff et al. 1985; Pätzold et al. 2007). The droplets that float in these surroundings are believed to contain ¾ water and ¼ sulfuric acid, which might suit acidophilic extremophiles (Young 1973; Cockell 2007). Clouds above this zone protect the planet from the deleterious UV radiation.

The remaining bodies, planets, moons, asteroids and comets can be divided into three groups when considering the possibility of life. Large planets (Jupiter, Saturn, Uranus and Neptune) do not have a solid rocky surface. The measurements by the Galileo mission revealed a narrow zone in the Jovian atmosphere where the pressure is a few bars, and the temperature is suitable for liquid water. The atmospheres of other giant planets have not been measured to large enough depths, but according to atmospheric models water clouds may be present at pressure levels of 10–100 bars. At these altitudes, light is a scarce commodity. The winds at these pressure levels in the Jovian atmosphere are quite fierce, 100–180 m/s. As it is hard to conceptualize how ventophile life could have formed and evolved for adaptation to windy conditions, we will not in the following discuss these planets.

The second large group of bodies are the ones that have surfaces exposed to the vacuum of space. The surface temperatures are generally such that water can be present in the form of solid ice. The conditions are so harsh that the possibility of life is very slim.

The last, most interesting group with respect to the existence of life are the celestial bodies that are covered by a shielding layer. This cover can be an atmosphere (Earth, Mars or Titan) or it could be ice (Europa, Enceladus, and possibly comets). We will consider these bodies in the following.

22.3 Mars

The possibility of life on Mars has been of high interest and imagination since the detection of channels on Mars by the Italian astronomer Giovanni Schiapparelli in 1877, using a modest-size telescope (Schiaparelli 1893). Due to a translation error these channels were thought to be canals, which would imply that they were built by intelligent beings. Later, Percival Lowell, an American astronomer, made a number of observations of Mars, and drew maps of the system of canals. Even seasonal changes were detected, which were explained by the growth and decay of vegetation. Around the same time spectroscopic observations of the Martian atmosphere indicated that the atmosphere contained no or very little water, in apparent conflict with the idea of Martian canals.

In 1965 the first photographs transmitted from the *Mariner 4* probe revealed a desolate and deserted Mars. The dream of a lush planet and a civilization building canals was buried into the wonderland of scientific wishes.

The *Viking* space missions in the 1970s were the first, and so far the only missions aimed to test for the presence of life on Mars. Their results were controversial (Stan-Lotter 2007). It was only after the *Viking* experiments in 1976 that a new domain of life Archaea was identified and defined, and most of these are known now out to be extremophiles (Woese and Fox 1977; Stan-Lotter 2007). Furthermore, the sensitivities of the Viking experiments were not very good. Recently, it has been realized that not even all Earth based environments would have given positive signatures in the *Viking* experiments. Chances for detecting more exotic hypothetical life forms such as hydrogen peroxide based life would have been even lower (Houtkooper and Schulze-Makuck 2007).

The conditions on Mars are quite harsh. The carbon dioxide atmosphere is thin. Average air pressure is 7 millibars, and varies seasonally between 5 and 11 millibars. The air pressure is highest during northern spring and autumn and lowest during the summer and winter, when part of the atmosphere condenses into carbon dioxide frost around the winter pole. Life protecting oxygen and ozone are almost completely missing from the atmosphere. The mean temperature on the surface of Mars is about $-60°C$, but varies over a large range. On winter nights it can plunge down to $-120°C$ but on a warm summer day on the equator it can reach a balmy $+20°C$. The amount of sunlight on Mars is about half of that on Earth.

The conditions on the Martian surface are deleterious to Earth-based life. The low temperature, the low air pressure, and the lack of liquid water make conditions difficult. For life, the most restrictive environmental feature on Martian surface is the high energy ultraviolet radiation (UVC, <280 nm), which directly destroys DNA. On Earth, the atmosphere absorbs this radiation so effectively that only one billionth of the incident amount reaches the surface. At Mars, our known life could survive on the surface for about 20 min, but just a few centimetres below the surface it would be protected from the harsh radiation. A few meters further down, pressure and temperature are higher, and at some point one would reach the conditions were water can remain in a liquid phase. In June 2008 the *Phoenix* space probe found water ice only a few centimeters below the surface dust (Mellon et al. 2009). The presence of subsurface ice within the top few meters has also been confirmed by copious amounts of hydrogen detected by the *Mars Odyssey* (Boynton et al. 2002).

There is no lack of water in Mars. The polar ice caps are formed of 3–4 km thick water ice (Zuber et al. 1998; Putzig et al. 2009). The combined water volume corresponds to the Greenland Ice shield. If this ice would melt it would cover Mars on average in a water 10 m deep. It is likely that close to polar regions and under the surface, there is a region which is sufficiently warm and humid for life. Due to lack of light the primary producers would have to be chemotrophic, which means that they harvest the necessary energy from chemical compounds.

A microbial ecosystem could thus be possible, but we can only have speculations of "higher" living beings.

Life could have emerged underground or in Martian oceans, which seem to have been present about three billion years ago. Mars and the Earth may have exchanged early forms of life contained in ejecta from asteroid impacting the planets. Indications of possible life have been found inside the stony meteorite ALH84001 and two other meteorites (Nakhla and Shergotty) (Gibson et al. 2001; Thomas-Keprta et al. 2009 and references therein). The interpretations of these findings are still somewhat controversial.

Methane has been found in the Martian atmosphere at levels of 10 ppb, and locally at 30 ppb (Formisano et al. 2004). The geographic positions of higher levels of methane correlate with locally higher concentrations of water. A natural explanation for this is the presence of methanogens, microbes that produce methane. Geologic processes have also been suggested such as active volcanism or the release of methanehydrate calthrates, but there are no direct indications for these (Geminale et al. 2011; Krasnopolsky et al. 2004; Krasnopolsky 2006; Bartoszek et al. 2011).

If life is ever found from Mars, it will have profound implications for the understanding of the origin and properties of life in general. If some of its properties differ from the Tellurian life, it will prove that life has had different origins on neighbouring planets. Some of the differences could be for instance in nucleic acid or protein structures or in their chiralities, or differences in the solvent used by the Martian life (water vs. hydrogen peroxide), or in a completely different strategy for the structure or chemistry of life. If such life is similar to life on Earth, then it would indicate that life was transferred between Mars and the Earth or that life can form only using similar chemistry and strategies. Detailed comparison of lives found on different planets would require extensive biological, biochemical and chemical analyses.

Martian environmental conditions have been recreated on micro scale in laboratories around Earth. Significant understanding of the survival of life in Martian conditions has been obtained. Due to the differences in the atmospheric conditions, natural conditions on Earth cannot fully emulate Martian conditions, but the tolerance of life to dry and cold conditions can still be investigated in deserts such as the Atacama desert in Chile or in the dry valleys of Antarctica, where the temperatures can fall close to levels found on Mars.

22.4 Europa

Jupiter's Moon Europa has been known to have a highly reflective surface since 1895 when Barnard published the first measurements of the diameter of the moon and compared it to the observed brightness (Barnard 1895). The photographs and other data transmitted from *Pioneer 10* and *11* and *Voyager 1* and *2* missions showed that Europa is an ice-covered body. No open holes or cracks have been detected in the ice cover. The young age of the surface is evident from the low number of larger craters, only half a dozen are present. Furthermore the surface shows features that resemble ice cracks that have been formed and refrozen at different epochs. Superficially the surface of Europa resembles the satellite images of the frozen Arctic Ocean and other Arctic seas and large lakes.

The *Galileo*-mission in the mid-1990s measured a deviation in the dipole magnetic field of Europa. This suggested that Europa has a layer of conductive matter under the ice cover. A natural explanation is a saline ocean with a depth of about 100 km (Zimmer and Khurana 2000). The thickness of the ice cover is not known, but it is most likely less than 20 km, and may be only a few hundred meters (e.g. Hand and Chyba 2007). Similar estimates for the thickness of the ice have been obtained from crater size distributions and their ejecta. The surface shows brownish areas that are composed of organic substance.

The surface of Europa is hostile for life. The cold surface is in almost complete vacuum, and is bombarded by charged particles carried and accelerated by Jupiter's intense magnetic field. It is exceedingly difficult to see how life could adapt to this environment at least in the form we know it. Under the ice cover, the environment may be much more suitable for forming and sustaining life. There the water is in liquid form, so in that respect at least one requirement is satisfied. But life also needs a source of energy. There are two options for this. First, the patterns reminiscent of frozen cracks on Europa's surface suggest that the ice cover breaks every now and then. During these times a habitat forms, where the light from the distant Sun can be harvested by phototrophic life forms. These could be dormant for most of the time and become active when the conditions become more suitable.

The ocean in Europa does not stay liquid without an input of external energy. The same energy source could be used by life. Europa orbits close to Jupiter. The tidal forces Europa experiences are about 1,000 times the tidal forces exerted by the Moon on the Earth. These forces pump energy into the stony core of Europa. It is conceivable that the bottom of Europa's ocean hosts structures similar to the black smokers found e.g. on the Atlantic mid-ridge. On Earth, the iron and sulfur compounds flowing out from these hot vents can be used for energy by chemotrophic microbes without the use of any sunlight. Simple ecosystems could be formed around communities of such primary producers, and they could also be connected to the subsurface ice crack communities.

Other environments analogous to the ocean on Europa can be found on Earth (Lorenz et al. 2011). About 150 subglacial lakes have been found in Antarctica. These are reminiscent of Europa in the sense that they have been deprived of sunlight for a long time. The largest of these lakes is situated below the Russian research station, Vostok. The lake resides below a 4 km thick ice cover and is 400,000 years old. Attempts to search for life in Lake Vostok have been made. Technical problems and ethical questions relevant to a Europa life expedition have turned up during the drilling of the ice. The final hole into the water of Lake Vostok has not been completed. The drill was stopped in February 2011 in the slush zone only 30 m above the liquid water (Schiermeier 2011). One of the most important problems is the question of protecting a unique environment. How can we avoid contaminating the possibly unique ecosystem of Lake Vostok with external microbes or with other organic matter, such as kerosene used in the drill. The sample of the lake will be obtained by conventional drilling and suitably balancing the 400 bar overpressure from the lake (Bulat, pers. comm). Two types of microbes have been found in the slush zone of Lake Vostok, most likely originating from the lake itself (Bulat et al. 2011). Small ice penetrating drills/pencil submarines have been planned for the drilling in Europa (Korablev et al. 2011; Weiss et al. 2011).

22.5 Titan

Titan was long considered to be the largest moon in the Solar System. It was known to have a dense atmosphere, but its composition remained a mystery, until the *Voyager* mission revealed that it consisted mainly of nitrogen. The difficulty of detecting this from ground based observations was understandable, because of our own nitrogen dominated atmosphere through which the observations were attempted. The *Huygens*-probe that landed in 2005 on the Titan surface and the *Cassini*-mission orbiting Saturn have opened a new window for studies of Titan and other Saturnian satellites.

Titan is a frigid place. The surface temperature is about $-179°C$. The conditions are so harsh that it appears that there is no place for life in that world. Light is also a scarce commodity on Titan. The amount of light that reaches the upper parts of the Titan's atmosphere is only 1/120 of the Earth average, or about the same as on a cloudy winter day in Southern Finland. Only part of this light is transmitted through the thick atmosphere. A couple of years before the arrival of the *Cassini-Huygens* mission to the neighbourhood of Titan, radio astronomers using radar techniques reported the detection of flat surfaces in Titan (Campbell et al. 2003). They suggested that these could be lakes or even seas. The space missions confirmed these findings. In July 2008 spectroscopic observations showed that these lakes are formed of light hydrocarbons, methane and ethane, as well as liquid nitrogen (Brown et al. 2008). The largest seas in the Northern Polar regions of Titan are similar in size to the Baltic Sea (Fig. 22.1).

The photographs taken during the descent of the *Huygens* probe show that the lakes are connected to features that appear to be river systems. *Cassini* has also observed methane clouds in the polar regions of Titan. Filling up and formation of lakes has also been detected (Jaumann et al. 2009 and references therein). In Titan, the light hydrocarbons have atmospheric and liquid cycles that resemble the water cycle on Earth (e.g. Turtle et al. 2011).

Huygens revealed that the surface of Titan contains "stones". Spectroscopy shows that the stones are made of water ice. It is not at all insignificant that liquid water, necessary for life as we know it, is found on Titan.

Could life thrive in the methane lakes? Life as we know it uses water as a liquid solvent, so it seems that life is not likely to be found in Titan. In any case, Titan is an interesting place. It resembles chemically the conditions found on the young Earth.

Fig. 22.1 Hydrocarbon lakes and seas in the Northen polar region of Titan. Over 400 lakes and seas have been found in Titan. The second largest sea, Ligeia Mare has an area of about 100,000 km^2, or the size of Gulf of Bothnia. The largest sea, Kranken Mare, shown in part at *right* and *down* from the *center* of is about the size of the Baltic Sea. Koitere Lacus, which is named by the Finnish Karelian Lake, Koiterejärvi, is located *left* and *down* from the *center* point of the image in the second slice from the *left*. The lake is drop shaped and shows a small island in its center (Credit: Cassini mission NASA/JPL/USGS)

The temperature is quite a bit lower, but for chemistry that means mainly slowed-down processes. Being lighter methane rises in the atmosphere above nitrogen. In the upper atmosphere the methane, under solar UV radiation, forms complex hydrocarbon molecules, tholines, which rain down to the surface and into the seas creating a rather rich hydrocarbon-based mixture of molecules. It is imaginable that processes that led to the emergence of life on Earth are still in operation in Titan. Recently a deficiency of hydrogen and acetylene has been reported (Clark et al. 2010; Strobel 2010). This might be explained by chemistry alone, but life is definitely an alternative for explaining the low quantities of these energy-rich compounds.

Recent models of Titan's interior bears some resemblance to Europa and other Jovian planets (Lorenz et al. 2008; Nimmo and Bills 2010). Under the surface ice cover there appears to be an ocean of an ammonium-water mixture. This environment and possible cryovolcanoes ejecting this liquid should be kept in mind when possible habitats for life are sought on Titan.

Titan analogues are difficult to find on Earth. The atmospheric pressures are of the same order, but the large concentration of oxygen found in Earth's atmosphere

Fig. 22.2 The Pitch Lake in Trinidad and Tobago is the largest of three natural asphalt lakes on Earth (© Harry J. Lehto)

is not present in Titan. The very low temperatures of Titan are not present in natural environments on Earth. Considering these limitations we can ask whether hydrocarbon lakes can be found on Earth. As surprising as it might sound the answer is positive. Three asfalt lakes are known in addition to some smaller pits. The first lake is in Trinidad (Fig. 22.2), the southernmost island of the Caribbean. The second one is in the deep jungles of Venezuela. It is located along the same plate boundary system as the Trinidad Pitch Lake. The third one is in Los Angeles, California.

The biological, chemical and geologic properties of Trinidad's Pitch lake are studied by an international group consisting of researchers from the United States, Canada, Trinidad and Tobago and Finland (Schulze-Makuch et al. 2011).

It provides a unique Titan analogue for studying the survival and presence of life in such a chemically extreme and rather hostile, dry and anoxic habitat. The main differences between Trinidadian and Titanian lakes are that in Trinidad, the temperature is above the freezing point of water, the hydrocarbons are long complex ones with high viscosity, although gaseous ethane and methane are present here too. In Titan, the temperature is very low, thus the small hydrocarbons, ethane and methane, are in liquid phase. There is some liquid nitrogen in the Titanian lakes also. The conditions only a few cm below the surface in the Pitch lake are anoxic, so it appears that the lake is indeed a good analogue for Titan's hydrocarbon lakes.

22.6 Enceladus

Like Jupiter's Europa, Saturnus's moon Enceladus has been known to be highly reflective. In 2005 the *Cassini* – mission detected "tiger stripes" that were 30° warmer than the surrounding ice cover. On subsequent flybys the probe looked on the shadow side of Enceladus and saw huge plumes of water being emitted by the moon (Fig. 22.3). More recently the molecular content of the plumes have been confirmed to be water with lesser amounts of organic molecules mixed in. The origin or the outflow has been localized to several spots in the tiger stripes (Fig. 22.4). This is a direct indication of a water reservoir that is hidden under the ice cover. It is not thought to be as extensive and deep as that of Europa, but since

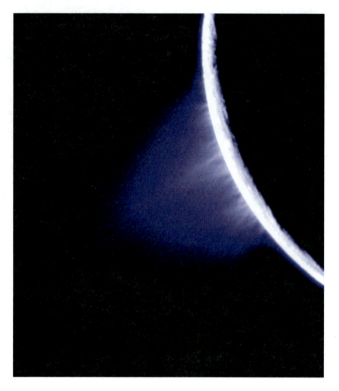

Fig. 22.3 Water geysirs on Enceladus, a moon of Saturn. Organic molecules were also found in the geysir water. This kind of images was analyzed carefully for planning the Enceladus flyby of August 2008 (Credit: Cassini mission NASA/JPL/USGS)

tidal forces exerted by Titan appear to provide energy, the necessary conditions for life, liquid water and energy, are fulfilled. Enceladus is indeed considered as one of the likely locations where extraterrestrial life could be found in our Solar system. For a review see Spencer et al. (2009).

22.7 Comets

Comets used to be called "stars with tails" in older Finnish Astronomy textbooks. They are not stars, but they do sometimes possess an impressive tail. The comets are a mixture of ices and rock. Many comets have significant amounts of water, carbon dioxide and carbon monoxide ice. A multitude of inorganic and organic molecules have been found (Sephton 2002; Schmitt-Kopplin et al. 2010). Many comets are on highly elongated eccentric orbits. As the comet enters the Earth's orbital distance from the Sun, their volatile ices start to evaporate creating the tail. Inside a comet the temperature can grow so high that water ice starts to melt. It has been speculated

Fig. 22.4 The tiger stripes of Enceladus. The stripes are visible below the *center* of the image near the South pole of Enceladus. During the flyby of August 2008, the locations of the eruption sites of the water geysirs were shown to originate in the sculci (Credit: Cassini mission NASA/JPL/USGS)

that because of the presence of liquid water and a large variety of organic molecules the conditions could be suitable for prebiotic chemistry. The Murchinson meteorite has shown us that the production of various organic compounds, e.g. amino acids is possible even on the smallest atmosphereless bodies of our Solar system.

22.8 Other Planetary Systems

Aleksander Wolszczan and Dale Frail (1992) were the first astronomers to report a planetary system around a star other than our Sun. The planets, around pulsar PSR 1257 + 12, were inferred from the cyclic variations in the time arrival of pulses from the pulsars. The first planetary system around a solar type star was discovered, when Michel Mayor and Didier Queloz (1995) noted that the spectral lines of the star 51 Pegasi shifted on a period of 4.2 days. They calculated that the mass of the planet to be at least half of Jupiter's mass, and that the orbital radius is 1/8 Mercury's orbit. Astronomers were surprised by the orbits of the first exoplanets discovered.

Recently, it has been revealed that correct signatures of planets were detected already prior to 1992, but the researchers did not dare to suggest the presence of planets (Campbell et al. 1988).

About 700 exoplanetary systems are known to date (January 13, 2012, http://exoplanet.eu/). The known number of planets is a few dozen larger, as some planetary systems have two, three and even five planets. A planetary system around HD 10180 with six or seven planets was reported in August 2010. So far nearly all confirmed planets differ significantly from our Solar planetary system.

Until 2010 planets were found most effectively by measuring the Doppler shifts of emission and absorption lines in spectra of bright stars. In our Solar system, Jupiter causes the Sun to wobble with an amplitude of 1 m/s and a period of 11 years, the orbital period of Jupiter. If the mass of the star can be calculated from its spectral type, observed brightness and distance, then this indirect method can be used to determine the properties of planets and their orbits. The period of the spectral variation gives directly the orbital period of the planet, but the mass cannot be calculated directly, unless the inclination of the orbit is known. The measured quantity, $m \cdot \sin i$, provides us with a lower limit of the mass, m, because of the unknown inclination, i.

If the plane of the orbit is along the line of sight ($i = 90°$), it will move across the stellar disk. The number of such planetary systems is at present about 130 and increasing rapidly because of on-going experiments specialised in detecting these. The dimming caused by the planetary eclipse is not complete as in a full Solar eclipse, but is more reminiscent of the Venus transit across the Solar disk. As the planet moves across the stellar disk the brightness of the star falls typically by about a tenth of a percent for a few hours. The information obtained from the planet is now significantly increased. Combined with spectroscopy, we can now measure the diameters of both the star and the planet, which in the case of the planet will allow us to calculate also its density.

In addition to these two methods, exoplanets have been detected by the gravitation lens effect. According to the general theory of relativity, a light beam traversing from a distant star to us will bend ever so slightly if it feels a gravitational field of, for example, an intervening star or a planet. Under favourable conditions, this bending and focusing of light is observed as a sharply peaked increase in brightness followed by subsequent fading to the initial level of brightness. The planet is detected as a small peak on the side of the gravitational lens event of the intervening star. Each star/planet gravitational lens event is unique and none of the planets have been re-observed with any other method. This is still an independent method for detecting planets and provides for additional statistics.

Some planets have been observed by direct imaging.

Could life thrive on the observed exoplanets? Not likely. First, with a few exceptions the planets are quite obvious gas giants – in principle similar to Jupiter and Saturn in our Solar system. Most of the planets are even larger than Jupiter. Some of these planets orbit the central star closer than Mercury orbits our Sun. The temperature on these planets is above the boiling point of water, and some of them are so hot that they are evaporating or "being consumed" by the star. For life they are not really cosy in any sense.

Most of the planets are on rather eccentric orbits, which means that at periastron the planet has very hot conditions. At apastron the planet ends up very far away from the central star in a cosmic deep freeze. These variable conditions are not very cosy for life either.

We can speculate whether the present planetary systems could hide Earth-like small rocky planets, which are beyond present observational capabilities. An Earth-like planet most likely does not survive on a giant planet-crossing orbit. When it encounters the giant planet's gravitational field it will be ejected onto a new eccentric orbit. In planetary systems where the giant planet's orbits are rather circular, an Earth-like planet could survive for a long time.

Planetary migration was used as an explanation for the unexpected orbit of the first planet 55 Pegasi B. The planet formed at a larger distance from the central star, and during the early phases of its formation it moved in the accretion disk closer to the central star. Migration of a giant planet in a planetary system sweeps clean the respective parts of the accretion disk of all Earth-like stony planets.

For a planet to have the possibility for Earth-like life, it should be orbiting the star in the "habitable zone", at a distance, where water can remain liquid. There are several variations of this definition depending on the exact temperature and time limits. These limits are largely determined by the properties of the central star and the orbits of the planets. About one tenth of the planets orbit their star in the habitable zone. Unfortunately, they are gaseous planets. Generally, it is expected that a living planet has a protecting atmosphere or an ice cover and a solid surface, which could be covered by a sea. Note that in all these cases the planet or moon should have a rocky core.

Although it appears that there is no sense in searching life from current exoplanets, all hope is not lost. We should keep in mind that the observational methods cause a strong bias towards detecting large planets on small orbits. Spectroscopically the detection of Earth-like planets is just becoming possible. It is also possible that in favourable situations a small planet could have survived the presence of a giant planet on an elliptical orbit or the migration of such a planet.

A giant planet in the habitable zone could also have moons of the size of a small planet. The definition of the habitable zone can be enlarged to encompass these environments. Conditions on these moons could be suitable for life.

Planetary systems similar to our own can abound, but we have not yet been able to detect them. With present detection methods and in the given time span we would have been able to detect Jupiter from a planetary system similar to our own. The situation is worse for detecting Saturn, Uranus and Neptune -like planets particularly because of their long orbital time scales. The inner planets could not have been detected because of their small mass. Improving observational capabilities push all these limits at present.

In 2007 the first small exoplanets were found around Solar type stars. Spectroscopic measurements revealed that the red dwarf star GL581 has four planets, two of which are small. The planet GL581c has a mass of five Earth Masses and GL581d has a mass of eight Earth masses. The former is particularly interesting because its mass is too small for a gaseous planet and because it orbits the central star at a distance that places it in the habitable zone. Based on the calculated temperature water could remain liquid at that distance. Unfortunately, at present we are unable to tell whether this planet has any life, or water, or even any atmosphere.

The smallest planet detected by Doppler techniques, GL581e, has a mass of about two Earth masses (Mayor et al. 2009). This low mass value was rivaled by Kepler 11-f detected in a six planet system (Lissauer et al. 2011).

22.9 Kepler

The *Kepler* spacecraft was launched on March 6, 2009. Science observations began on May 2, 2009. Kepler is looking at individual stars in the direction of the constellations of Cygnus and Lyra. In February 2011 the *Kepler* mission released data of over 150,000 stars (Borucki et al. 2011a, b). During the period of about 1.5 years transit-like signatures of 1,235 planetary candidates were detected (Fig. 22.5). Planets with Earth-size radii ($<1.25R_{Earth}$) number 68. Within the habitable zone of the star 54 planets are found, six of these are less than two Earth masses. The team

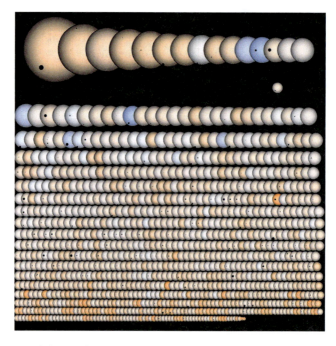

Fig. 22.5 A portrait image of the host stars of 1,235 candidate planets. The planet candidates are shown as *black* spots against the star. The Sun is shown *below* the end of the *top row*. Both Jupiter and the Earth are shown against the Sun. The colors of the stars indicate the spectral type with *blue* stars being warmer and *orange* stars cooler than the Sun. All sizes reflect the true relative size of the stars and planets (Credit: Kepler mission, Jason Rowe, NASA Ames Research Center and SETI Institute)

estimates statistically that about every third star should have a planetary candidate. *Kepler* is expected to be in orbit at least for 3.5 years, with a possible extension to 6 years. The present statistics on confirmend exoplanets are expcted to change in the very near future.

The Keppler team in collaboration with other astronomers reported in December 2011 the discovery of a two Earth diameter planet, Keppler 20b, orbiting in the habitable zone of the star. Within a month, in January 2012, three sub-Earth size planets were reported, the smallest of which has a radius of about 0.6 Earth radius close to the size of Mars. These planets are in the system with a catalogue name KOI-961.

22.10   SuperWASP, MOST, COROT, GAIA, JWST

Other collaborations are also looking for exoplanets. Some of the most noteworthy are *SuperWASP*, *COROT* and *MOST*. The *SuperWASP* measures the brightness of millions of stars and searches for eclipses caused by planets. The *COROT* satellite was launched in Earth orbit in 2007. It searches for planets in the vicinity of nearby stars. *MOST* satellite launched in 2003 has produced a wealth of data on stellar variability. The ground based SuperWASP, and the space missions *COROT* and *MOST* have all produced exoplanet detections.

The GAIA mission will measure radial and proper motions of a billion stars. These stars are located in our own milky way and in the local group of galaxies. It will provide the most accurate map of our galactic neighbourhood. As a side product we will get information of thousands of planetary systems.

The follow up to the Hubble Space Telescope will be the James Webb Space Telesope. It will measure spectra of exoplanets. The infrared spectrum will be searched for signs of water, carbon dioxide and ozone. A substantial amount of oxygen or ozone tells that the planet's atmosphere is in an unstable state maintained by life.

22.11   SETI

By "listening", astronomers are trying to make a major leap and find a planet with life that is capable to technical communication. This method would not reveal a planet filled with simple single cell organisms. The Search for Extra Terrestrial Intelligence (SETI) uses two strategies.

Radio frequencies are monitored in millions of narrow band frequencies, and amongst these one is searching for radio signals that would reveal artificial radio emission, such as radio transmitters or radar-like instruments. The Doppler shift of this emission should indicate that the planet is rotating, and the measurement should be repeatable after one rotation period. This method is heavily based on the assumption that an advanced civilization "leaks" radio emission into outer space. This assumption could also be its Achilles' heal, as it is quite possible that radio emission is not used for communication in highly advanced civilizations.

The general public can join this search by using a screen saver seti@home. This screen saver searches for these signals when the computer would otherwise be running idle.

The second strategy for finding civilizations is based on searches in optical wavelengths. Stars are monitored with optical telescopes for nanosecond pulses. Such short pulses are not known to occur naturally and their presence would indicate a civilization that is using nanosecond pulses for communication. The idea is to intercept the communication e.g. between two civilizations or between host planets and e.g. an interstellar space mission. Optical pulses are more effective than radio pulses, because more information can be packed per unit time and because they can be aimed at a much higher accuracy. Furthermore, the transmitter/receiver system is smaller and cheaper to build than the radio system with the same information transmission capability.

Because of the uniqueness and short duration of the signals, observations are done with three telescopes simultaneously with the hope of detecting the same event at three monitoring stations at least.

22.12 The Consequences of Finding Extraterrestrial Life

The news that life has been found outside Earth is going to cause a state of general confusion in the general public, and after that, depending on the person, either a full ignorance or even a certain level of denial or alternatively great interest. What happens after this will depend on what kind of life has been found and where it was found.

When life is found from our solar system, it will most likely be microbial. It is quite unlikely that highly complicated life will ever be found even from the waters of Europa or Enceladus. The detection of life would cause an imposition of quarantine between the newly detected life and Earth. A mutual contamination risk is a possibility. Strict standards for a minimal contamination risk for example in Mars missions have been imposed. A similar kind of situation is encountered when samples are returned from Mars or other bodies in our Solar system. Of course, the characterization of life is an academic problem. What are its basic properties, and how does it differ from our life and in what ways is it similar? Depending on the results the whole planet could become a protected zone. On the other hand understanding of another kind of life could result in new applications and discoveries, which could help us to save the Earth from the man made eco-catastrophy. Life on another Solar system body would be the only possibility for us to study and see directly a different kind of life.

When a living planet is found outside our solar system, for example, on a planet orbiting within the habitable zone of a central star, and having an atmosphere containing water and oxygen or other gases that appear to be out of equilibrium, then the problem will have a different nature. The planet is now so distant, that it makes no sense to send probes using present technology. At a speed of one tenth of the speed of light, it would take 330 years to reach a planet at a distance of ten parsecs. The probe should be smart, as the communication would in practice be unidirectional.

Obtaining further information after the first signs of life on a planet will be very challenging. One of the aims will be to investigate the level of complexity that life has reached. The task will be formidable.

If the first indication of extraterrestrial life will come through SETI, it would have the deepest impact of the alternatives above. The presence of a second civilization would be for some people a relief and for some people a great mental shock. If the discovery is made at radio wavelengths, then one can sort out the planets rotation period and the orbital timescale. If the central star can be identified, then the planet's orbital radius and temperature can be calculated. Combining the information with the second generation missions we could get information about the composition of the atmosphere and continents. Correlating this information with the radio signal we could determine whether the main radio emission regions are close to the coastline, such as on Earth!

Philosophically, the discovery of a second civilization would be the most earth-shattering news. How could we get in contact with "them"? Could we learn something new about natural sciences and about our position in the universe? The deep influence of detecting a new civilisation is reflected by the announcement given by the Vatican in May 2008, which resolves a theological issue that has been mentioned already in the fifteenth century. The Vatican announced that it is possible that the living beings on other planets are sinless (Fuenes 2008).

References

Barnard EE (1895) Filar micrometer measures of the diameters of the Four Bright Satellites of Jupiter, made with the 36-inch Equatorial of Lick Observatory. Mon Not R Astron Soc 55:382

Bartoszek M, Wecks M, Jakobs G, Möhlmann D (2011) Photochemically induced formation of Mars relevant oxygenates and methane from carbon dioxide and water. Planet Space Sci 59:259–262

Borucki WJ et al (2011a) Characteristics of *Kepler* planetary candidates based on the first data set. Astrophys J 728:117. doi:10.1088/0004-637X/728/2/117

Borucki WJ et al (2011b) Characteristics of planetary candidates observed by Kepler, II: analysis of the first four months of data. Astrophys J arXiv:1102.0541

Boynton WV et al (2002) Distribution of hydrogen in the near surface of Mars: evidence for subsurface ice deposits. Science 297:81–85

Brown RH, Soderblom LA, Soderblom JM, Clark RN, Jaumann R, Barnes JW, Sotin C, Burratti B, Baines KH, Nicholson PD (2008) The identification of liquid ethane in Titan's Ontario Lacus. Nature 454:607–610

Bulat SA, Alekhina IA, Marie M, Martins J, Petit JR (2011) Searching for life in extreme environments relevant to Jovian's Europa: lessons from subglacial ice studies at Lake Vostok (East Antarctica). J Adv Space Res. doi:doi:10.1016/j.asr.2010.11.024

Campbell B, Walker GAH, Yang S (1988) A search for substellar companions to solar-type stars. Astrophys J 331:902–921

Campbell DB, Black GJ, Carter LM, Ostro SJ (2003) Radar evidence for liquid surfaces on Titan. Science 302:431–434

Clark RN, Curchin JM, Barnes JW, Jaumann R, Soderblom L, Cruikshank DP, Brown RH, Rodriguez S, Lunine J, Stephan K, Hoefen TM, Le Mouélic S, Sotin C, Baines KH, Buratti BJ, Nicholson PD (2010) Detection and mapping of hydrocarbon deposits on Titan. J Geophys Res 115:E10005. doi:10.1029/2009JE003369

Cockell CS (2007) Habitability. In: Horneck G, Rettberg P (eds) Complete course in astrobiology. Wiley-VCH, Weinheim, pp 151–177

Formisano V, Atreya S, Encrenaz T, Ignatiev N, Giuranna M (2004) Detection of methane in the atmosphere of Mars. Science 306:1758–1761

Fuenes (2008) Il rapporto tra astronomia e fede in un'intervista a padre Fuenes, dirretore della Specola Vaticana: L'extraterrestre è mio fratello (14/05/2008)

Geminale A, Formisano V, Sindoni V (2011) Mapping methane in Martian atmosphere with PFS-MEX data. Planet Space Sci 59:137–148

Gibson EK Jr, McKay DS, Thomas-Keprta KL, Wentworth SJ, Westall F, Steele A, Romanek CS, Bell MS, Toporski J (2001) Life on Mars: evaluation of the evidence within Martian meteorites ALH84001, Nakhla, and Shergotty. Precambrian Res 106:15–34

Hand KP, Chyba CF (2007) Empirical constraints on the salinity of the European ocean and implications for a thin ice shell. Icarus 189:424–438

Houtkooper JM, Schulze-Makuck D (2007) A possible biogenic origin for hyrdogen peroxide on Mars: the Viking results reinterpreted. Int J Astrobiol 6:147–152

Jaumann R, Kirk RL, Lorenz RD, Lopes RMC, Stofan E, Turtle EP, Keller HU, Wood CA, Sotin C, Soderholm LA, Tomasko MG (2009) Geology and surface processes on Titan. In: Brown RH, Lebreton J-P, Waite JH (eds) Titan from Cassini-Huygens. Springer, Dordrecht, pp 75–140

Kanervo E, Lehto K, Ståhle K, Lehto H, Mäenpää P (2005) Characterization of growth and photosynthesis of Synechocystis sp. PCC 6803 cultures under reduced atmospheric pressures and enhanced CO_2 levels. Int J Astrobiol 4:95–98

Korablev O, Gerasimov M, Dalton JB, Hand K, Lebreton J-P, Webster C (2011) Methods and measurements to assess physical and geochemical conditions at the surface of Europa. J Adv Space Res. doi:doi:10.1016/j.asr.2010.12.010

Krasnopolsky VA, Maillard JP, Owen TC (2004) Detection of methane in the Martian atmosphere: evidence for life? Icarus 172:537–547

Krasnopolsky VA (2006) Some problems related to the origin of methane on Mars. Icarus 180:359–367

Lissauer JJ et al (2011) A closely packed system of low-mass, low-density planets transiting Kepler-11. Nature 470:53–58

Lorenz RD, Stiles BW, Kirk RL, Allison MD, del Marmo PP, Iess L, Lunine JL, Ostro SJ, Hensley S (2008) Titan's rotation reveals an internal ocean and changing zonal winds. Science 391:1649–1651

Lorenz RD, Gleeson D, Prieto-Ballesteros O, Gomez F, Hand K, Bulat S (2011) Analog environments for a Europa lander mission. J Adv Space Res. doi:doi:10.1016/j.asr.2010.05.006

Mayor M, Queloz D (1995) A Jupiter-mass companion to a solar-type star. Nature 378:355–359

Mayor M, Bonfils X, Forveille T, Delfosse X, Udry S, Bertaux J-L, Beust H, Bouchy F, Lovis C, Pepe F, Perrier C, Queloz D, Santos NC (2009) The HAPRS search for southern extra-solar planets. Astron Astrophys 507:487–494

Mellon MT et al (2009) Ground ice at the Phoenix landing site: stability state and origin. J Geophys Res 114:1–15

Nimmo F, Bills BG (2010) Shell thickness variations and long-wavelength topography of Titan. Icarus 208:896–904

Pätzold M, Häusler B, Bird MK, Tellmann S, Mattei R, Asmar SW, Dehant V, Eidel W, Imamura T, Simpson RA, Tyler GL (2007) The structure of Venus' middle atmosphere and ionosphere. Nature 450:657–660

Putzig NE, Phillips RJ, Campbell BA, Holt JW, Plaut JJ, Carter LM, Egan AF, Bernardini F, Safaeinili A, Seu R (2009) Subsurface structure of Planum Boreum from Mars reconnaisance orbiter shallow radar soundings. Icarus 204:443

Roadcap GS, Sanford RA, Qusheng J, Pardinas JR, Bethke CM (2006) Extremely alkaline (pH > 12) ground water hosts diverse microbial community. Ground Water 44:511–517

Schiaparelli G (1893) Il pianeta Marte. Estratto dai fascicoli N.º 5 e 6, 1 e 15 febbraio 1893 della Rivista Natura ed Arte

Schulze-Makuch D, Haque S, de Sousa Antonin MR, Ali D, Hosein R, Song YC, Yang J, Zaikova E, Beckles DM, Guinan E, Lehto HJ, Hallam SJ (2011) Microbial life in a liquid asphalt desert. Astrobiology 11:241–258

Schiermeier Q (2011) Race against time for raiders of the lost lake. Nature 469:275

Schmitt-Kopplin P, Gabelica Z, Gougeon RD, Fekete A, Kanawati B, Harir M, Gebefuegi I, Eckel G, Hertkorn N (2010) High molecular diversity of extraterrestrial organic matter in Murchinson meteorite revealed 40 years after its fall. Publ Natl Acad Sci 107:2763–2768

Seiff A, Schofield JT, Kliore AJ, Taylor FW, Limaye SS, Revercomb HE, Sromovsky LA, Kerzhanovich VV, Moroz VI, Marov MY (1985) Models for the structure of the atmosphere of Venus from the surface to 100 km altitude. Adv Space Res 5(11):3–58

Sephton MA (2002) Organic compounds in carbonaceous meteorites. Nat Prod Rep 19:292–311

Spencer JR, Barr AC, Esposito LW, Helfenstein P, Ingersoll AP, Jaumann R, McKay CP, Nimmo F, Porco CC, Waite JH (2009) Enceladus: an active Cryovolcanic satellite. In: Dougherty MK, Esposito LW, Krimigis SM (eds) Saturn from Cassini-Huygens. Springer, Dordrecht, pp 683–724

Stan-Lotter H (2007) Extremophiles, the physiochemical limits of life (growth and survival). In: Horneck G, Rettberg P (eds) Complete course in astrobiology. Wiley-VCH, Weinheim, p 121

Strobel DF (2010) Molecular hydrogen in Titan's atmosphere: implications of the measured tropospheric and thermospheric mole fractions. Icarus 208:878–886

Thomas-Keprta KL, Clemett SJ, McKay DS, Gibson EK, Wentworth SJ (2009) Origins of magnetite nanocrystals in Martian meteorite ALH84001. Geochim Cosmochim Acta 73:6631–6677

Turtle EP, Del Genio AD, Barbara JM, Perry JE, Schaller EL, McEwen AS, West RA, Ray TL (2011) Seasonal changes in Titan's meteorology. Geophys Res Lett 38:L03203. doi:10.1029/2010GL046266

Weiss P, Yung KL, Kömle N, Ko SM, Kaufmann E, Kargl G (2011) Thermal drill sampling system onboard high-velocity impactors for exploring the subsurface of Europa. Adv Space Res 58:743–754. doi:10.1016/j.asr.2010.01.015

Woese CR, Fox GF (1977) Phylogenetic structure of the prokaryotic domain: the primary kingdoms. Proc Natl Acad Sci USA 74:5088–5090

Wolszczan A, Frail DA (1992) A planetary system around the millisecond pulsar PSR1257+12. Nature 355:145–147

Young AT (1973) Are the clouds of Venus sulphuric acid. Icarus 18:564–582

Zimmer C, Khurana KK (2000) Subsurface oceans on Europa and Callisto: constraints from Galileo magnetometer observations. Icarus 147:329–347

Zuber MT et al (1998) Observations of the north polar region of Mars from the Mars orbiter laser altimeter. Science 282:2053–2060

Further Reading

Gerda Horneck, Petra Rettberg (eds) (2007) Complete course in astrobiology. Wiley-VCH, Weinheim

Teerikorpi P, Valtonen M, Lehto K, Lehto H, Byrd G, Chernin A (2008) The evolving universe and the origin of life – the search for our cosmic roots. Springer-Verlag, New York

http://sci.esa.int/

http://www.nasa.gov/missions/index.html

Appendix 1: Geological Time (in Million Years, Ma). Simplified from the International Stratigraphic Chart (International Commission of Stratigraphy, 09/2010)

Eon	Era	Period	Epoch	Age (Ma)	Eon	Era	Age (Ma)
Phanerozoic	Cenozoic	Quaternary	Holocene				542
			Pleistocene	2.6	Precambrian Proterozoic	Neoproterozoic	
		Neogene	Pliocene				1,000
			Miocene	23		Mesoproterozoic	
		Paleogene	Oligocene				1,600
			Eocene			Paleoproterozoic	
			Paleogene	65			2,500
	Mesozoic	Cretaceous	Upper		Archean	Neoarchean	
			Lower	146			2,800
		Jurassic	Upper			Mesoarchean	
			Middle				3,200
			Lower	200		Paleoarchean	
		Triassic	Upper				3,600
			Middle			Eoarchean	
			Lower	251			4,000
	Paleozoic	Permian	Lopingian			Hadean	
			Guadalupian			(informal)	4,600
			Cisuralian	299			
		Carboniferous	Pennsylvanian				
			Mississippian	359			
		Devonian	Upper				
			Middle				
			Lower	416			
		Silurian	Pridoli				
			Ludlov				
			Wenlock				
			Llandovery	444			
		Ordovician	Upper				
			Middle				
			Lower	485			
		Cambrian	Furongian				
			Series 3				
			Series 2				
			Terreneuvian	542			

Appendix 2: Layered Structure of Earth's Interior

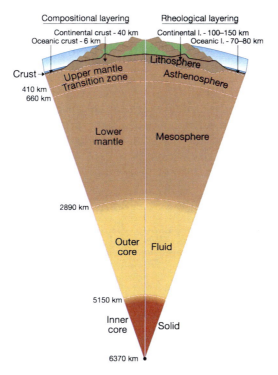

The interior of the Earth is layered both in terms of composition and rheological properties, as depicted in attached picture (modified from Fig. 2.39 in Kearey P, Klepeis K and Vine F 2009 Global tectonics 3rd edition, Wiley-Blackwell, Chichester, UK). The boundaries between the **crust** and **mantle** (Mohorovičić discontinuity or Moho), mantle and core (Gutenberg discontinuity) as well as liquid core and solid core are defined by discontinuities in seismic wave velocities.

The thickness of the outer layers varies markedly depending on the geotectonic environment. The oceanic crust is in average 6 km thick and consists mainly of basaltic rocks and sedimentary interlayers, while the continental crust is in average about 40 km thick but may reach in collisional orogenic belts a depth of 75 km and has variable composition, silicic rocks prevailing in the upper crust and more mafic rocks in the lower crust. The **lithosphere** consists of crust and solid ultramafic uppermost mantle. Its average thickness is in oceanic areas about 70–80 km, in continental areas 100–150 km, and beneath Archean cratons the lithosphere often extends to more than 300 km. The strong, rigid lithosphere is broken into plates that float on less viscous asthenosphere. The **asthenosphere** deforms by creep. It is essentially solid, but may contain in upper parts small amounts of interconnected melt along grain boundaries. A seismic low-velocity zone (LVZ) occurs in the upper parts of the asthenosphere (beneath the lithosphere), but it is poorly developed beneath Proterozoic shields and may be absent beneath Archean shields. The **transitional zone** between the seismic discontinuities at 410 and 660 km is an inhomogeneous zone characterized by transformation of upper mantle minerals like olivine to more densely packed phases. Although the mineral composition of the upper mantle and lower mantle (**mesosphere**) vary markedly, the chemical composition probably remains nearly similar, corresponding common stone meteorites. At 2890 km the solid silicate rock of the lower mantle changes abruptly to a liquid iron-nickel alloy of the **outer core**, and the density jumps from about 6 to 10 g/cm^3. Convections in the outer core stir up electric currents in conducting iron to create a geodynamo which is responsible of the Earth's magnetic field. The **inner core** is solid iron-nickel alloy. The temperature at Earth's center is about 5,000°C and the density of the inner core about 13 g/cm^3.

Appendix 3: Layers of Earth's Atmosphere

Layer	Distance from Earth's surface	Comments
Exosphere	690–10,000 km	Free moving hydrogen and helium atoms
Thermosphere	85–690 km	Aurora in lower parts
Mesosphere	50–85 km	Average temperature −85°C, most meteors burn upon entering the atmosphere
Stratosphere	6–20 – 50 km	Contains the ozone layer, pressure 1/10,000 of the atmospheric pressure on the sea level
Troposphere	Earth's surface – 6–20 km	Contains about 80% of the mass of the atmosphere

Ionosphere is the part of the atmosphere that is ionized by solar radiation. It extends from about 50 to 1000 km and overlaps both the exosphere and thermosphere and is responsible for the auroras. It forms the inner edge of magnetosphere.

Magnetosphere is formed when a stream of charged particles (the solar wind) from the Sun interacts with and is deflected by the Earth's magnetic field.

Index

A

Acidification, 167–169, 176
Aerosol particle, 219–225, 229
Albedo, 148, 220, 225
Anaerobic microbe, 196
Anorogenic, 28, 51, 53, 56
Anoxic basin, 5, 160
Anthropogenic climate change, 199, 211, 230, 237
Archean, 2, 4, 12, 15, 20, 22–24, 40, 47, 49–55, 57, 61–70, 77, 85, 107–114, 116, 119, 121, 329, 332
Arkona basin, 153, 155–157
Asthenosphere, 11, 65, 69, 332
Atmosphere, 2, 3, 5, 6, 140, 145, 147, 152, 159, 163–167, 187–196, 199, 200, 206, 209, 213, 220, 221, 224, 225, 227–229, 270–272, 276, 295, 297, 298, 301, 303, 304, 309–313, 315, 316, 321–325, 333
Aulacogen, 52
Aurora Borealis (northern lights), 6, 265
Aviation, 272–273

B

Bacteria, 142, 174, 181, 196, 310
Baltic Sea, 3–5, 38, 134, 139–149, 151–160, 168, 169, 171–182, 203, 207, 208, 284, 315, 316
Banded iron formation (BIF), 21, 71, 195, 196
Belomorian province, 50, 52, 53
Bimodal magmatism, 56, 57
Biogenic volatile organic compound (BVOC), 223–225
Biological pump, 165–166
Biology, 5, 171–182
Bornholm basin, 155, 157–159, 175

Bottom layer, 151, 154, 156–160
Bouguer anomaly, 131, 133

C

Carbon
  cycle, 5, 53, 164, 189–193, 220
  isotope excursion, 193, 195
  isotopes, 189, 192, 193, 195
  reservoir, 164
Carbon dioxide cycle, 164
Chemical pump, 165
Collision, 4, 23, 30, 31, 42, 50, 52, 54–57, 77, 78, 90, 276
Comet, 2, 269, 311, 318–319
Composite Svecofennian orogen, 54
  Fennia orogen, 54
  Lapland-Savo orogen, 54
  Nordic orogen, 54
  Svecobaltic orogen, 54
Concordia diagram, 105, 109, 110, 114, 122
Convection currents, 121
Craton, 4, 12–24, 26–28, 31, 47, 50, 53–55, 61–78, 112–114, 195, 332
Cratonic mantle, 61–78
Cretaceous, 121, 329
Crust
  continental, 4, 11, 37, 42, 62, 94, 107, 112, 120, 121, 292, 332
  oceanic, 11, 37, 42, 53, 61, 68, 94, 114, 332
Cyanobacteria, 2, 3, 157, 174, 180, 195

D

Desalination, 175, 182, 243, 251, 252
Devonian, 194, 329
Diabase, 115

Diamond, 29, 61–78, 121
Discordia, 105, 109, 121, 122
Drainage basin, 151, 152
Drift ice, 141–144, 147, 148

E
Earth, 1–7, 121–13, 21, 37–45, 49, 61–65, 68, 82–85, 90–92, 94, 96, 97, 104, 106, 111, 112, 114, 130–132, 134, 135, 163, 164, 187–196, 199–202, 206, 220, 225, 227, 241–243, 252, 257, 262, 263, 266–270, 272, 274–276, 279–281, 283–286, 288, 291, 292, 295, 296, 299, 301–307, 309–318, 321–325, 331–333
Earth cross section, 61, 62
Earth's atmosphere, 5, 164, 187–196, 316, 333
Earth's interior, 331–332
East-European platform, 47
Ecosystem, 143, 144, 149, 158, 171, 173, 178, 182, 222–224, 242, 309, 313–315
Eemian, 203–205
EIS. *See* Eurasian ice sheets (EIS)
Enceladus, 311, 317–319, 324
  Cassini, 317
ESA. *See* European Space Agency (ESA)
Eukaryote, 196
Eurasia, 5, 202–205
Eurasian ice sheets (EIS), 203, 204
Europa
  Galileo, 314
  Pioneer 12, 13, 313
  Voyager 1, 2, 313
Europe, 5–7, 12, 47, 71, 81–99, 130, 141, 144, 145, 147, 154, 199–214, 223, 227–238, 245, 246, 268, 272, 273, 277, 280, 284, 285, 288, 296, 304
European Space Agency (ESA), 268, 274, 277, 284, 285, 301, 302, 304, 306
Eutrophication, 145, 149, 174, 180–182

F
Fennoscandia, 4, 37, 42, 49, 52, 54, 56, 81–99, 104, 127–135, 141, 201, 203, 205–207, 209–212, 291
Fennoscandian shield, 42, 45, 47, 50, 52, 57, 70, 71, 77, 103, 195
Finland, 4–6, 37–45, 47–57, 61–78, 81–99, 103, 104, 106–110, 112–122, 127–130, 133, 140–144, 148, 149, 152, 155–158, 160, 171, 175, 180, 201, 203–205, 209–212, 222–224, 233–236, 241–246, 248–252, 258, 265, 266, 283, 284, 291, 298, 299, 302–304, 315, 317

Fish, 158, 174–176, 178, 181, 182, 241
Forest, 205, 220, 221, 223–225, 242, 249

G
Galileo satellite, 277
Geoid, 130–134, 284, 286, 291, 292
Geological time, 13, 329
Geomagnetic activity, 259, 260, 262, 263, 269
Glacier, 135, 199, 200, 205, 210, 211, 220, 246–247, 284–286, 291, 292
Global climate model (GCM), 221, 229, 230
Global Earth Observation System of Systems (GEOSS), 290
Global Geodetic Observing System (GGOS), 290
Global Positioning System (GPS), 6, 277, 283, 284, 287–292
Global warming, 135, 147, 171, 213, 230, 237, 291
GLONASS (Russian Global Navigation Satellite System), 288
Gotland Basin, 155, 156, 159, 182
GPS. *See* Global Positioning System (GPS)
Gravity satellites, 284, 286, 291, 292
Greenhouse gases, 5, 147, 187–191, 195, 196, 206, 221, 227–230, 233–235, 237, 249
Gross primary production (GPP), 223–225
Group of Earth Observation (GEO), 290
Gulf Stream, 209

H
Hadean, 61, 63–64, 111, 329
Halocline, 139, 152, 153, 155, 157, 158, 160, 178–180, 182
HELCOM (Helsinki Commission, Baltic Marine environmental Protection Commission), 168
Holocene, 4, 5, 200, 201, 203, 205–210, 329
Holocene Thermal Maximum (HTM), 208–210
Hubble space telescope, 323
Human population, 211, 229, 252
Hummocked ice, 142
Hydrogen sulfide, 2, 151, 157–160
Hydropower plant, 250
Hydrosphere, 2, 3, 5, 190, 199

I
Ice age, 128, 151, 201–203, 206, 230, 251, 291
Ice cap, 246, 295, 312
Ice conditions, 4, 5, 141–146, 148, 211
Ice season, 139–149, 177

Ice sheet, 4, 134, 135, 141, 142, 151, 188, 199, 202–206, 208, 209
Igneous rock, 48, 52, 92, 109, 110, 116
IMAGE (magnetometer chain from Estonia to Spitzbergen), 266
Inner core, 62, 332
Interglacial, 5, 188, 199, 200, 202–205, 207, 210
International Association of Geodesy (IAG), 289, 290
International Panel on Climate Change (IPCC), 147, 148, 175–178, 243
Interstadial, 204
Ionosphere, 6, 257, 272, 273, 276, 333
Island arc, 42, 45, 53–55, 66, 84, 92, 94
Isotope, 53, 63, 64, 70, 103–122, 189, 190, 192–195, 202, 204, 210, 309
Isotopic age, 16
Isotopic method, 112, 119, 121
Isotopic microanalysis
 laser ablation mass spectrometry (LAMS), 106, 111
 Lu-Hf, 105, 106, 111, 116, 117, 120
 Pb-Pb, 120
 Rb-Sr, 64, 103, 120
 secondary ion mass spectrometry (SIMS), 105–107, 110, 111, 120, 122
 Sm-Nd, 64, 112, 113, 120
 U-Pb, 71, 73, 74, 77, 103–107, 111, 113, 115–119, 121

J

James Webb telescope, 323
Jormua-Outokumpu ophiolite, 53, 57
Jupiter, 2, 311, 313, 314, 317, 320–322
Jurassic, 61, 329

K

Karelian craton, 4, 55, 63, 65, 69–78, 113
Karelian province, 47, 50–53, 57, 107–114, 119, 121
Kepler spacecraft, 322
Killer electrons, 270, 274–276
Kimberlite, 4, 29, 31, 61–78, 121
Kola craton, 15, 23, 71
Kola province, 50, 52, 53, 112
Komatiite, 49, 53, 69, 77
Kyoto protocol, 237

L

Lamproite, 4, 69, 71–74, 78, 121
Lamprophyre, 72, 121

Landfast ice, 141–144, 148, 177
Land uplift, 4, 127–135, 145
Lapland-Kola orogen, 4, 23, 42, 51, 53, 54, 57
Large dam, 249–251
Laser ablation mass spectrometry (LAMS), 106, 111
Late heavy bombardment (LHB), 63
Lava ocean, 2
Layered intrusion, 52, 53, 57, 83, 91
Levering, 135
Life
 characteristics, 309–310
 limits, 309–310
 origin, 313
 requirements, 309–310
Lithosphere
 continental, 61–64, 77, 332
 oceanic, 61, 66–68, 114, 332
Little Ice Age (LIA), 5, 211–213
Lunar eclipse, 280, 281

M

Magnetic indices, 258, 263
Magnetic record, 258–260, 263
Magnetic storm, 259, 262, 267, 270, 273, 275
Magnetosphere, 6, 257, 263, 269, 270, 274, 275, 333
Major Baltic Inflow, 153–157, 160
Mantle, 2–4, 11, 20, 31, 37, 38, 40, 41, 43, 44, 49, 50, 52, 53, 55, 56, 83, 92, 94, 111, 112, 114, 120, 121, 132, 135, 291, 292, 331, 332
Mantle formation, 61–78
Mantle plume, 20, 56, 68–69
Mars
 Mariner 4, 297, 298, 312
 Viking, 297, 312
Mercury, 2, 310, 311, 320
Mesoarchean, 4, 111, 329
Mesoproterozoic, 121-31, 47, 49, 121, 329
Metallic mine, 4, 97, 98
Metallogenic areas, 87, 88
Metamorphic rock, 110
Metasomatism, 69, 75, 77, 78
Meteorite, 116, 313, 319, 333
Meteorite impact, 2, 3
Methane, 2, 5, 164, 188, 191, 195, 196, 202, 227, 301, 304, 313, 315–317
Microcontinent, 14, 23, 42, 45, 49, 54, 55, 57
Mid ocean ridge basalt (MORB), 50, 54, 66
Migmatite, 4, 50, 51, 57, 110, 115, 117
MilanKovitch cycle, 199, 200, 202

Mineral deposit, 4, 85, 90, 91, 96, 97
Mineral resources, 81–99
Minette, 72
MIR, 271, 307
Missions to Mars
  Beagle, 301, 302
  Curiosity rover, 303
  ExoMars mission plans, 304
  Mariner, 297, 298, 312
  Mars, 297–298
  Mars Climate orbiter, 299
  Mars Express (MEX), 299, 301, 305
  Mars Global Surveyor orbiter, 299
  Mars Observer, 299
  Mars Odyssey (MO), 300, 312
  Mars Pathfinder lander, 299
  Mars Polar lander, 299
  Mars Reconnaissance Orbiter (MRO), 301, 305
  Nozomi, 299, 300
  Opportunity rover, 301
  Phobos-Grunt, 303
  Phobos Sample Return Mission, 303
  Phoenix lander, 302
  Sojourner rover, 299
  Spirit rover, 301
Mohorovičić discontinuity (MOHO), 37, 40–43, 331
Moon, 63, 82, 270, 277, 279, 298, 303, 305–307, 310, 311, 313–315, 317, 318, 321
  formation of the Moon, 2

N
National Aeronautics and Space Administration (NASA), 268, 274, 298, 300–304, 307, 316, 318, 319, 322
National Oceanic and Atmospheric Administration (NOAA), 259, 260, 274, 275
Navigation satellite, 276, 284, 286, 288
Neoarchean, 50, 103, 107, 108, 111, 112, 114, 115, 118, 119, 329
Neoproterozoic, 21, 111, 121, 193, 201, 329
North Atlantic Oscillation (NAO), 145, 148, 154, 206, 207, 212, 213
Northern Europe, 5, 91, 145, 147, 154, 199–214, 227–238
Northern lights, 257, 265–267
Nutrient, 143, 145, 158, 160, 166, 172, 174, 176–182

O
Orbital forcing, 5, 206, 208
Orbiting Geophysical Observatory (OGO), 268
Ore deposit
  in Fennoscandia, 85, 86, 88, 89
  in Finland, 4, 84, 87, 88, 92, 93
  genetic types, 88, 90, 92
Ore resources, 96
Orthogneiss, 71
Outer core, 61, 62, 332
Oxygen content, 5, 159, 188, 193, 194

P
Paleocene, 202
Paleomagnetic reconstruction, 15, 26
Paleomagnetism, 13
Paleoproterozoic, 12, 14, 20, 21, 47, 54–57, 71, 95, 103, 107–109, 112–119, 121, 193, 329
Paleoproterozoic rifting, 4, 49, 50, 52–54
Paragneiss, 50, 107, 110–112
pH, 165, 167, 176, 189, 310
Phanerozoic, 47, 50, 65, 104, 188–190, 193–194, 196, 329
pH change, 168, 169
Photosynthesis, 3, 164, 178, 191, 192, 223–225
Physical pump, 165
Phytoplankton, 166, 174, 175, 177–179, 181, 189
Planetary systems, 190, 309, 319–323
Planetesimal, 2, 63
Plankton, 157, 158
Plate tectonics, 3, 11, 45, 49, 83–85, 94, 108, 112, 284
Pliocene, 329
Plutonic rock, 4, 47, 49–51, 53–56, 112
Polar Year, 266
Postglacial rebound, 132, 291
Precambrian continents
  Amazonia, 15
  Australia, 15
  Baltica, 15
  Congo, 15
  India, 15
  Kalahari, 15
  Laurentia, 15
  North China, 15
  Sao Francisco, 15
  Siberia, 15
Precipitation, 94, 95, 148, 152, 165, 169, 171, 179, 182, 200, 201, 207, 209–211, 220, 230–235, 245, 246, 252, 257, 291
Pressure ridge, 142, 143

Proterozoic, 14, 15, 31, 40–43, 47, 49, 65, 66, 70, 75, 77, 85, 113, 120, 121, 201, 329, 332
Protoplanet, 2
Pulsar, 319

Q
Quaternary, 5, 199–214, 329
Quaternary Environment of the Eurasian North project (QUEEN), 203

R
Radiation flux, 187
Radiative balance, 6, 220–221
Rainmaking, 251, 252
Rainwater harvesting, 251, 252
Rapakivi granite, 4, 24, 26, 27, 45, 47, 49, 56, 57, 120
Rift, 28, 37, 52, 53, 55, 57, 84, 92, 93
Rifting, 4, 15, 21, 23, 28, 31, 49, 50, 52–57, 90

S
Saalian, 203
Salinity, 5, 139–141, 145, 151–159, 168, 169, 171–176, 178, 180–182, 310
Satellite positioning, 272–273
Search for Extraterrestrial Intelligence (SETI), 322–325
Secondary ion mass spectrometry (SIMS), 105–107, 110, 111, 120, 122
Seismic survey
 reflection profile, 38
 refraction profile, 39, 40, 43, 44
 tomography, 44, 65
Shield, 4, 12, 15, 37, 42, 45, 47, 50, 52, 57, 70, 71, 77, 85, 103, 107, 191, 194, 195, 269, 270, 311, 312, 332
Snowball Earth, 21, 201
Sodankylä Geophysical Observatory, 258, 267
Solar activity, 6, 209, 212, 214, 257–259, 262, 263, 267–270, 299
Solar and Heliospheric Observatory (SOHO), 268, 274, 275
Solar eclipse, 268, 281–283, 320
Solar eruption
 coronal mass ejection, 6, 257, 267, 268, 270, 276
 flare, 267, 270
Solar forcing, 5, 206, 260
Solar storm, 265–277
Solar wind, 6, 257, 259, 261–263, 269, 270, 275, 276, 298, 333

Space geodesy
 GLONASS, 288
 satellite laser ranging (SLR), 286, 288, 289, 292
 techniques, 282, 285–290
 very long baseline interferometry (VLBI), 286, 288, 289, 292
Space weather, 6, 257–263, 265–277
Spreading, 12, 84
Stagnation, 158–160, 179, 180
Stellar triangulation, 283
Stratification, 140, 143, 145, 147, 148, 151, 152, 154, 157, 158, 178–180, 182
Struve Geodetic Arc, 281
Subduction, 4, 25, 26, 28, 42, 43, 50–56, 65–68, 71, 77
Sulfate, 157, 158
Sulfide, 2, 64, 84, 91, 93–95, 151, 157–160
Sulfur, 2, 84, 91, 93–95, 213, 311, 314
Sulfur isotopes, 193, 194
Sunspot, 257–260, 262, 263, 267–269, 276
Sunspot cycle, 257, 259, 260, 262, 263, 267, 276
Supercontinents
 Arctica, 12, 15, 30, 135, 200, 202, 223, 291, 313, 314
 Atlantica, 12, 22
 Columbia (or Hudsonland), 12, 247
 Gondwana, 12, 30
 Laurussia, 12
 Nena, 12, 16, 31
 Nuna, 12, 22, 24, 27, 28, 31
 Pangea, 12, 13, 31
 Rodinia, 12, 28–31
 Ur, 12, 16, 20, 22, 25
Supracrustal, 21, 47, 50–52, 56, 104, 111, 112, 114, 117–119, 121
Surface layer, 152, 153, 157, 158, 164, 166–168, 172, 174, 175, 178, 179
Svecofennian orogen, 24, 40, 42, 54, 57, 71, 94, 96, 112, 113, 115, 118, 119, 121

T
Telecommunication, 272–274, 306
Thermal ionization mass spectrometry (TIMS), 103, 105, 107, 110
Thermohaline circulation (THC), 205
Titan
 Cassini-Huygens, 315
 Huygens, 315
 hydrocarbon lakes and seas, 316
 Voyager, 315

Tonalite-trondhjemite-granodiorite association (TTG), 50, 52, 107, 111
Total ecosystem respiration (TER), 223, 225
Towing of icebergs, 252
Triangulation, 266, 281, 283
Triassic, 327

U
Ultraviolet radiation, 187, 194, 269, 272, 312
United Nations Framework Convention on Climate Change (UNFCCC), 252
United Nations Framework Convention on Water Solidarity (UNFCWS), 252

V
Van Allen radiation belt, 271, 277

Venus, 2, 269, 311, 320
Virtual water, 249, 250
Volcanic activity, 5, 202, 206, 212, 228, 302
Volcanic rock, 49–55, 57, 90, 92–93, 119

W
Water consumption, 242, 243, 245, 248, 252
Water plume, 317
Water resources, 241–245, 247–249, 251, 252
Water transfer, 135, 243, 244, 248–252
Weather satellite, 284
Weichselian, 147, 202–204, 208

X
Xenocryst, 61–78
Xenolith, 4, 61–78, 121